Microbiology Monographs

Volume 27

Series Editor
Alexander Steinbüchel
Münster, Germany

More information about this series at http://www.springer.com/series/7171

Min-Tze Liong
Editor

Beneficial Microorganisms in Food and Nutraceuticals

Springer

Editor
Min-Tze Liong
Universiti Sains Malaysia, School of Industrial Technology
Penang, Malaysia

Series Editor
Alexander Steinbüchel
Institut für Molekulare Mikrobiologie und Biotechnologie
Westfälische Wilhelms-Universität
Münster
Germany

ISSN 1862-5576 ISSN 1862-5584 (electronic)
Microbiology Monographs
ISBN 978-3-319-23176-1 ISBN 978-3-319-23177-8 (eBook)
DOI 10.1007/978-3-319-23177-8

Library of Congress Control Number: 2015957590

Springer Cham Heidelberg New York Dordrecht London

Printed on acid-free paper

Springer International Publishing AG Switzerland is part of Springer Science+Business Media (www.springer.com)

Preface

This volume communicates aspects of beneficial microorganisms in relation to food and nutraceuticals. The conventional roles of microorganisms in foods typically emphasize on bio-preservation, extended shelf life, and production of higher digestible nutrients via natural fermentation processes. However, the recent introduction of nutraceuticals has broadened the potentials of microorganisms to transform normal foods to nutraceutical products, many with well-characterized health claims. In combination with such increasing demands for nutraceuticals, various new food technology techniques have been developed, conventional technologies have been re-innovated, and various new beneficial microorganisms have also been identified.

Penang, Malaysia Min-Tze Liong

Contents

From Traditional Knowledge to an Innovative Approach for Bio-preservation in Food by Using Lactic Acid Bacteria

Cristina Stewart Bogsan, Luis Augusto Nero, and Svetoslav Dimitrov Todorov

Contents

Abstract Fermentation processes have been part of the human food preparation for centuries. Empirical knowledge of these processes has been transmitted from one generation to another and has survived over the years. However, with the establishment of the scientific basis of microbiology, all food fermentation processes have been re-evaluated from the new perspective—the physiological characteristics of lactic acid bacteria. The growth of lactic acid bacteria and production of different

C.S. Bogsan
Department of Biochemical-Pharmaceutical Technology, Faculty of Pharmaceutical Sciences, University of São Paulo, 05508-000 São Paulo, São Paulo State, Brazil

L.A. Nero
Veterinary Department, Universidade Federal de Viçosa, Campus UFV, 36570-900 Viçosa, Minas Gerais State, Brazil

S.D. Todorov (✉)
Veterinary Department, Universidade Federal de Viçosa, Campus UFV, 36570-900 Viçosa, Minas Gerais State, Brazil

Laboratory of Food Microbiology, Department of Food and Experimental Nutrition, Faculty of Pharmaceutical Sciences, University of São Paulo, Av. Prof. Lineu Prestes, 580, Bloco 14, 05508-900 Sao Paulo, São Paulo State, Brazil
e-mail: slavi310570@abv.bg

M.-T. Liong (ed.), *Beneficial Microorganisms in Food and Nutraceuticals*, Microbiology Monographs 27, DOI 10.1007/978-3-319-23177-8_1

metabolites play an essential role not only on the sensorial characteristics of the fermented food products but also in terms of the safety concern. Several antimicrobial metabolites produced by lactic acid bacteria have been described in the literature. Some of them, known as bacteriocins, not only have been explored intensively over the last few decades by the food industry but also have been acknowledged as promising antibacterial compounds with pharmaceutical applications. In this chapter, we will review some of the traditional applications of lactic acid bacteria, showing the importance of antimicrobial metabolites with special focus on antimicrobial proteins (bacteriocins), and discuss some specific cases on its applications.

1 From the Past to the Future

An overview of the human history of the nutrition showed that the traditional empirical knowledge has been used for the preservation and preparation of food products. Furthermore, nutrition has been established as an object of the science only in beginning of twentieth century. However, only in last two centuries the systematic basis of the preservation and conservation of the food products has been established, following after the development of science, especially microbiology, physiology and biochemistry and more recently genetics and nutrition. If we re-evaluate the history of the preparation of the fermented food products, we can clearly acknowledge that from the ancient times, they have been part of the nutritional habits of the human civilizations. Acknowledgment of the good food was a part of the lifestyle of Ancient Greeks and Romans. Parts of their diet were several fermented food products including wine, olives, meat, milk and fish products. We can address the preparation of alcoholic beverages to the Egyptians. Nomadic peoples from central Asia were preparing yogurt, kumis and kefir. In North Europe, fermented meat products were a part of the nutritional habits of Germanic tribes, and Eskimos were known by the preparation and preservation of fish. The ancient Persians were preparing boza and other fermented products based on cereals, milk, fruits and vegetables. Fermented products based on corn (maize) were part of the nutritional and spiritual habits of the native tribes in pre-Columbian America. All these products were prepared based on the empirical knowledge. Only in the nineteenth century, after the establishment of the modern microbiology by Leeuwenhoek, Louis Pasteur, Metchnikoff and other pioneers, the basis of fermentation process and preservation was established.

By the end of nineteenth century, scientific basis of fermentation processes and modern microbiology was established by Louis Pasteur. However, Ilja Metchnikoff and his collaborator, Stamen Grigorov, were the first to suggest the basic concept of the modern understanding of functional food products and probiotics. Again, at this time, they did not know very well yet how and why probiotics worked and all observation and conclusions were mostly based on the observations and empirical conclusions. In time of Metchnikoff, even the term probiotics was not yet defined and used. Only much later, Fuller and Gibson (1997) defined this term as live microorganisms which, when administrated in adequate amounts, confer health benefits to the host such as reduction of gastrointestinal infections and inflammatory bowel disease and modulation of the immune system (Fuller and Gibson 1997).

2 Role of Lactic Acid Bacteria in the Most Traditional Way of Application: Preparation of Food Products

The use of microorganisms with antagonistic properties and/or their antimicrobial metabolites that can increase the shelf life is known as bio-preservation process (Voulgari et al. 2010; Liua et al. 2014). Fermentation is the oldest bio-preservation process (Kabak and Dobson 2011; Metha et al. 2012), and it is applied to dairy products, vegetables, meat and fish. Fermentation processes are the transformation of raw materials into products using microorganisms such as yeasts, moulds and bacteria (Voulgari et al. 2010).

These microorganisms have different morphological, metabolic, physiological and taxonomic properties (Metha et al. 2012). The primary bacteria found in traditional fermented food are the lactic acid bacteria (LAB). LAB can be represented by different Gram-positive and catalase-negative microorganisms, with the production of lactic acid as a final metabolite of sugar fermentation (Voulgari et al. 2010).

LAB can be detected in different ecological niches ranging from plant surfaces to the gastrointestinal tract of animals. The most representative genera of LAB are *Lactobacillus*, *Lactococcus*, *Enterococcus*, *Leuconostoc*, *Pediococcus*, *Oenococcus*, *Carnobacterium*, *Weisella* and *Streptococcus* (Stiles and Holzapfel 1997; Ross et al. 2002). LAB are aerotolerant anaerobes; are acid resistant, non-spore forming, and catalase negative; and have a DNA base composition G+C less than 54 % (Stiles and Holzapfel 1997; Myoshi et al. 2009).

LAB are extensively used as start cultures in fermented food products. They are traditionally used for food bio-preservation to prolong storage, enhance nutritive and therapeutic values and promote changes in flavour, texture, and tastes (Aguirre and Collins 1993; Kabak and Dobson 2011). These microorganisms produce organic acids, hydrogen peroxide, diacetyl and bacteriocins with antagonistic microbiological properties to suppress the growth of undesirable microbiota (Ross et al. 2002; Liua et al. 2014). Besides, LAB increase presence in the fermented food products of the essential amino acids and vitamins such as group B vitamins (Kabak and Dobson 2011). Moreover, most of the LAB are recognized as generally recognized as safe (GRAS) status and are considered as safe for human consumption (Aguirre and Collins 1993; Myoshi et al. 2009).

After the industrial revolution in the nineteenth century, with the concern of possible shortage of food, there has been an increased interest in the production of traditional fermented food products (Kabak and Dobson 2011). Most of the traditional fermented food products are originated from Asia, Africa, India, Europe and the Middle Eastern countries (Blandino et al. 2003). Although the process of bio-preservation by fermentation dates back more than 10,000 years ago, new fermented milks, cereals, fruits, vegetables, fish and meat are still emerging until nowadays. However, many food products can contain toxins or anti-nutritional compounds, such as aflatoxins, tannins and phytates, that could be removed by the action of LAB. LAB can inhibit the growth of pathogenic bacteria via the primary metabolites, such as lactic and acetic acids, ethanol and carbon dioxide or antimicrobial compounds such as formic, benzoic and acids, hydrogen peroxide, diacetyl and acetoin besides bacteriocin (Kabak and Dobson 2011; Metha et al. 2012; Elsanhoty et al. 2014).

Traditional fermented foods are dependent on the microbiota characteristics of a specific raw material or food source that are directly influenced by the region geography, topography, climate, agricultural factors and animals that are native to the area (Ross et al. 2002; Metha et al. 2012). Nowadays, genetic, biochemical and physiological studies are the key to understand the bio-preservation mechanism of LAB (Shiby and Mishra 2013). Some examples of fermented food products from milk, cereals, vegetables and meat and fish origin, along with the importance of LAB in the preparation, are included in the following.

Milk is a complex food matrix, an emulsion containing proteins, fats, lactose and various vitamins and minerals. It is a food secreted from the mammary glands of mammals and serves as nourishment for their young (Oliveira 2009). The quality of milk depends on the health of the milk-producing animal and also the hygiene level of the environment (Metha et al. 2012). The traditional dairy products are obtained from spontaneous acidification by LAB that are naturally present in milk as adventitious contaminants, accompanied by yeasts or moulds (Wouters et al. 2002).

The traditional dairy fermented products are produced by the "back -slopping method", that is, the reintroduction of inoculum from the previous fermentation into new product (Oliveira 2009; Kabak and Dobson 2011). The most common pathogenic microorganisms found in raw milk and inhibited by LAB are *Staphylococcus aureus* and *Escherichia coli*. However, *Streptococcus agalactiae*, *Streptococcus uberis*, *Mycobacterium* spp., *Brucella* spp., *Bacillus* spp. and *Listeria* spp. have also been reported (Metha et al. 2012).

The production of the first fermented milks products can be addressed to the Middle East communities (Kabak and Dobson 2011). Prepared from the raw milk, all of them share the lactic acid as a result of lactose fermentation from LAB growth, which promotes the coagulation of the milk proteins (Oliveira 2009). The production of dairy products is dependent on the use of defined LAB starters. These strains have been replaced from the undefined strain mixtures traditionally used (Ross et al. 2002; Oetterer 2014). Traditional fermented milks originated from different countries and include more than 500 different milk products, such as yogurt, fermented milk, acidophilus milk, kefir, kumis, and buttermilk, among others described herein (Oliveira 2009); however, the exact mechanisms of bacterial interaction with food matrices are not fully clarified yet.

Acidophilus milk originated from Europe and the milk is inoculated by selected *Lactobacillus acidophilus* strains. The incubation temperature is at 37 to 40 °C for 18–20 h until coagulation of milk. The final products are characterized by its acidity and contain about 1 % of lactic acid (Oliveira 2009; Shiby and Mishra 2013).

Amasi is a traditional product from Africa (Mokua 2004). Traditionally, the fresh milk is placed in a gourd, added a bit of amasi from a previous batch ("back-slopping method"), and then allowed to ferment for 2–3 days at ambient temperature, until reaching the pH of 4.0. The microbiota is dominated by different LAB, including *Lactobacillus* spp., *Lactococcus* spp. *and Leuconostoc* spp., which originated from the environment (Todorov 2008).

Buttermilk, also known as chaa/lassi, originated from the Balkans region of Europe, the Middle East, and Central Asia (Sharma et al. 2013). It is a fresh dairy beverage prepared from *Streptococcus*, *Leuconostoc* and *Lactococcus lactis* subsp. *lactis/cremoris* inoculation into a liquid leftover resulting from butter churning, fermented at 20–25 °C until reaching the pH of 4.5 (Oliveira 2009). Buttermilk presents the average composition about 47 % lactose (w/w), 29 % protein (w/w), 3.1 % moisture (w/w), 12 % fat (w/w) and 7.4 % ash (w/w) (Peteán et al. 2014).

Dahi is a fermented dairy product originated from India. It is a yogurt-like product made from cow or buffalo milk or mixture of both, in addition to *Lactobacillus acidophilus*, *Streptococcus lactis*, *Lactococcus lactis* and *Lactococcus lactis* subsp. *diacetylactis*. These strains are added to a boiled and cooled milk and fermented for 7 h at 30 °C. The final product presents a clean acid taste and a yellowish cream white colour (Shiby and Mishra 2013). Dahi presents the average composition about 15 % total solids (w/w), 1.8 % protein (w/w), 5 % lipid (w/w) and 12 % sugar (w/w) (Shiby and Mishra 2013).

Kefir, originated from the Caucasus region of Asia, is made from the inoculation of the "kefir grain" in fresh milk. It is an acidic, middle alcoholic fermented dairy beverage (Kabak and Dobson 2011; de Oliveira et al. 2013; Nielsen et al. 2014). Kefir has a refreshing taste, is lightly carbonated and is slightly acidic (Beshkova et al. 2002). Kefir grains are composed of an insoluble protein and polysaccharide matrix, gelatinous, and yellowish and vary in size from 0.3 to 3.5 cm in diameter (Kabak and Dobson 2011; Nielsen et al. 2014). The grains range from white to yellow in colour. The beneficial properties and composition of kefir are highly dependent on the type of milk that was fermented and may include the high concentration of vitamin B12.

The kefir grains are washed with water and stored at a low temperature, prior to serving as inoculums in a subsequent fermentation (Nielsen et al. 2014). The complex matrix includes a mixture of bacteria and yeasts such as *Lactobacillus kefiranofaciens* and *Lactobacillus kefir* and *Lactobacillus* spp., *Lactococcus* spp., *Streptococcus* spp., *Leuconostoc* spp. (Powell et al. 2007), *Kluyveromyces marxianus*, *Candida inconspicua*, *Candida maris*, *Torulopsis kefir* and *Saccharomyces cerevisiae* (Simova et al. 2002) and differs from that found in the fermented beverage: it occurs because of the complex symbiotic interactions between the organisms in the kefir grains during the kefir production. The ratio of lactic acid bacteria to yeast in kefir grains is 1000:1 (Nielsen et al. 2014). Kefir beverage can be made from cow, goat, sheep, camel and buffalo milk. Also, kefir could be made from milk substitutes such as soy milk, rice milk and coconut milk (Nielsen et al. 2014) resulting in end products with different compositions. The main contents are about 1.5 % alcohol (w/w) and a pH around 4.5 (Kabak and Dobson 2011).

During the process of fermentation of kefir, several changes in composition of nutrients and other ingredients occur. Lactic acid produced by the LAB and propionic acid by propionibacteria are final metabolites usually find in kefir. Other substances that can also be found in kefir and contribute to the flavour are pyruvic acid, acetic acid, diacetyl and acetoin, citric acid, acetaldehyde and various amino acids as a result of the metabolism of milk proteins.

As a result of the fermentation process conducted using LAB and yeasts, only a little concentration of lactose remains in the final product. This is an important advantage for the lactose-intolerant individuals to be able to tolerate kefir (de Oliveira et al. 2013).

Kumis, originated from India, the Middle East, Mongolia and Turkey, is made from mare milk, involving acid and alcoholic fermentation using microorganisms *Lactobacillus delbrueckii* subsp. *bulgaricus*, *Lactobacillus plantarum* and *Kluyveromyces marxianus*, *Saccharomyces lactis* and *Torula koumiss* (Tamime et al. 1999; Danova et al. 2005; Wang et al. 2008; Oliveira 2009). The metabolites are lactic acid, ethanol and carbon dioxide in the ratio of 2:3:1 (Tamime et al. 1999). The fermentation process is similar to the production of kefir, using the "back-slopping method" as starter culture, producing a grey-coloured liquid milk, lightly carbonated with sharp alcoholic and acidic taste (Tamime et al. 1999; Kabak and Dobson 2011). Kumis presents the average composition about 11 % total solids (w/w), 1.8 % protein (w/w), 1 % lipid (w/w), 4.1 % lactose (w/w), 0.5 % ash (w/w) and a pH value of 4.0 (Kabak and Dobson 2011).

Yogurt, one of the most popular fermented milk products worldwide, originated from the Balkans and Middle East. Yogurt is the result of the fermentation of cow, goat, sheep or buffalo milk using *Streptococcus thermophilus* and *Lactobacillus delbrueckii* subsp. *bulgaricus*, both present naturally in milk or added as starter culture using the "back-slopping method". Yogurt has a viscosity and a distinctive acidic, sharp flavour (Tamime and Robson 1999; Oliveira 2009). The LAB used in yogurt production ferment in a symbiotic growth pattern where *Streptococcus thermophilus* grows faster than *Lactobacillus delbrueckii* subsp. *bulgaricus* and then ferments the lactose in the presence of dissolved oxygen and releases more acid lactic and formic acid. In the presence of formic acid, *Lactobacillus delbrueckii* subsp. *bulgaricus* grows producing amino acids and small peptides. The fermentation process occurs at 37–42 °C until pH reaches 4.5 after 4 h of fermentation (Tamime and Robson 1999; Oliveira 2009; Kabak and Dobson 2011; Metha et al. 2012).

Labneh, originated from the Middle East, is made from either cow, goat or sheep milk strained from the traditional yogurt in a special cloth bag for 10–14 h to remove the whey, representing approximately 65 % of the total weight. Additionally, some salt could be added to improve the shelf life (Kabak and Dobson 2011).

Labneh presents the average composition of 80 % moisture, 9 % protein, 9 % fat and 8 % lactose (Kabak and Dobson 2011).

Vili originated from Nordic countries, and it is made from the inoculation of milk with *Lactococcus lactis* subsp. *cremoris*, *Lactococcus lactis* subsp. *lactis* biovar. *diacetylactis* and *Leuconostoc mesenteroides* subsp. *cremoris* associated with yeasts *Kluyveromyces marxianus* and *Pichia fermentans* producing a viscous, gelatinous and almost ropey consistency (Tamime 2007).

Cheese production dates back 8000 years ago and includes more than 1000 varieties (Beresford et al. 2001). Farms and shepherds developed the cheese from spontaneous fermentation of LAB into cow, goat, sheep and buffalo milk on a small scale, incubating the milk from the previous day under specific conditions. These traditional processes are still used to produce the "artisanal products" (Wouters et al. 2002).

The primary function of LAB in cheese technology is to produce acid during fermentation process; however the strains used will differentiate the cheese ripening and flavour compounds obtained by proteolysis of milk amino acids (Fox and Wallace 1997; Beresford et al. 2001). LAB used as starter bacteria to cheese curd are divided into thermophilic bacteria or mesophilic bacteria. The mesophilic bacteria are used for the production of cheese which temperature of curd is under 40 °C during fermentation, and thermophilic bacteria are used to that in which the acidification process is made at a temperature higher than 40 °C (Fleet 1999; Wouters et al. 2002). In cheese production, the starter culture are associated as well with other microorganisms, including non-LAB, moulds and yeasts (Metha et al. 2012); however, the main LAB found as starter cultures are *Lactococcus*, *Lactobacillus*, *Streptococcus*, *Leuconostoc* and *Enterococcus, Lactococcus lactis* subsp. *cremoris* and *Lactococcus lactis* subsp. *lactis* (Beresford et al. 2001).

The traditional way of preservation, mainly for vegetables, was to keep them in the presence of solution with high salt concentration or brine. These processes are used until today for the manufacture of pickles, but only in the last century, it was discovered that this conservation process is related to the action of LAB, which lowers pH and controls with high osmotic pressure caused by the salt and the action of spoilage microorganisms. Cucumber, sauerkraut, kimchi, olives, peppers, tomatoes and carrots are the most traditional fermented vegetable foods (Metha et al. 2012).

Boza, originated from the Balkans region of Europe and the Middle East, is a traditional low alcoholic beverage and highly viscous made from rice and wheat flours (Kabak and Dobson 2011). This cereal fermented drink is traditionally prepared from cooked grains with the addition of water, cooled and sweetened with the addition of 20 % sugar (w/w). Then the product is fermented addition of starter from a previous boza beverage in "back-slopping method" for 24 h at 30 °C until reaching around pH 4.0 (Kabak and Dobson 2011). The LAB responsible for the fermentation of traditional boza are a mixture of different *Lactobacillus* spp. and *Leuconostoc* spp. (Von Mollendorff et al. 2006; Todorov et al. 2008; Kabak and Dobson 2011). Boza presents the average composition about 1.5 % protein, 12 % carbohydrate and 80 % moisture.

LAB and yeast are the principal microbiota of boza and are responsible for the production of a number of vitamins and increase the nutritional value of the product (Von Mollendorff et al. 2006). Traditional medicine recommends the application of boza in several cases as a parallel treatment for the control of diarrhoea in kids and adults and to reduce symptoms of inflammation of upper respiratory tract and even tuberculosis (LeBlanc and Todorov 2011; Todorov et al. 2014).

Kimchi, originated from Korea, is made from cabbage and various other vegetables, salted overnight, washed and drained. Then the vegetables were fermented in a jar for 24 h with the presence of natural LAB microbiota at 15 °C until reaching the pH of 4.2. The predominant bacteria are *Leuconostoc*, *Weissella* and *Lactobacillus* (Metha et al. 2012).

Pickles originated in Persia. In the preservation process of this food, cucumbers are brined, where the salt concentration controls the spoilage microorganisms until the LAB became the dominant microbiota (Metha et al. 2012). The process of preparation of fermented cucumbers is conducted using *Leuconostoc*, *Pediococcus cerevisiae* and *Lactobacillus plantarum* and yeasts *Torulopsis holmii* and *Brettanomyces versatilis* in the presence of salt (solution of 10–15 % NaCl in water) for 4 weeks at 20 °C, until the pH reaches 3.5 (Caplice and Fitzgerald 1999; Kabak and Dobson 2011; Metha et al. 2012).

Sauerkraut, originated from North Europe, is made from finely cut cabbage added to 2 % NaCl (w/w) and let to ferment into a closed crock for 3 weeks using *Leuconostoc mesenteroides*, *Lactobacillus lactis* and *Pediococcus* spp. at 18 °C until reaching the pH of 4.0 (Metha et al. 2012).

Tarhana, originated from the Middle East, is prepared by mixing cracked wheat of flour, yogurt, yeast and a variety of cooked vegetables and spices and has an acidic and sour taste with a strong yeast flavour (Kabak and Dobson 2011). These food products are fermented using *Lactococcus lactis*, *Lactococcus diacetylactis*, *Lactobacillus acidophilus*, *Lactobacillus casei* and *Leuconostoc cremoris*, *Lactobacillus delbrueckii* subsp. *bulgaricus* and yeast *Streptococcus cerevisiae* for 7 days until reaching the pH of 4.0 (Campbell-Platt 1994). Tarhana presents the average composition about 20 % protein, 60 % carbohydrate, 15 % fat, 2 % fibre, 8 % salt and 10 % ash (Kabak and Dobson 2011).

Olives are an essential part of the Mediterranean area food habits. Olives are harvest by hand and washed to immerse in a brine solution to ferment for 10 days at 20 °C. The fermentation occurred using natural microbiota that include *Lactobacillus mesenteroides*, *Lactobacillus brevis*, *Lactobacillus plantarum and Lactobacillus pentosus* (Metha et al. 2012).

The traditional meat product has its origin in the Babylonian age. The preservation of meat is based on dry curing, salting and fermentation processes (Metha et al. 2012). The traditional meat product is used to be dry cured by salt, nitrate, nitrite and sugars and then fermented using LAB. The LAB are essential agents to improve the hygienic and sensory quality of the final sausage (Fadda et al. 2010). The fermentation using LAB during the storage of product promotes the decrease of pH and reduction of nitrates and nitrites to nitric oxide, development of red colour, dehydration, lipolysis, fat autoxidation and proteolysis (Fadda et al. 2010).

In Europe, fermented sausages are traditionally prepared from seasoned raw meat stuffed in casings and allowed to ferment and mature. The curing process must contain a fermentation stage using *different Lactobacillus spp.*, which leads to the competitive expulsion of pathogenic organisms (Moretti et al. 2004; Metha et al. 2012).

Lukanka is Balkan Peninsula (Europe) traditional dry-fermented sausage made from pork meat and predominantly fermented using *L. plantarum*. However, other bacteria as *Lactobacillus alimentarius*, *Lactobacillus casei*, *Lactobacillus delbrueckii*, *Lactobacillus farciminis*, *Lactobacillus paraplantarum*, *Lactobacillus pentosus* and *Lactobacillus sharpeae* are also found in the autochthonous microbiota of this product (Talon et al. 2007).

Sucuk is originated from Turkey. This is a dry-fermented sausage made from sheep and/or beef meat or goat meat or buffalo meat. The mixture is then stuffed into the small intestines of cattle and allowed to ferment and dry for several weeks at ambient temperature. Sucuk presents the average composition about 40 % moisture (w/w) and 50 % fat (w/w) with a pH of 5.0. The fermentation occurs spontaneously using the innate microbiota. The main LAB found in the traditional products are *Lactobacillus plantarum*, *Lactobacillus curvatus*, *Lactobacillus fermentum*, *Pediococcus pentosaceus*, *Pediococcus acidilactici* and *Lactobacillus sake* (Kabak and Dobson 2011).

The spoilage of fish is a very rapid process: before refrigeration process, fermentation using LAB increases the palatability and nutritional value, to extend the shelf life. Preservation of fish can be traced back to the ancient Greek civilization (Metha et al. 2012). The natural microbiota present in the fish is salt sensitive, and without salt, the putrefactive microbiota grows fast. Under anaerobic conditions, *Lactobacillus* spp. became the dominant bacteria. The final product may be preserved as the entire fish, such as anchovies, or in the form of pastes like bagoong or in the form of sauces like nuoc-mam (Taira et al. 2007; Oetterer 2014; Chuon et al. 2014).

Bagoong originated from the Philippines and Indonesia. It is a condiment made of partially or completely fermented fish or shrimp and salt (Oetterer 2014). The fermentation process takes several months and is made through the fish indigenous microbiota at 45 °C which results in fish sauce (Taira et al. 2007; Metha et al. 2012).

Nuoc-mam, originated from Southeast Asia, is made from various low-quality fish that are mixed with salt and packed in wooden pots to create an anaerobic condition. The fermentation process takes several months and uses the "back-slopping method" as a starter culture from previous fish sauce (Metha et al. 2012; Oetterer 2014).

3 Bio-preservation and Antimicrobial Peptides

The application of the bio-preservation and antimicrobial compounds has been part of the human life since the rise of the human civilization. We have sufficient evidences to believe that even ancient Egyptians have profited from bio-preservation in preparation of different fermented food products. However, all these application have been related to the empirical knowledge and been transferred from generation to the next generation without any scientific knowledge. Only after the discovery of antibiotics by Fleming and a bit later detection of nisin, various antimicrobial peptides have been subject of intensive research and scientific interests. Almost 100 years later, we can say that antimicrobial peptides can be detected in all life forms, including not only microorganisms but also animals, insects, fishes, birds and plants. Very frequently antimicrobial peptides produced by microorganisms are referred to as "bacteriocins". In last 50 years, bacteriocins have been intensively studied by different research groups working in the area of biochemistry, food science and recently medical-related research groups. According to Cotter et al. (2005), bacteriocins can be produced by almost all bacterial species as a part of the defending molecules. However, special attention has been paid to the bacteriocins produced by LAB, as potential candidates in the preservation of fermented food products (Heng et al. 2007). Several research groups have focused on the inhibitory effect of bacteriocins from LAB against foodborne pathogens and spoilage bacteria (De Vuyst and Vandamme 1994; García et al. 2010). But in the last decade, the research focus in bacteriocins involved in their potential application in the treatment of human and animal diseases and fight against antibiotic resistance problems was increased.

Microorganisms, understanding LAB as well, can produce various antimicrobial compounds, including organic acids (lactic, acetic, formic, propionic acids) that normally intensify their action by reducing the pH of the media. Other substances, like fatty acids, acetoin, hydrogen peroxide, diacetyl, antifungal compounds (propionate, phenyl-lactate, hydroxyphenyl-lactate, cyclic dipeptides and 3-hydroxy fatty acids), bacteriocins (nisin, reuterin, reutericyclin, pediocin, lacticin, enterocin and others) and bacteriocin-like inhibitory substances (BLIS) (Reis et al. 2012), can be actively involved in the antimicrobial action. Thus, bacteriocins are just one specific share of a big range of bio-inhibitor compounds produced by LAB. Another antimicrobial entourage of LAB may include different biopolymers, sugars, sweeteners, nutraceuticals, aromatic compounds and various enzymes, in this way indicating that LAB have higher flexibility and wider application than just as starter cultures. Research for the new antimicrobial compounds is crucial in order to provide an alternative to the chemical additives, in this way offering to the market new and more natural food products. The specific spectrum of activity of bacteriocins against certain emerging foodborne pathogens and spoilage microorganisms, their resistance to thermal proceeding and low pH, combined with their sensitivity to human proteolytic enzymes, are important positive characteristics in the application of these compounds in food preservation (Masuda et al. 2011).

4 Bacteriocins

By definition, bacteriocins are ribosomally synthesized antimicrobial proteins (polypeptide or small proteins), usually active against genetically related species (Cotter et al. 2005). In the last decade, based on the intensive research in the area of bacteriocins, we have sufficient examples for bacteriocins that may have application in controlling Gram-negative bacteria, some yeast, *Mycobacterium* spp. and even viruses (Todorov et al. 2010a, b; Schirru et al. 2012). However, amino acid sequences of only few of these unusual bacteriocins are provided (Todorov et al. 2010b). These reports need to be carefully screened, since only a few work on explaining the mechanisms of mode of action for this "unusual" bacteriocinogenic activity.

Many classifications of bacteriocins have already been proposed (Cotter et al. 2005), but according to the most recent (Heng et al. 2007), bacteriocins of Gram-positive bacteria are grouped into four classes, based on their structure and function: class I, lantibiotic peptides; class II, small non-modified peptides with <10 kDa; class III, large proteins with >10 kDa; and class IV, cyclic proteins.

Class I bacteriocins are subdivided in three subgroups: type A corresponds to linear peptides, type B to globular peptides and type C to multicomponent bacteriocins. Type A bacteriocins are further divided in two subtypes: subtype AI comprises nisin-like bacteriocins (including various natural mutations of nisin, streptin) and subtype AII comprises SA-FF22-like bacteriocins (SA-FF22, lacticin 481, salvaricin A, sublancin 168). Mersacidin and cinnamycin, for instance, belong to the class I type B globular peptides. Class I type C examples are lacticin 3147 and cytolysin, both being formed by more than one component, all necessary for biological activity (Favaro et al. 2015).

Class II bacteriocins are also subdivided in three subgroups: type IIa corresponds to pediocin-like bacteriocins (pediocin Pa-1, carnobacteriocin B2, listerocin 743A and ubericin A), type IIb are multicomponent bacteriocins (lactococcin G, thermophilin 13, lactacin F and lactocin 705) and type IIc are miscellaneous bacteriocins, a diverse group that includes sakacins Q, T and X and aureocin A53 (Favaro et al. 2015).

Class III bacteriocins include lysins (class IIIa) and non-lytic bacteriocins (class IIIb). Class IV bacteriocins include circularly inhibitory peptides and the prototype is enterocin AS-48 (Abriouel et al. 2005; Favaro et al. 2015).

5 The Short Story of Nisin

One of the important challenges faced by food industry is to preserve the qualities (nutritive and sensorial) of the raw material and to provide safe and spoilage-free and pathogenic bacteria-free products to the consumers. However, in last few decades, consumers are requesting not only safe but also healthier food. More and more frequently, chemical preservatives and additives are rejected by consumers, and in this way, the use of bacteriocins is a promising alternative in reducing the use and adding of chemical preservatives in foods. In addition, the combined application of bacteriocins, essential oils and other natural antibacterial metabolites could result in food products which are more naturally preserved and have better sensorial and nutritional characteristics (Reis et al. 2012).

According to www.scopus.com, till end of 2014, more than 6200 scientific papers have been published dealing with bacteriocins. The biggest parts of these studies are related to bacteriocins from LAB, but even very extensively studied, their usage as food preservatives is still very limited, related to several technological or legislation barriers. A high number of bacteriocins may have potential applications in food bio-preservation, but they are not currently approved as antimicrobial food additives (Favaro et al. 2015).

One of the most studied bacteriocins by numerous research groups are nisin and pediocin PA-1, and both of them have commercial applications in the food industry and in veterinary and human medicine (Cotter et al. 2005). The commercial application of nisin dates back to 1953, when it was marketed in England. Nowadays, nisin has been approved in more than 50 countries worldwide. According to European Union food safety regulation authorities, nisin is licensed as a food preservative (E234), and it is recognized as a safe additive in food by the Joint Food and Agriculture Organization/World Health Organization (FAO/WHO) Expert Committee on Food Additives in 1969 (Favaro et al. 2015). Applications of nisin as a natural food preservative have been explored in dairy products and especially in processed cheese. According to FAO/WHO Codex Committee on Milk and Milk Products, nisin is allowed as a food additive for processed cheese products at a maximal concentration of 12.5 mg (applied as pure nisin) per kilogram of product (Reis et al. 2012). However, national legislations may differ concerning the levels of nisin in various food products. In the United Kingdom, the application of nisin can be added to cheese without limit, while in Spain, Brazil, Argentina, Italy and Mexico, 12.5 mg/kg or 500 IU/g is a maximum concentration that is allowed (Cleveland et al. 2001; Sobrino-López and Martín-Belloso 2008). In the United States, a much higher level of 10,000 IU/g is permitted (Cleveland et al. 2001). The application of nisin is not facing serious safety issues, since nisin is not toxic with cytotoxicity similar to that of NaCl (Jozala et al. 2007), is sensitive to digestive proteases and does not influence sensory properties of the food products (Pongtharangkul and Demirci 2004).

Nisin serves as a model in explaining the mode of action for most of bacteriocins. According to Wiedemann et al. (2001), lipid II (the main transporter of peptidoglycan subunits from the cytoplasm to the cell wall) acts as a docking station (receptor) for the adhesion of nisin. After interaction with lipid II, nisin wedges itself into the cell membrane to form short-lived pores, which disturb the integrity of the cytoplasmic membrane and cause the efflux of intracellular components (Wiedemann et al. 2001). However, at high concentrations of nisin, pore formation may occur even when lipid II is not presented on cell surface or been blocked by other molecules. In this case, nisin can interact with highly presented negatively charged phospholipids on cell membrane (Wiedemann et al. 2001). The positively charged C-terminus of nisin is important for initial adhesion to the binding to negatively charged phospholipids. Similar mode of action was described for mersacidin and the antibiotic vancomycin, which can also bind to lipid II, but to a different part of the molecule (Cotter et al. 2005).

6 Application of Bacteriocins in Bio-preservation

The application of bacteriocin(s) produced by LAB can be as such "as simple", as applying the producer strains as a part of the starter culture in case that this strains can have well-proven technological properties, been part of the starter cultures mixture or been applied as an adjustment culture. However, when we apply/introduce bacteriocin producers into the food matrix, it is very important to pay attention to the physiological and biochemical characteristics of the LAB. LAB are normally microorganisms that can ferment glucose and several other carbohydrates, and they can grow at temperatures from around 15 °C to 45 °C and in the presence or absence of oxygen (Holt et al. 1994). The problem is that very frequently fermentation time in the preparation of certain fermented food products is very short, and after that, products are kept at refrigeration condition. In this situation, bacteriocin producers are not able to produce and express the bacteriocin(s), based on the very simple reason that they are in "not growing" condition. Most probably, LAB will survive at the refrigeration and freezing conditions, but definitely we cannot count on the production of bacteriocin(s) in order to have bio-preservation characteristics (Favaro et al. 2015). Another potential problem is the fact that very frequent bacteriocin(s) production can be dependent on the needs of the specific nutrients. There are several research papers pointing on the effect of the type of medium and medium components themselves on the production of bacteriocin (s) (Parente and Ricciardi 1994; Parente et al. 1994; Todorov 2008; Furtado et al. 2014; Schirru et al. 2014). Very frequently bacteriocin production can be detected in liquid growth medium and not in solid medium or opposite. Bacteriocin (s) can be produced and expressed in laboratory commercial mediums, but not in the milk or meat substrates (Todorov 2008; Furtado et al. 2014; Schirru et al. 2014). Bacteriocin production can be regulated by the level of pH, temperature and presence of different proteins and be a part of quorum senses (Parente and Ricciardi 1994; Parente et al. 1994; Cotter et al. 2005; Von Mollendorff et al. 2009; Todorov et al. 2010a, b; Favaro et al. 2015). If the fermentation process is very short, substrate environments, temperature or pH and various other factors will be not sufficient or good for the production and expression of bacteriocins, and then another option will be to apply the semi-purified or in some cases pure bacteriocin(s). In this scenario, we need to be aware of the biochemical characteristics of the bacteriocin and to be sure that they will be applicable for our specific way of application. Most of the bacteriocins are stable and express their antimicrobial spectrum of activity in neutral or low pH, and they are thermostable and are not affected by several surfactants and chemicals (Biscola et al. 2014; Paula et al. 2014; Ribeiro et al. 2014). However, the presence of proteolytic enzymes and high lipids composition can be a factor reducing the effect of bacteriocin(s) in the bio-preservation processes (Cotter et al. 2005; Todorov et al. 2007b; Favaro et al. 2015).

The application of bacteriocinogenic strains or bacteriocins themselves can be a way to control spoilage and even pathogenic bacteria (and even some other microorganism groups), but some innovative ideas need to be applied in order to have better effect.

7 Bacteriocin Produced by *Lactobacillus plantarum* 423: A Multiple Application of the Same Peptide

Lactobacillus plantarum 423 is a strain isolated from sorghum beer, identified as a bacteriocin producer (Van Reenen et al. 1998); its bacteriocin production has been optimized (Verellen et al. 1998; Todorov et al. 2007c), its mode of action has been investigated (Todorov and Dicks 2006a), and its probiotic potential was characterized (Brink et al. 2003). Previously, Van Reenen et al. (1998; 2003) reported that plantaricin 423, produced by *Lactobacillus plantarum* 423, is a small heat-stable antimicrobial protein and belongs to class IIa (anti-listerial antimicrobial peptide). Plantaricin 423 shares a high similarity to pediocin PA-1 and coagulin operons; in fact *pla*C and *pla*D genes are identical to *ped*C and *ped*D of the pediocin PA-1 operon, as well as coaC and coaD of the coagulin operon (Van Reenen et al. 2003). In study of Van Reenen et al. (2003), plantaricin 423 was successfully cloned on a shuttle vector under the control of a yeast promoter and heterologously produced by *Saccharomyces cerevisiae*. However, plantaricin 423 production was recorded in the recombinant *Saccharomyces cerevisiae* cells by overlay approach only in 2-, 3- and 4-day-old colonies using soft agar containing *Listeria monocytogenes* cells as target microorganisms. Based on this academic exercise, Van Reenen et al. (2003) suggested that it will be possible to construct GMO, in this case yeast with application in food industry with bactericidal properties. However, similar work has been previously reported by Schoeman et al. (1999) related to pediocin PA-1.

The problem in this study was that no inhibitory activity was detected in the supernatant of the recombinant yeast without concentration (Van Reenen et al. 2003). Schoeman et al. (1999) also faced similar problem and detected inhibition zones around colonies but found very low bacteriocin activity in concentrated supernatant and speculated that the bacteriocin remained cell wall associated.

Even if the previous work were just academic exercises and not practically applied, based on the low levels of expression of the produced bacteriocins, these studies present an opportunity for the application of biological control of food spoilage microorganisms during food fermentation processes, including wine production. In addition, the usage of GMO must be considered as a concern, once international safety requirements determine a number of specifications in using these organisms in foods (there are several ethical and technological limitations for the use of GMO).

In a different study, the same bacteriocinogenic strain of *Lactobacillus plantarum* 423 (Van Reenen et al. 2003) has been used in bio-preservation of salami produced by different types of meat: beef, horse, mutton, blesbok (*Damaliscus dorcas phillipsi*) and springbok (*Antidorcas marsupialis*). In parallel, *Lactobacillus curvatus* DF38, producer of curvacin DF38, and a bacteriocin-negative mutant of *Lactobacillus plantarum* 423 (423m) were evaluated as starter cultures in the production of salami (Todorov et al. 2007b). According to the results, the best growth of *Lactobacillus plantarum* 423 and *Lactobacillus curvatus* DF38 and the highest level of bacteriocin production were recorded in blesbok salami. *Listeria innocua* was successfully inhibited in the experimentally contaminated salami. However, no inhibition of *Listeria innocua* was recorded in horse salami, suggesting that the meat contained compounds that inhibited bacteriocin activity (Todorov et al. 2007b). This observation is in agreement with a finding pointed already in the same study that the bacteriocin activity can be interfered by several factors from temperature, pH to the type and composition of the food matrix. Negative point in this study is that they did not investigate the influence of other antimicrobial substances that may be involved in the inhibition of *Listeria innocua*. For example, it was reported that *Lactobacillus rhamnosus* LC-705 produces 2-pyrrolidone-5-carboxylic acid, which inhibits the growth of a number of bacteria (Huttunen et al. 1995; Mäyrä-Mäkinen and Suomalainen 1995, 1996; Yang et al. 1997). We need to acknowledge that most probably the final inhibitory effect on *Listeria innocua* resulted from the complex action of all antimicrobial metabolites potentially produced by LAB. However, differences in the observed inhibitory effect of salami prepared with *Lactobacillus plantarum* 423 (bacteriocin producer) and *Lactobacillus plantarum* 423m (bacteriocin gene-cured mutant) are good arguments to address part of the reduction of *Listeria innocua* as a result of the produced and expressed bacteriocin (plantaricin 423).

An important point in the application of LAB or inhibitor substances produced by LAB is that it is not going to interfere with the organoleptic characteristics of the final products. In other words, the consumers need to be satisfied with the quality of the final products. In an experiment where *Lactobacillus plantarum* 423 was used in the preparation of salami from beef, horse, mutton, blesbok and springbok, the colour of salami did not differ statistically from the controls; the end pH of salami ranged from 4.4 to 4.7. No significant differences were recorded among the three starter cultures regarding colour and venison-like aroma (Todorov et al. 2007b).

Maybe one of the most studied food products for a bio-preservation was dairy products. Several examples for dairy bio-preservation, including discussion of the potential problems, were presented by Favaro et al. (2015). Most probably dairy products gained this special attention based on the fact that they are one of the more popular food products between consumers from all ages; very frequently, they are used a vehicles for the delivery of probiotic bacteria, and there are numerous reported applications of nisin as a natural food preservative especially for dairy products and processed cheese.

Based on our previous experiment working with the application of LAB and bacteriocin(s) in the preservation of cheeses (Pingitore et al. 2012; Furtado et al. 2015), we can conclude that in most of the experiments, the inhibitory effect needs to be addressed to the combined effect of bacteriocin production and expression and the changes in the pH, water activity and interaction between LAB and *Listeria monocytogenes*.

Furtado et al. (2015) evaluated the application of bacteriocinogenic LAB (*Lactococcus lactis* subsp. *lactis*) with proven in vitro anti-*Listeria monocytogenes* activity (Furtado et al. 2014) to be an innovative technological approach for the control of spoilages in fresh cheeses. Furtado et al. (2015) observed a reduction of *Listeria monocytogenes* cell counts after 10 days that is around 3 log units lower than in control cheeses with no added LAB. The problem faced in this study was that this effect did not differ significantly from that obtained with a non-bacteriocinogenic *Lactococcus lactis* strain. When nisin (12.5 mg/kg) was added, a rapid decrease in the number of viable *Listeria monocytogenes* in the cheeses was recorded. As previously mentioned, we can conclude that the inhibition of *Listeria monocytogenes* is a more complex process. However, in another study by Furtado et al. (2014), the probiotic potential of *Lactococcus lactis* DF04Mi was investigated. In relation to the latter, creating a cheese product containing a bio-preservative culture with probiotic properties can be considered as an innovative idea with potential beneficial and health characteristics.

However, we need to acknowledge that not all bacteriocinogenic strains can be considered as ideal bio-preservatives for every type of products. It is well recorded that nisin works well in several dairy products, but in relation to the high lipid content, nisin is not well adapted for application in meat products (Favaro et al. 2015). An important point is the level of produced bacteriocin and if the technological condition of the production of the food products will support the production and expression of these bacteriocins in the sufficient level to perform their mode of action against pathogenic bacteria. Another potential problem with hard phase fermented products is that bacteriocin can be produced in one part of the product, while the pathogenic bacteria can be at the other part. Very frequently postproduction contaminations are located on the surface of the products.

In the study conducted by Pingitore et al. (2012), the first step was focused on comparing the in vitro characteristics of two bacteriocinogenic strains isolated from cheeses—*Enterococcus mundtii* CRL35 and *Enterococcus faecium* ST88Ch, including the level of bacteriocin production, effect to different *Listeria monocytogenes* strains, adsorption of bacteriocin to the pathogen and interference of the media components to this process. In the second part of the study, Pingitore et al. (2012) evaluated *Enterococcus mundtii* CRL35 and *Enterococcus faecium* ST88Ch in situ for their capability to control growth of *Listeria monocytogenes* in experimentally contaminated fresh Minas cheese during refrigerated storage. The growth of *Listeria monocytogenes* was inhibited in a more significant way in cheeses containing *Enterococcus mundtii* CRL35 up to 12 days. In comparison, cheese prepared with bioprotective culture *Enterococcus faecium* ST88Ch was less effective in controlling the growth of *Listeria monocytogenes*, as the bacteriostatic effect occurred only after 6 days but with the similar inhibitory profile as non-bacteriocinogenic strain of *Enterococcus faecalis* ATCC 19443. Based on the in vitro tests reported by Pingitore et al. (2012) about the bio-preservation potential of *Enterococcus mundtii* CRL35 and *Enterococcus faecium* ST88Ch, it was clearly shown that *Enterococcus mundtii* CRL35 has a better potential as compared with *Enterococcus faecium* ST88Ch. In the same study, in cheeses containing nisin (12.5 mg/kg), less than one log reduction was observed. This research underlines the potential application of *Enterococcus mundtii* CRL35 in the control of *Listeria monocytogenes* in fresh Minas cheeses (Pingitore et al. 2012).

One of the intensively studied product regarding options for the control of *Listeria monocytogenes* contaminations is smoked salmon. Every year, large number of products from salmon industry was discarded because of microbial contaminations. Different options were reported in the area of bio-preservation and food safety: protamies (Johansen et al. 1996), monoglycerides (Wang and Johnsonm 1997) and bacteriocins (Leisner et al. 1995; Stecchini et al. 1995; Goff et al. 1996; McMuller and Stiles 1996; Rodriguez et al. 1997; Duffes 1999; Duffes et al. 1999, 2000; Vaz-Velho et al. 2005).

The application of *Carnobacterium divergens* V41 and its bacteriocin was used to control *Listeria innocua* during salmon-trout cold smoking process. Culture of *Carnobacterium divergens* V41 and bacteriocin-containing cell-free supernatant were applied on raw fish fillets (Vaz-Velho et al. 2005). Fillets were artificially contaminated with equal level (2 %, v/v) of *Listeria innocua* and *Carnobacterium divergens* or treated with 2 % or 5 % of bacteriocin-containing cell-free supernatant. Fillets were subject on treatment simulating industrial procedure for the preparation of smoked salmon including cold smoking and stored for 3 weeks in vacuum packs. Both treatments, with bacterial culture or bacteriocin-containing cell-free supernatant, showed significant inhibition of *Listeria innocua* in contaminated samples compared with the controls (Vaz-Velho et al. 2005). An important point of this work is that based on the sensorial test performed on fillets treated with culture of *Carnobacterium divergens* or bacteriocin-containing cell-free supernatant, trained team of panelists cannot detect any specific differences in the product.

The bioprotective properties of LAB strains isolated from fermented meat products from Portugal (Todorov et al. 2013) have been investigated as potential bio-preservative cultures in the preparation of spray on *chouriço*, produced at industrial scale and packed under modified atmosphere packaging (MAP) (8 % or 12 % v/v CO_2) and stored at 5 °C. In addition, bacteriocinogenic strains have been tested related to their safety properties (Todorov et al. 2014). Jacome et al. (2014) reported on the effect of two different LAB cultures (*Lactobacillus sakei* ST153 (Todorov et al. 2013) and BLC35 (presenting commercial mixed starter, including strains of *Lactobacillus curvatus*, *Staphylococcus xylosus* and *Pediococcus acidilactici*; CHR Hansen). In this study, two different methods of application and two different MAP on potential growth control of *Listeria monocytogenes* and on sensory properties of *chouriço* were studied. *Lactobacillus sakei* ST153 and BLC35, both with previously shown bacteriocinogenic activity against *Listeria monocytogenes*, were applied by immersion or spray on smoked pork slices, packed under MAP (8 % or 12 % v/v CO_2) and stored at 5 °C. Meat products were contaminated by *Listeria monocytogenes* using a sterile cotton swab. The microbiological control was performed during the storage period, together with a quantitative descriptive sensory test, performed by a sensory trained panel at 30, 90 and 120 days. Special attention has been given to meat colour, greasiness, characteristic odour, off odour, hardness, succulence, characteristic taste, acid taste and bitter taste. Jacome et al. (2014) reported the decrease of *Listeria monocytogenes* to values <100 CFU/g (below detection level). The application of used bio-preservation method had no significant effect on any of the analysed sensorial attributes. However, authors reported that BLC35 used as bio-preservative culture resulted in a less succulent *chouriço* when compared to the product produced with *Lactobacillus sakei* ST153 (Jacome et al. 2014).

Microbiological study of boza showed that this product is a very rich source of LAB (Todorov and Dicks 2006b; Todorov 2010; LeBlanc and Todorov 2011; Todorov and Holzapfel 2015) and some of them can be considered as strong bacteriocin producers (Todorov and Dicks 2006b; Todorov 2010; LeBlanc and Todorov 2011; Todorov and Holzapfel 2015) and potential probiotics (LeBlanc and Todorov 2011; Todorov et al. 2008). In addition, it has been shown that some of the expressed bacteriocins can inhibit several foodborne and human pathogens and reduce the growth and viability of *Mycobacterium tuberculosis* and some viruses (LeBlanc and Todorov 2011; Todorov et al. 2008). Even more, cultures of some LAB isolated from boza can be responsible for inhibiting the growth of some fungus based on cell to cell interaction.

In a different experiment, Todorov et al. (2009) investigated the possibilities to apply bacteriocinogenic strain of *Enterococcus mundtii* ST4SA. *Enterococcus mundtii* ST4SA was isolated from soybeans, and its bacteriocin has been described as an active substance against several foodborne and human pathogens, including pathogens isolated from human middle ear and some viruses (Todorov et al. 2005). At the same time, this strain was evaluated for its probiotic potential (Botes et al. 2008). *Enterococcus mundtii* ST4SA was applied as a potential bioprotective culture for control of contaminants during the production of boza (Todorov et al. 2009). The production process was simulated according to the recipe for the production of boza with difference that *Enterococcus mundtii* ST4SA was added to the product in combination with *Lactobacillus sakei* as a sensitive strain, representing possible contamination. Dynamics of the bacterial population was monitored by classical microbiological approach and by biomolecular methods, including DGGE and species-specific PCR taking into consideration to not target DNA from dead cells. Results supported the previously raised hypothesis for possible control of spoilage bacteria by applying the bacteriocinogenic culture (Todorov et al. 2009). Botes et al. (2008) demonstrated the probiotic potential for *Enterococcus mundtii* ST4SA, and combined with the fact that this strain can survive well in boza and can be very effective in the control of spoilage bacteria, it opened new ways for the application of bacteriocinogenic and probiotic strains in food industry. In addition, sensorial analysis of boza prepared with *Enterococcus mundtii* ST4SA as starter culture showed no differences from the boza prepared using the traditional way (Todorov et al. 2009).

Antimicrobial properties of kefir have been intensively studied. This activity can be addressed to the presence of the LAB and yeast in the kefir (Powell et al. 2006, 2007). It has been shown that kefir possesses antimicrobial activity against various bacteria, including some pathogens as *Staphylococcus*, *Clostridium*, *Salmonella* and *Listeria*, and even can reduce growth of *Mycobacterium tuberculosis* (Rodrigues et al. 2005; van Wyk et al. 2011; de Oliveira et al. 2013; Ahmed et al. 2013; Miao et al. 2014).

Lactobacillus plantarum ST8KF, a strain isolated from kefir grains, has been characterized as a bacteriocin producer (Powell et al. 2007). This strain produces a 3.5 kDa bacteriocin with activity against *Enterococcus mundtii*. The experimental design proposed by Powell et al. (2006) was to reincorporate this strain (*Lactobacillus plantarum* ST8KF and its plasmid-cured and bacteriocin-negative variant *Lactobacillus plantarum* ST8KF^{-}) in kefir grains and to follow the growth of *Enterococcus mundtii* using fluorescence in situ hybridization (FISH) technique. Kefir produced with grains containing *Lactobacillus plantarum* ST8KF prevented the growth of *Enterococcus mundtii* in situ. However, no evidences of inhibition of *Enterococcus mundtii* were recorded when kefir was produced from grains containing a bacteriocin-negative variant (ST8KF^{-}) of the previously described strain (Powell et al. 2006). It is important to underline that authors used appropriate controls in order to estimate the effect of bacteriocin, but we need to acknowledge the possible effect in the inhibition of *Enterococcus mundtii* as an effect of complex various inhibitors produced during fermentation of kefir.

8 Application of LAB in the Control of Moulds

It is important to point out the antagonistic effect of some LAB against fungi. This application of LAB is well known from the time of empirical application and knowledge, but even nowadays, it is still uncommonly applied. The control of filamentous moulds and yeast, common spoilage organisms of food products, is important since they may also produce health-damaging mycotoxins (Legan 1993). However, antifungal activities were frequently observed only as an effect of interactions between LAB and tested yeast or moulds and not as an effect of bacteriocins produced by LAB on tested yeast or moulds (Schnürer and Magnusson 2005; Smaoui et al. 2010; Stoyanova et al. 2010; Todorov 2010; Belguesmia et al. 2012, 2014). In food industry, the addition of propionic acid and its salts, modified atmosphere packaging, irradiation and pasteurization are widely applied in order to minimize fungal spoilage of fermented food products (Legan 1993). More interesting from scientific and ecological points of view is bio-preservation (the control of one organism by another). LAB are one of the extensively studied potential bio-preservatives, since a big part of them has a well-accepted GRAS status. Most of the LAB are producers of the organic acids that have antifungal activity (Röcken and Voysey 1995; Röcken 1996; Stiles 1996) as well as a wide range of low molecular weight compounds (Niku-Paavola et al. 1999), peptides (Okkers et al. 1999) and proteins (Magnusson and Schnürer 2001) with antifungal activity. Ström et al. (2002) described the antifungal cyclic dipeptides: cyclo(L-Phe-L-Pro) and cyclo(L-Phe-trans-4-OH-L-Pro) produced by *Lactobacillus plantarum* MiLAB 393 (Ström et al. 2002). *Lactobacillus plantarum* MiLAB 393 has been described as a potential strain for the control of *Aspergillus nidulans*. Cyclic dipeptides have been previously explored as potential antibacterial and antifungal inhibitors (Gratz et al. 2004) and most probably these substances, previously related only to the metabolism of strains of *Lactobacillus plantarum* (Lindgren and Dobrogosz 1990), can be also produced by other LAB, e.g. *Pediococcus pentosaceus* and *Lactobacillus sakei* (Magnusson et al. 2003).

In literature, only a limited number of reports related to the LAB have been presented as presenting activity against yeast and moulds: *Lactobacillus coryniformis* subsp. *coryniformis* Si3 (Magnusson and Schnürer 2001), *Lactobacillus plantarum* TN635 (Smaoui et al. 2010), *Lactococcus lactis* subsp. *lactis* (Stoyanova et al. 2010), *Lactobacillus plantarum* ST69BZ and *Leuconostoc lactis* ST612BZ (Todorov 2010), *Lactobacillus harbinensis* K.V9.3.1Np (Belguesmia et al. 2014) and *Enterococcus durans* A5-11 (Belguesmia et al. 2012). In addition, Schnürer and Magnusson (2005) suggested that the mechanism of antimicrobial action is difficult to elucidate due to complex and commonly synergistic interactions between different compounds.

One of the biggest concerns of the presence of food spoilage mould is the occurrence of their mycotoxins. Mycotoxins constitute a serious health hazard, and they are related with a high lost of food products. Due to this, it is important to develop biological control strategy that should help to improve the safety of products by controlling mycotoxin producers or mycotoxin contaminations. We have sufficient scientific data to believe that many LAB can inhibit mould growth and that some of them have the potential to interact with mycotoxins and contribute to their deactivation.

The actions of the antifungal properties of LAB on some mycotoxin-producing moulds have been extensively studied but have been reported only by a few authors (Dalié et al. 2010). According to Dalié et al. (2010), the main LAB recognized for their ability to prevent or limit mycotoxin-producing mould growth belong to the genera *Lactococcus* and *Lactobacillus*, as well as *Pediococcus* and *Leuconostoc* (Dalié et al. 2010). In study reported by Roy et al. (1996), a collection of 2100 isolates of LAB was screened for the production of substances actively inhibiting several types of moulds. However, only six isolates were identified for their antifungal activity against *Aspergillus flavus*, and only one of them (*Lactococcus lactis* subsp. *lactis* CHD 28.3) showed a broad spectrum of antifungal activity against *Aspergillus flavus*, *Aspergillus parasiticus* and *Fusarium* spp. However, *Aspergillus* was the most sensitive indicator of the antifungal metabolite produced by this lactic strain. Some other *Lactococcus* spp. strains [*Lactococcus lactis* (Wiseman and Marth 1981; Coallier-Ascah and Idziak 1985; Luchese and Harrigan 1990), *Lactococcus lactis* subsp. *diacetylactis* DRCI (Batish et al. 1989), *Lactococcus lactis* subsp. *cremoris* (Florianowicz 2001)] were reported as potential candidates for the control of mycotoxinogenic mould growth.

Strains from the genus *Lactobacillus* were frequently described as possessing antifungal activity. The antifungal strains have been isolated from different environments such as sourdough (Corsetti et al. 1998; Hassan and Bullerman 2008), grass silage (Magnusson and Schnürer 2001; Magnusson et al. 2003) and vegetable products (Sathe et al. 2007). A cell-free supernatant of *Lactobacillus plantarum* 21B isolated from sourdough and grown in wheat flour hydrolysate was shown to possess an efficient antifungal activity against *Penicillium* spp., *Aspergillus* app. and *Fusarium* spp. (Lavermicocca et al. 2000). Lavermicocca et al. (2000) demonstrated that part of the antifungal activity of *Lactobacillus plantarum* 21B can be addressed to the production of phenyllactic and 4-hydroxy-phenyllactic acids. Niku-Paavola et al. (1999) described the ability of *Lactobacillus plantarum* VTTE-78076 to suppress the growth of *Fusarium* spp. In this case, antifungal activity was described to the occurrence of effect of benzoic acid, an imidazolidinedione derivative and a piperazinedione derivative.

Lactobacillus coryniformis subsp. *coryniformis* Si3, isolated from grass silage, was able to inhibit the growth of a great number of mycotoxinogenic moulds including *Aspergillus* spp., *Penicillium* spp. and *Fusarium* spp. (Magnusson and Schnürer 2001). However, in liquid medium, the production of antifungal metabolites by *Lactobacillus coryniformis* subsp. *coryniformis* Si3 was shown to be a growth phase-dependent process, and after a partial purification, the molecular mass of the potent antifungal compound was estimated to be close to 3 kDa and to be heat stable, sensitive to proteolytic enzymes and active within a narrow pH range. It is interesting to underline that these characteristics are in accordance with those of subclass II bacteriocins (Klaenhammer 1993).

Other species of *Lactobacillus*, including *Lactobacillus casei* (Gourama 1997; Suzuki et al. 1991; Mäyrä-Mäkinen et al. 1994; Florianowicz 2001), *Lactobacillus sanfrancisco* (Corsetti et al. 1998) and *Lactobacillus rhamnosus* (Stiles et al. 2002), have also been described as potential producers of inhibitors of toxinogenic mould growth. Moreover, some papers have reported the ability of the genus *Pediococcus* to control growth of mycotoxinogenic mould (Effat et al. 2001; Mandal et al. 2007; Rouse et al. 2008). In study of Mandal et al. (2007), antifungal lactic strain, identified as *Pediococcus acidilactici* LAB 5, was isolated from vacuum-packed fermented meat and has been shown to exhibit varying degrees of antifungal activity against *Aspergillus fumigatus*, *Aspergillus parasiticus*, *Fusarium oxysporum* and *Penicillium* sp.

9 Food or Medicine?

Bacteriocins have been of interest to the medical industry, based on simple fact that they are produced by nonpathogenic bacteria, most of them with GRAS status, that are normally present in the human gastrointestinal tract (GIT) and several fermented food products. In the last decades, bacteriocins have also been suggested to be candidates in the cancer treatment (Farkas-Himsley and Yu 1985; Baumal et al. 1982). Some promising studies have been suggesting them as diagnostic agents for some cancers (Farkas-Himsley et al. 1995; Musclow et al. 1987; Saito et al. 1979; Cruz-Chamorro et al. 2006; Sand et al. 2007); however, most of these studies still remain experimental and outside of the main thread of cancer research. Most probably, this is because of several questions about their mechanism of action and, by definition, the presumption that antibacterial agents (including bacteriocins) have no obvious connection to killing or inhibiting mammalian tumor cells.

In the last decades, some bacteriocins were tested as potential AIDS drugs, but the studies did not progress beyond in vitro tests on cell lines (Farkas-Himsley et al. 1991). However, antiviral activity of bacteriocins has been studied by several research groups and shows promising results as a potential treatment in some cases as a single or accompanying therapy (Todorov et al. 2005, 2010b; Wachsman et al. 1999; Kassaa et al. 2014). Wachsman et al. (2003) studied the inhibition of replication of herpes simplex virus (HSV) types 1 and 2 in model of Vero cells in the presence of enterocin CRL35, a class IIa bacteriocin produced by *Enterococcus faecium* CRL35 previously described by Farias et al. (1996). It was interesting to observe that virus adsorption and penetration are not affected by the presence of enterocin CRL35. However, a late step of virus multiplication is slowed down when 100 μg/ml enterocin CRL35 was added at 8 h postinfection, and it caused a 90 % inhibition of virus release (Wachsman et al. 2003). The effect of enterocin CRL35 on HSV antigen expression was determined by immunofluorescence using a polyclonal serum and a monoclonal antibody against glycoprotein D (γ protein). Wachsman et al. (2003) indicated that enterocin CRL35 impeded the second round of infection, apparently because of the inhibition of glycoprotein D expression. Studies on the effect of enterocin CRL35 on viral protein synthesis showed that in the presence of enterocin CRL35, HSV late γ proteins were not synthesized. Authors suggested that inhibition of HSV spreading by enterocin CRL35 is due to the prevention of mainly late glycoprotein synthesis (Wachsman et al. 2003).

Férir et al. (2013) reported on labyrinthopeptin A1 (LabyA1), a prototype peptide of a novel class of carbacyclic lantibiotics. Authors evaluated its broad-spectrum activity against HIV and HSV in vitro and studied its mechanism of action. LabyA1 also demonstrated additive to synergistic effects in its anti-HIV-1 and anti-HSV-2 activity with anti(retro)viral drugs in dual combinations such as tenofovir, acyclovir, saquinavir, raltegravir and enfuvirtide (Férir et al. 2013). LabyA1 also did not affect the growth of vaginal *Lactobacilli* populations (Férir et al. 2013). Based on the lack of toxicity on the vaginal *Lactobacillus* strains and its synergistic/additive profile in combination with clinically approved anti(retro) virals, this lantibiotic has a potential and deserves further attention as a candidate in the prevention of sexual transmitted diseases (Férir et al. 2013).

Montalbán-López et al. (2011) reviewed possible applications of the bacteriocins in therapeutical practice in human and veterinary medicine in addition to earlier review focusing only on two peptides lantibiotics (part of the big family of bacteriocins) in the medical practices by Lawton et al. (2007). One of the best known lantibiotic is a nisin; however, in the last decade, special attentions have been granted to numerous two-peptide lantibiotics, i.e. lantibiotics that function optimally as a consequence of the synergistic activity of two peptides (Lawton et al. 2007). Based on the study of the genetic determinants and protein structure, lacticin 3147, staphylococcin C55, plantaricin W, Smb, BHT-A and haloduracin have been shown to be closely related. These antimicrobial peptides are extremely potent tools in nanomolar concentrations with activity against a number of microorganisms, including activity against multidrug-resistant nosocomial pathogens such as methicillin-resistant *Staphylococcus aureus* (MRSA) and vancomycin-resistant enterococcus (VRE). And even more important, to date, the development of significant levels of resistance has not been apparent. Based on their study on the application of lacticin 3147 and previously mentioned related two-peptide lantibiotics, Lawton et al. (2007) suggested a possible potential medical and veterinary application for these bacteriocins.

Nisin, one of the first and most extensively studied lantibiotics, has a long history of application in the food industry, and it has been approved for use as a food preservative/additive in over 50 countries worldwide (Delves-Broughton et al. 1996). In addition, nisin also showed a promising application in the control of clinically significant bacteria such as methicillin-resistant *Staphylococcus aureus* (MRSA), vancomycin-resistant *Enterococcus* (VRE), *Clostridium difficile* and *Helicobacter pylori* (Lawton et al. 2007). Also, veterinary applications of nisin based on its activity against bovine mastitis-related pathogens have been proposed (Broadbent et al. 1989). Nisin has been applied as an effective contraceptive in rat, rabbit and monkey models, but it also has the advantage of being an antimicrobial and may also help in the prevention of the spread of sexually transmitted diseases (Lawton et al. 2007).

In addition to nisin, the application of several other bacteriocins has been described as an effective treatment against clinical bacterial pathogens. Mersacidin produced by *Bacillus subtilis* has attracted significant interest due to its wide spectrum of activity, including activity against MRSA (Sahl et al. 1995). Its efficacy against MRSA using a murine model has been investigated (Kruszewska et al. 2004).

Epidermin, a lantibiotic produced by *Staphylococcus epidermidis* Tü3298, and a number of natural variants of epidermin have been reported (Lawton et al. 2007). Gallidermin, produced by *Staphylococcus gallinarum*, differs from epidermin by a single amino acid and was observed to be more effective against Gram-positive pathogens such as *Staphylococcus aureus* and *Propionibacterium acne* (Kellner et al. 1988). Nascimento et al. (2006) proposed the application of epidermin against a range of MRSA strains including preventing coagulase-negative *Staphylococci*

(CNS) adhering to catheters. Mutacin 1140, an epidermin family peptide, is a licensed and patented antimicrobial peptide produced by *Streptococcus mutans*. Mutacin 1140 was a very potent agent with activity against clinically isolated Gram-positive bacteria and even some clinically significant Gram-negative strains. This lantibiotic has been applied in the control of pathogenic bacterial strains, such as those responsible for gastric ulcers, pneumonia, listeriosis and "strep" throat (Lawton et al. 2007).

Salivaricin A2 and salivaricin B, produced by *Streptococcus salivarius* K12, were applied in the control of oral bacteria responsible for bad breath (Wescombe et al. 2006). In addition, two salvaricin A-producing strains (*Streptococcus salivarius* 20P3 and 5) were incorporated into a milk product for children, and it has been indicated that these probiotics strains were capable of colonization of and persistence in the oral cavity and to be involved in the control of *Streptococcus pyogenes* (Dierksen et al. 2007). *Streptococcus salivarius* has also been shown to be producing antimicrobial peptide active against *Propionibacterium acne*, strain related to acne of the skin (Filip et al. 2006).

Lacticin 3147, produced by *Lactococcus lactis* subsp. *lactis* DPC3147, is of particular interest from an application point of view. This lantibiotic had activity against MRSA, VRE, penicillin-resistant *Pneumococcus*, *Propionibacterium acne* and *Streptococcus mutans*, all of which are significant human pathogens (Lawton et al. 2007). Lacticin 3147 showed effective role in the treatment of mastitis in cattle related to *Streptococcus dysgalactiae* (Ryan et al. 1999). Based on efficacy of lacticin 3147 in treatment of cattle *Streptococcus dysgalactiae*, most probably, this lantibiotic will be effective in humans as well. *Streptococcus dysgalactiae* has been related to acute pharyngitis in children, bacteraemia in a patient with pyomyositis and reactive arthritis, vertebral osteomyelitis, perinatal morbidity, neonatal mortality and postpartum endometritis in women (Torres et al. 2002; Zaoutis et al. 2004; Kumar et al. 2005). Lacticin 3147 has been shown to have the ability to inhibit *Streptococcus agalactiae* and *Streptococcus dysgalactiae* and may be a prospective agent for the control of these microorganisms (Twomey et al. 2002). It has been shown that lacticin 3147 has a potential in the control of cariogenic strains and may be exploited in the future for the prevention of dental caries (O'Connor et al. 2006).

Macedocin ST91KM, produced by *Streptococcus gallolyticus* subsp. *macedonicus* ST91KM, showed to be bactericidal to *Streptococcus agalactiae*, *Streptococcus dysgalactiae*, *Streptococcus uberis* and *Staphylococcus aureus* associated with mastitis infections, including strains resistant to methicillin and oxacillin (Pieterse et al. 2010). A teat seal preparation containing macedocin ST91KM effectively released the peptide and inhibited the growth of *Streptococcus agalactiae* in the model system. Pieterse et al. (2010) proposed that macedocin ST91KM could form the basis for alternative dry cow therapy to prevent mastitis infections in dairy cows as it is effective against pathogens that display resistance to conventional antibiotic therapy.

Bacteriocins produced by different strains were proposed as potential agents for the treatment of bacterial vaginosis. Kaur et al. (2013) reported on bacteriocin HV6b, produced by *Lactobacillus fermentum* active against *Gardnerella vaginalis*. Todorov et al. (2006) characterized bacteriocin produced by *Lactococcus lactis* subsp. *lactis* HV219 isolated from vaginal secretion to be active against vaginal pathogens and to be presenting a probiotic potential (Todorov et al. 2006, 2007a).

Bacteriocins with different structures (nisin A, lacticin Q, and nukacin ISK-1) were investigated for the efficacy against MRSA and biofilm formation compared to vancomycin, a glycopeptide antibiotic used in the treatment of MRSA infections, usually with bactericidal activity against only planktonic cells (Okuda et al. 2013). Nisin A and lacticin Q showed the bactericidal activity against MRSA, both planktonic cells and biofilm cells. However, nukacin ISK-1 showed bacteriostatic activity against planktonic cells and did not show bactericidal activity against biofilm cells. Okuda et al. (2013) suggest the application of bacteriocins that form stable pores on biofilm cells that may be an option for the treatment of MRSA biofilm infections in medical practices.

A big advantage of bacteriocins is that they can target individual bacterial species (narrow spectrum of activity) or provide broad-spectrum killing of many microbes. Maybe bacteriocins can be looked as potential single agents or coagents in the treatment of multidrug-resistant pathogens. In addition, they could even be produced in the body of humans or other animals by intentionally introduced beneficial bacteria, as some probiotics do.

Since the beneficial effects of yogurt have been mentioned, Metchnikoff can be considered as a father of the modern probiotics science. However, a long way has been taken since the first observation of the beneficial effect of fermented food products and empirical application of LAB to the scientific proofs of positive effect of LAB to human and animal health. Speaking of the various beneficial LAB, most of the time, we describe them as probiotics. The Food and Agriculture Organization/World Health Organization in 2001 states that probiotics are "live microorganisms which when administered in adequate amounts confer a health benefit to the host" (FAO/WHO 2001). According to the International Scientific Association for Probiotics and Prebiotics (ISAPP), probiotics must (1) be alive when administered, (2) have undergone controlled evaluation to document health benefits in the target host, (3) be a taxonomically defined microbe or a combination of microbes (genus, species, and strain level), and (4) be safe for its intended use (ISAPP 2009; LeBlanc et al. 2015).

The modern human and veterinary medicines are based on the application of different approaches in the treatment of diseases related to well establish therapeutically practices and new technologies. Complexity of the treatment of pathogenic bacteria includes not only the application of antibiotics (and other antimicrobial preparations). A modern treatment helps protect and reestablishes the integrity of the natural GIT microbiota. Probiotics play an important role in these processes. In the past, physicians were prescribed to the patience on intensive antibiotic treatment to have a rich on LAB products as yogurt, fermented milk products or boza in addition to vitamin B2 and vitamin B complex. The beneficial effects of LAB from fermented food products can be addressed to fact that some of them can be involved in the recolonization of the GIT. However, some of these LAB can be producing bacteriocins, taking part in the inhibition of pathogenic bacteria.

It is important to mention that the synergetic effect between antibiotics and bacteriocins produced by LAB has been shown (Todorov et al. 2009, 2010a, b; Salvucci et al. 2010).

Salvucci et al. (2010) evaluated possible inhibition effectiveness of synthetic enterocin CRL35 in combination with cell wall, membrane-acting antibiotics (monensin, bacitracin and gramicidin) and muranolytic enzymes (mutanolysin and lysozyme) using *Listeria innocua* and *Listeria monocytogenes* as a model strains. In this study, tested antibiotics showed positive interactions with enterocin CRL35 in the inhibition of both tested *Listeria* strains. However, when mutanolysin and enterocin CRL35 were added to not actively growing cells in a buffer system, the lytic effect of mutanolysin was enhanced. The addition of mutanolysin as a single agent showed no inhibitory effect on the growth of *Listeria innocua* cells in a tested culture medium. In contrast, the combination of lysozyme and enterocin CRL35 resulted in a 50 % inhibition of the *Listeria innocua* growth. Salvucci et al. (2010) suggested that the combination of synthetic enterocin CRL35 and some antibiotics is more effective against *Listeria innocua* and *Listeria monocytogenes* cultures and even more importantly the amount of these antibiotics can be used at lower quantities. The effectiveness of the combination of synthetic enterocin CRL35 with mutanolysin and lysozyme most probably depends on complex environments, and the authors suggested more detailed studies to be performed to elucidate this issue. Enterocin CRL35 represents a promising agent that not only can ensure the quality and safety of food (as has been reported earlier in this chapter), but it can also be combined with several antimicrobial agents important in the medical field.

Minahk et al. (2004) have also reported about the effect of sublethal concentrations of enterocin CRL35, combined to erythromycin, chloramphenicol and tetracycline, and observed that the peptide induced a significant membrane gradient dissipation without appreciable cell death. Most probably, this membrane depolarization is necessary, but not sufficient to provoke cell death, and another concentration-dependent step may be required. It has been described that pleurocidin and derivatives, which are antimicrobial peptides from eukaryotic organisms, lost their ability to damage cell membranes at sublethal concentrations while maintaining their capacities to inhibit macromolecular synthesis (Patrzykat et al. 2002; Todorov et al. 2010b).

When sublethal doses of ciprofloxacin were combined with bacteriocin ST5Ha, produced by *Enterococcus faecium* ST5Ha, a strain isolated from smoked salmon, a strong enhancement of the bioactivity was observed (Todorov et al. 2010b). The authors suggested that bacteriocin ST5Ha increased the effectiveness of the ciprofloxacin through dissipation of the proton gradient, responsible for the extrusion of these antibacterial compounds, and resulted in a well-observed synergistic effect. For a tested time period, the combination of ciprofloxacin and bacteriocin ST5Ha resulted in a stronger inhibition of *Listeria ivanovii* subsp. *ivanovii* ATCC19119, compared with the use of ciprofloxacin or bacteriocin ST5Ha as single antibacterial agents (Todorov et al. 2010b).

Todorov and Dicks (2009) also reported on synergism between bacteriocin ST44AM produced by *Pediococcus pentosaceus* (isolated from marula) and ciprofloxacin. Growth inhibition of *Listeria ivanovii* subsp. *ivanovii* ATC19119 was recorded during the first 12 h in the presence of ciprofloxacin and bacteriocin ST44AM and continued for the duration of the experiment. Authors observed that cells treated with either bacteriocin ST44AM or ciprofloxacin developed resistance after 24 h. It may be that in this cases, bacteriocin ST44AM was degraded by proteolytic enzymes or inactivated due to aggregation (Todorov and Dicks 2009). However, similar effects of interaction between ciprofloxacin and bacteriocins have been recorded even with other five bacteriocins (produced by *Lactobacillus plantarum* ST69BZ, *Enterococcus faecium* ST62BZ and *Leuconostoc lactis* ST63BZ, ST611BZ and ST612BZ, all isolated from boza), showing that most probably the effect of synergism between these studied antibiotics and bacteriocins is more general (Todorov 2010). These results indicate that the mechanism by which the cationic peptide increases the effectiveness of ciprofloxacin is similar.

The combined use of antibiotics, particularly ciprofloxacin, and bacteriocins is a promising approach to reduce the amounts of antibiotics required for the treatment of infectious diseases in human and veterinary medicine, overcoming the development of resistant strains.

Antimicrobial peptides, including bacteriocins, are very promising antimicrobial agents; however, delivery systems are still a challenge. Arthur et al. (2014) reviewed several bacteriocin delivery systems and their potential application, not only for food bio-preservation but focusing on the potential human and veterinary medicine. As previously discussed, bacteriocins can act as antibiotic synergists or even as alternatives to enhance the therapeutic effects of current infection treatments and especially to decrease the prevalence of resistant strains. The introduction of bacteriocins into the biomedical industry in appropriate vehicles and further development of food applications were still highly dependent on the slow development of reliable delivery systems. Silver- or carbohydrate-based nanoparticles, nanofiber scaffolds, nanospheres, implant impregnation, catheter coating, hydrogel, oral tablet, gum technology, livestock feed supplementation and aquaculture dry spray, in addition to food products with bacteriocin or bacteriocin producer, incorporated into food packaging are just a few options for the application of bacteriocins discussed by Arthur et al. (2014).

Antibiotic drug resistance is an emerging issue in the food industry and human and veterinary medicine. The development of new and more effective antimicrobial alternative drug therapies will most probably help in the reduction of our dependence on classical antibiotic drugs. From our perspective, antimicrobial peptides, including bacteriocins, have highly promising potential as antibiotic alternatives or synergist treatment. A high number of bacteriocins have proven their efficacy in the laboratory conditions, and even some of them have been successfully applied in industrial processes. However, more in vivo studies need to be conducted before bacteriocins can be incorporated into control of bacterial and viral infections.

And this is only beginning of the future of the bacteriocins.

References

Abriouel H, Lucas R, Ben Omar N, Valdivia E, Maqueda M, Martínez-Cañamero M, Gálvez A (2005) Enterocin AS-48RJ: a variant of enterocin AS-48 chromosomally encoded by *Enterococcus faecium* RJ16 isolated from food. Syst Appl Microbiol 28(5):383–397

Aguirre M, Collins MD (1993) Lactic acid bacteria and human clinical infection. J Appl Bacteriol 75(2):95–107

Ahmed Z, Wang Y, Ahmad A, Khan ST, Nisa M, Ahmad H, Afreen A (2013) Kefir and health: a contemporary perspective. Crit Rev Food Sci Nutr 53:422–434

Arthur TD, Cavera VL, Chikindas ML (2014) On bacteriocin delivery system and potential applications. Future Microbiol 9:235–248

Batish VK, Lal R, Grover S (1989) Screening lactic acid starter cultures for antifungal activity. J Cultur Dairy Prod 24:21–25

Baumal R, Musclow E, Farkas-Himsley H, Marks A (1982) Variants of an interspecies hybridoma with altered tumorigenicity and protective ability against mouse myeloma tumors. Cancer Res 42(5):1904–1908

Belguesmia Y, Choiset Y, Rabesona H, Baudy-Floch M, Le Blay G, Haertlé T, Chobert J-M (2012) Antifungal properties of durancins isolated from *Enterococcus durans* A5-11 and of its synthetic fragments. Lett Appl Microbiol 56:237–244

Belguesmia Y, Rabesona H, Mounier J, Pawtowsky A, Le Blay G, Barbier G, Haertlé T, Chobert JM (2014) Characterization of antifungal organic acids produced by *Lactobacillus harbinensis* K.V9.3.1Np immobilized in gellan-xanthan beads during batch fermentation. Food Control 36:205–211

Beresford TP, Fitzsimons NA, Brennan NL, Cogan TM (2001) Recent advances in cheese microbiology. Int Dairy J 11:259–274

Beshkova DM, Simova ED, Simov ZI, Frengova GI, Spasov ZN (2002) Pure cultures for making kefir. Food Microbiol 19:537–544

Biscola V, Abriouel H, Todorov SD, Capuano VSC, Gálvez A, Franco BDGM (2014) Effect of autochthonous bacteriocin-producing *Lactococcus lactis* on bacterial population dynamics and growth of halotolerant bacteria in Brazilian charqui. Food Microbiol 44(1):296–301

Blandino A, Al-Aseeri ME, Pandiella SS, Cantero D, Webb C (2003) Cereal-based fermented foods and beverages. Food Res Int 36:527–543

Botes M, van Reenen CA, Dicks LMT (2008) Evaluation of *Enterococcus mundtii* ST4ST and *Lactobacillus plantarum* 423 as probiotics by using a gastro-intestinal model with infant milk formulation as substrate. Int J Food Microbiol 128:362–370

Brink M, Fraser T, Todorov SD, Vaz-Velho M, Senekal M, Dicks LMT (2003) A combined use of probiotics and prebiotics in a soymilk-based food supplement aimed at improving the gastrointestinal flora of children with HIV AIDS. Electron J Environ Agric Food Chem 2(4):504–509

Broadbent JR, Chou YC, Gillies K, Kondo JK (1989) Nisin inhibits several Gram-positive, mastitis-causing pathogens. J Dairy Sci 72:3342–3345

Campbell-Platt G (1994) Fermented foods: a world perspective. Food Res Int 27:253–257

Caplice E, Fitzgerald GF (1999) Food fermentations: role of microorganisms in food production and preservation. Int J Food Microbiol 50:131–149

Chuon MR, Shiomoto M, Koyanagi T, Sasaki T, Michihata T, Chan S, Mao S, Enomoto T (2014) Microbial and chemical properties of Cambodian traditional fermented fish products. J Sci Food Agric 94:1124–1131

Cleveland J, Montville TJ, Nes IF, Chikindas ML (2001) Bacteriocins: safe, natural antimicrobials for food preservation. Int J Food Microbiol 71(1):1–20

Coallier-Ascah J, Idziak E (1985) Interaction between *Streptococcus lactis* and *Aspergillus flavus* on production of aflatoxin. Appl Environ Microbiol 49:163–167

Corsetti A, Gobetti M, Rossi J, Damiani P (1998) Antimould activity of sourdough lactic acid bacteria: identification of a mixture of organic acids produced by *Lactobacillus sanfrancisco* CB1. Appl Microbiol Biotechnol 50:253–256

Cotter PD, Hill C, Ross RP (2005) Bacteriocins: developing innate immunity for food. Nat Rev Microbiol 3(10):777–788

Cruz-Chamorro L, Puertollano MA, Puertollano E, de Cienfuegos GA, de Pablo MA (2006) In vitro biological activities of magainin alone or in combination with nisin. Peptides 27(6): 1201–1209

Dalié DKD, Deschamps AM, Richard-Forget F (2010) Lactic acid bacteria – potential for control of mould growth and mycotoxins: a review. Food Control 21:370–380

Danova S, Petrov K, Pavlov P, Petrova P (2005) Isolation and characterization of *Lactobacillus* strains involved in koumiss fermentation. Int J Dairy Technol 58:100–105

De Oliveira Leite AM, Miguel MA, Peixoto RS, Rosado AS, Silva JT, Paschoalin VMI (2013) Microbiological, technological and therapeutic properties of kefir: a natural probiotic beverage. Braz J Microbiol 44(2):341–349

De Vuyst LD, Vandamme EJ (1994) Bacteriocins of lactic acid bacteria: microbiology, genetics and applications. Blackie Academic & Professional, London

Delves-Broughton J, Blackburn P, Evans RJ, Hugenholtz J (1996) Application of bacteriocin, nisin. Antonie Van Leeuwenhoek 69:193–202

Dierksen KP, Moore CJ, Inglis M, Wescombe PA, Tagg JR (2007) The effect of ingestion of milk supplemented with salivaricin a-producing *Streptococcus salivarius* on the bacteriocin-like inhibitory activity of streptococcal population on the tongue. FEMS Microbiol Ecol 59: 584–591

Duffes F (1999) Efficacité de *Carnobacterium* spp. a inhiber *Listeria monocytogenes* dans le saumon fumé – Identification de deux proteins impliques dans le résistance de *Listeria monocytogenes* a la resistance de divercin V41. Ph.D. theisis, Ecole National Superieure Agronomique de Rennes, France

Duffes F, Leroi F, Boyaval P, Dousset X (1999) Inhibition of *Listeria monocytogenes* by *Carnobacterium* spp. strains in a simulated cold-smocked fish system stored at 4°C. Int J Food Microbiol 47:33–42

Duffes F, Corre C, Leroi F, Dousset X, Boyaval P (2000) Inhibition of *Listeria monocytogenes* by *in situ* produced and semipurified bacteriocins from *Carnobacterium* spp. on vacuum-packed refrigerated cold-smoked salmon. J Food Prot 62:1394–1403

Effat BA, Ibrahim GA, Tawfik NF, Sharaf OM (2001) Comparison of antifungal activity of metabolites from *Lactobacillus rhamnosus*, *Pediococcus acidilactici* and *Propionibacterium thoenii*. Egypt J Dairy Sci 29:251–262

Elsanhoty RM, Salam SA, Ramadan MF, Badr FH (2014) Detoxification of aflatoxin M1 in yoghurt using probiotics and lactic acid bacteria. Food Control 43:129–134

Fadda S, López C, Vignolo G (2010) Role of lactic acid bacteria during meat conditioning and fermentation: peptides generated as sensorial and hygienic biomarkers. Meat Sci 86:66–79

FAO/WHO (2001) Report of a Joint FAO/WHO expert consultation on evaluation of health and nutritional properties of probiotics in food including powder milk with live lactic acid bacteria

Farias ME, Farias RN, de Ruiz Holgado AP, Sesma F (1996) Purification and N-terminal amino acid sequence of enterocin CRL35, a 'pediocin-like' bacteriocin produced by *Enterococcus faecium* CRL35. Lett Appl Microbiol 22:417–419

Farkas-Himsley H, Yu H (1985) Purified colicin as cytotoxic agent of neoplasia: comparative study with crude colicin. Cytobios 42(167–168):193–207

Farkas-Himsley H, Freedman J, Read SE, Asad S, Kardish M (1991) Bacterial proteins cytotoxic to HIV-1-infected cells. AIDS 5(7):905–907

Farkas-Himsley H, Hill R, Rosen B, Arab S, Lingwood CA (1995) The bacterial colicin active against tumor cells *in vitro* and *in vivo* is verotoxin 1. Proc Natl Acad Sci USA 92(15): 6996–7000

Favaro L, Penna ALB, Todorov SD (2015) Bacteriocinogenic LAB from cheeses - application in biopreservation? Trends Food Sci Technol 41(1):37–48

Férir G, Petrova MI, Andrei G, Huskens D, Hoorelbeke B, Snoeck R, Vanderleyden J, Balzarini J, Bartoschek S, Brönstrup M, Süssmuth RD, Schols D (2013) The lantibiotic peptide

labyrinthopeptin A1 demonstrates broad anti-HIV and anti-HSV activity with potential for microbicidal applications. PLoS One 8(5):e64010. doi:10.1371/journal.pone.0064010
Filip JC, Bowe WP, DiRienzo JM, Volgina A, Margolis DJ (2006) Inhibition of *Propionibacterium acnes* by bacteriocin-like inhibitory substances (BLIS) produced by *Streptococcus salivarius*. J Drugs Dermatol 5:868–871
Fleet GH (1999) Microorganisms in food ecosystems. Int J Food Microbiol 50:101–117
Florianowicz T (2001) Antifungal activity of some microorganisms against *Penicillium expansum*. Eur Food Res Technol 212:282–286
Fox PF, Wallace JM (1997) Formation of flavor compounds in cheese. Adv Appl Microbiol 45: 17–85
Fuller R, Gibson GR (1997) Modification of the intestinal microflora using probiotics and prebiotics. Scand J Gastroenterol 32:28–31
Furtado DN, Todorov SD, Landgraf M, Destro MT, Franco BDGM (2014) Bacteriocinogenic *Lactococcus lactis* subsp. *lactis* DF04Mi isolated from goat milk: characterization of the bacteriocin. Braz J Microbiol 45(4):1541–1550
Furtado DN, Todorov SD, Landgraf M, Destro MT, Franco BDGM (2015) Bacteriocinogenic *Lactococcus lactis* subsp. *lactis* DF04Mi isolated from goat milk: application in the control of *Listeria monocytogenes* in fresh Minas-type goat cheese. Braz J Microbiol. 41(1):201-206.
García P, Rodríguez L, Rodríguez A, Martínez B (2010) Food biopreservation: promising strategies using bacteriocins, bacteriophages and endolysins. Trends Food Sci Technol 21(8): 373–382
Goff JH, Bhunia AK, Johnson MG (1996) Complete inhibition of low levels of *Listeria monocytogenes* on refrigerated chicken meat with pediocin AcH bound to heat-killed *Pediococcus acidilactici* cells. J Food Prot 59:1187–1192
Gourama H (1997) Inhibition of growth and mycotoxin production of *Penicillium* by *Lactobacillus* species. Lebensmittel-Wissenchaft Technol 30:279–283
Gratz S, Mykkänen H, Ouwehand AC, Juvonen R, Salminen S, El-Nezami H (2004) Intestinal mucus alters the ability of probiotic bacteria to bind aflatoxin B1 in vitro. Appl Environ Microbiol 70:6306–6308
Hassan YI, Bullerman LB (2008) Antifungal activity of *Lactobacillus paracasei* ssp. *tolerans* isolated from a sourdough bread culture. Int J Food Microbiol 121(1):112–115
Heng NC, Wescombe PA, Burton JP, Jack RW, Tagg JR (2007) The diversity of bacteriocins in Gram-positive bacteria. In: Bacteriocins. Springer, Berlin, pp 45–92
Holt JG, Krieg NR, Sneath PHA, Stanely JT, Williams ST (1994) Bergey's manual of determinative bacteriology, 9th edn. Williams & Wilkins, Baltimore, MD
Huttunen E, Noro K, Yang Z (1995) Purification and identification of antimicrobial substance produced by two *Lactobacillus casei* strains. Int Dairy J 5:503–513
ISAPP (2009) Clarification of the definition of a probiotic of the International Scientific Association for Probiotics and Prebiotics (online). http://www.isapp.net/Portals/0/docs/ProbioticDefinitionClarification.pdf. Accessed 14 Dec 2014
Jacome SL, Fonseca S, Pinheiro R, Todorov S, Noronha L, Silva J, Gomes A, Pintado M, Morais AMMB, Teixeira P, Vaz-Velho M (2014) Effect of lactic acid bacteria on quality and safety of ready-to-eat sliced cured/smoked meat products. Chem Eng Trans 38:403–408
Johansen C, Gill T, Gram L (1996) Changes in cell morphology of *Listeria monocytogenes* and *Shewanella putrefaciens* resulting from the action of protamine. Appl Environ Microbiol 62: 1058–1064
Jozala AF, De Andrade MS, De Arauz LJ, Oessoa A Jr, Penna TCV (2007) Nisin production utilizing skimmed milk aiming to reduce process cost. Appl Biochem Biotechnol 137–140: 515–528
Kabak B, Dobson ADW (2011) An introduction to the traditional fermented foods and beverages of Turkey. Crit Rev Food Sci Nutr 51:248–260

Kassaa IA, Hober D, Hamze M, Chihib NE, Drider D (2014) Antiviral potential of lactic acid bacteria and their bacteriocins. Probiot Antimicrob Proteins 6:177–185

Kaur B, Balgir P, Mittu B, Chauhan A, Kumar B (2013) Purification and physicochemical characterization of anti-*Gardnerella vaginalis* bacteriocin HV6b produced by *Lactobacillus fermentum* isolate from human vaginal ecosystem. Am J Biochem Mol Biol 3:91–100

Kellner R, Jung G, Horner T, Zahner H, Schnell N, Entian KD, Gotz F (1988) Gallidermin: a new lanthionine-containing polypeptide antibiotic. Eur J Biochem 177:53–59

Klaenhammer TR (1993) Genetics of bacteriocin produced by lactic acid bacteria. FEMS Microbiol Rev 12:39–85

Kruszewska D, Sahl HG, Bierbaum G, Pag U, Hynes SO, Ljungh A (2004) Mersacidin eradicates methicillin-resistant *Staphylococcus aureus* (MRSA) in a mouse rhinitis model. J Antimicrob Chemother 54:648–653

Kumar A, Sandoe J, Kumar N (2005) Three cases of vertebral osteomyelitis caused by *Streptococcus dysgalactiae* subsp. *equisimilis*. J Med Microbiol 54:1103–1105

Lavermicocca P, Valerio F, Evidente A, Lazzaroni S, Corsetti A, Gobetti M (2000) Purification and characterization of novel antifungal compounds from sourdough *Lactobacillus plantarum* strain 21 B. Appl Environ Microbiol 66:4084–4090

Lawton EM, Ross RP, Hill C, Cotter PD (2007) Two-peptide lantibiotics: a medical perspective. Minirev Med Chem 7:1236–1247

LeBlanc JG, Todorov SD (2011) Bacteriocin producing lactic acid bacteria isolated from Boza, a traditional fermented beverage from Balkan Peninsula – from isolation to application. In: Méndez-Vilas A (ed) Science against microbial pathogens: communicating current research and technological advances. Microbiology Book Series, no 3, vol 2. Formatex Research Center, Badajos, pp 1311–1320

LeBlanc JG, Todorov SD, de LeBlanc AM (2015) Beneficial effects of microorganisms isolated from papaya. In: Todorov SD, Ivanova IV (eds) Tropical fruits – from cultivation to consumption and health benefits: papaya. Nova Publisher, New York, pp 105–118

Legan JD (1993) Mould spoilage of bread: the problem and some solutions. Int Biodeterior Biodegrad 32:33–53

Leisner JJ, Greer GG, Dilts BD, Stiles ME (1995) Effect of growth of selected lactic acid bacteria on storage life of beef stored under vacuum and in air. Int J Food Microbiol 26:231–243

Lindgren S, Dobrogosz W (1990) Antagonistic activities of lactic acid bacteria in food and feed fermentations. FEMS Microbiol Rev 87:149–163

Liua W, Zhanga L, Yia H, Shib J, Xuea C, Lia H, Jiaoa Y, Shigwedhaa N, Dua M, Hana X (2014) Qualitative detection of class IIa bacteriocinogenic lactic acid bacteria from traditional Chinese fermented food using a YGNGV-motif-based assay. J Microbiol Methods 100(1): 121–127

Luchese RH, Harrigan WF (1990) Growth of and aflatoxin production by *Aspergillus parasiticus* when in the presence of either *Lactococcus lactis* or lactic acid and at different initial pH values. J Appl Bacteriol 69:512–519

Magnusson J, Schnürer J (2001) *Lactobacillus coryniformis* subsp. *coryniformis* strain Si3 produces a broad-spectrum proteinaceous antifungal compound. Appl Environ Microbiol 67:1–5

Magnusson J, Ström K, Roos S, Sjögren J, Schnürer J (2003) Broad and complex antifungal activity among environmental isolates of lactic acid bacteria. FEMS Microbiol Lett 219: 129–135

Mandal V, Sen SK, Mandal NC (2007) Detection, isolation and partial characterization of antifungal compound(s) produced by *Pediococcus acidilactici* LAB 5. Nat Prod Commun 2: 671–674

Masuda Y, Ono H, Kitagawa H, Ito H, Mu F, Sawa N et al (2011) Identification and characterization of leucocyclicin Q, a novel cyclic bacteriocin produced by *Leuconostoc mesenteroides* TK41401. Appl Environ Microbiol 77(22):8164–8170

Mäyrä-Mäkinen A, Suomalainen T (1995) *Lactobacillus casei* subsp. *rhamnosus*, bacterial preparations comprising said strain, and use of said strain and preparations for the controlling of yeast and moulds. US Patent 5378458. http://fi.espacenet.com

Mäyrä-Mäkinen A, Suomalainen T (1996) Inhibition of clostridium with lactic acid bacteria. EP Patent 0698347. http://fi.espacenet.com

Mäyrä-Mäkinen AK, Kristianinkatu A, Suomalainen TV (1994) A novel microorganism strain, bacterial preparations comprising said strain, and use of said strain preparations for the controlling of yeasts and moulds. European Patent Application, 0,576,780, A2

McMuller LM, Stiles ME (1996) Potential use of bacteriocin producing lactic acid bacteria in the preservation of meats. J Food Prot (Suppl) 64–71

Metha B, Kamal-Eldin A, Iwanski RZ (2012) Fermentation effects on food properties. CRC, Boca Raton, FL

Miao J, Guo H, Ou Y, Liu G, Fang X, Liao Z, Ke C, Chen Y, Zhao L, Cao Y (2014) Purification and characterization of bacteriocin F1, a novel bacteriocin produced by *Lactobacillus paracasei* subsp. *tolerans* FX-6 from Tibetan kefir, a traditional fermented milk from Tibet, China. Food Control 42:48–53

Minahk CJ, Dupuy F, Morero RD (2004) Enhancement of antibiotic activity by sub-lethal concentrations of enterocin CRL35. J Antimicrob Chemother 53:240–246

Mokua RA (2004) Effect of Kenyan fermented milk on *Escherichia coli*. The Graduation College University of Wisconsin-Stout, Menomonie, WI, 51p

Montalbán-López M, Sánchez-Hidalgo M, Valdivia E, Martínez-Bueno M, Maqueda M (2011) Are bacteriocins underexploited? Novel applications for old antimicrobials. Curr Pharm Biotechnol 12:1205–1220

Moretti VA, Madonia G, Diaferia C, Mentasti T, Paleari MA, Panseri S, Pirone G, Gandini G (2004) Chemical and microbiological parameters and sensory attributes of a typical Sicilian salami ripened in different conditions. Meat Sci 66:845–854

Musclow CE, Farkas-Himsley H, Weitzman SS, Herridge M (1987) Acute lymphoblastic leukemia of childhood monitored by bacteriocin and flow cytometry. Eur J Cancer Clin Oncol 23(4): 411–418

Myoshi A, Florence ACR, Silva RC, Pontes D, Chatel JM, Serror P, Langella P, Azevedo V, Oliveira MN (2009) Uso Biotecnológico De Bactérias Lácticas Como Probióticos, Para A Produção De Proteínas Heterólogas E Como Vetores De DNA. Microbiologia in Foco 1:7

Nascimento JS, Ceotto H, Nascimento SB, Giambiagi-Demarval M, Santos KR, Bastos MC (2006) Bacteriocins as alternative agents for control of multiresistant staphylococcal strains. Lett Appl Microbiol 42:215–221

Nielsen B, Gürakan GG, Ünlü G (2014) Kefir: a multifaceted fermented dairy product. Probiotics Antimicrob Proteins 6:123–135

Niku-Paavola ML, Laitila A, Mattila-Sandholm T, Haikara A (1999) New types of antimicrobial compound produced by *Lactobacillus plantarum*. J Appl Microbiol 86:29–35

O'Connor EB, O'Riordan B, Morgan SM, Whelton H, O'Mullane DM, Ross RP, Hill C (2006) A lactacin 3147 enriched food ingredient reduces *Streptococcus mutans* isolated from the human oral cavity in saliva. J Appl Microbiol 100:1251–1260

Oetterer M (2014) USP/ESALQ/LAN-662. The process of seafood brew (Anchovamento). www.esalq.usp.br/departamentos/.../Fermentacao%20do%20pescado.pdf. Accessed 15 Dec 2014

Okkers DJ, Dicks LM, Silvester M, Joubert JJ, Odendaal HJ (1999) Characterization of pentocin TV35b, a bacteriocin-like peptide isolated from *Lactobacillus pentosus* with a fungistatic effect on *Candida albicans*. J Appl Microbiol 87:726–734

Okuda K, Zendo T, Sugimoto S, Iwase T, Tajima A, Yamada S, Sonomoto K, Mizunoe Y (2013) Effects of bacteriocins on methicillin-resistant Staphylococcus aureus biofilm. Antimicrob Agents Chemother 57:5572–5579

Oliveira MN (2009) Tecnologia de produtos lácteos funcionais. Ed Atheneu, Sao Paulo

Parente E, Ricciardi A (1994) Influence of pH on the production of enterocin 1146 during batch fermentation. Lett Appl Microbiol 19:12–15

Parente E, Ricciardi A, Addario G (1994) Influence of pH on growth and bacteriocin production by *Lactococcus lactis* subsp. *lactis* 140VWC during batch fermentation. Appl Microbiol Biotechnol 41:388–394

Patrzykat A, Friedrich CL, Zhang LJ, Mendoza V, Hancock REW (2002) Sublethal concentrations of pleurocidin-derived antimicrobial peptides inhibit macromolecular synthesis in *Escherichia coli*. Antimicrob Agents Chemother 46:605–614

Paula AT, Jeronymo-Ceneviva AB, Silva LF, Todorov SD, Franco BDGM, Choiset Y, Chobert J-M, Haertlé T, Xavier D, Penna ALB (2014) *Leuconostoc mesenteroides*: bacteriocinogenic strain isolated from Brazilian water-buffalo mozzarella cheese. Probiot Antimicrob Proteins 6(3-4):186–197

Peteán M, Beccarí A, Burns AP, Sihufe G, Zacarías MF, Binetti A, Reinheimer J, Vinderola G (2014) Influence of technological variables on the functionality of the cell-free fraction of fermented buttermilk. Int J Dairy Technol 67:39–46

Pieterse R, Todorov SD, Dicks LMT (2010) Mode of action and *in vitro* susceptibility of mastitis pathogens to macedocin ST91KM and preparation of a teat seal containing the bacteriocin. Braz J Microbiol 41(1):133–145

Pingitore EV, Todorov SD, Sesma F, Franco BDGM (2012) Application of bacteriocinogenic *Enterococcus mundtii* CRL35 and *Enterococcus faecium* ST88Ch in the control of *Listeria monocytogenes* in fresh Minas cheese. Food Microbiol 32(1):38–47

Pongtharangkul T, Demirci A (2004) Evaluation of agar diffusion bioassay for nisin quantification. Appl Microbiol Biotechnol 65(3):268–272

Powell JE, Todorov SD, Van Reenen CA, Dicks LMT, Witthuhn RC (2006) Growth inhibition of *Enterococcus mundtii* in Kefir by in situ production of bacteriocin ST8KF. LeLait 86:1–5

Powell JE, Witthuhn RC, Todorov SD, Dicks LMT (2007) Characterization of bacteriocin ST8KF produced by a kefir isolate *Lactobacillus plantarum* ST8KF. Int Dairy J 17:190–198

Reis JA, Paula AT, Casarotti SN, Penna ALB (2012) Lactic acid bacteria antimicrobial compounds: characteristics and applications. Food Eng Rev 4(2):124–140

Ribeiro SC, Costa MC, Todorov SD, Franco BDGM, Dapkevicius MLE, Silva CCG (2014) Characterization of bacteriocin-producing bacteria isolated from Pico Cheese an artisanal cow´s milk cheese. J Appl Microbiol 116(3):573–585

Röcken W (1996) Applied aspects of sourdough fermentation. Adv Food Sci 18:212–216

Röcken W, Voysey PA (1995) Sourdough fermentation in bread making. J Appl Bacteriol 79: 38S–48S

Rodrigues KL, Caputo LRG, Carvalho JCT, Evangelista J, Schneedorf JM (2005) Antimicrobial and healing activity of kefir and kefiran extract. Int J Antimicrob Agents 25:404–408

Rodriguez JL, Gaya P, Medina M, Nunez M (1997) Bactericidal effect of enterocin 4 on *Listeria monocytogenes* in a model dairy system. J Food Prot 60:28–32

Ross RP, Morgan S, Hill C (2002) Preservation a fermentation: past, present and future. Int J Food Microbiol 79:3–16

Rouse S, Harnett D, Vaughan A, van Sinderen D (2008) Lactic acid bacteria with potential to eliminate fungal spoilage in foods. J Appl Microbiol 104:915–923

Roy U, Batish VK, Grover S, Neelakantan S (1996) Production of antifungal substance by Lactococcus lactis subsp. lactis CHD-28.3. Int J Food Microbiol 32:27–34

Ryan MP, Jack RW, Josten M, Sahl HG, Jung G, Ross RP, Hill C (1999) Extensive post-translational modifications, including serine to D-alanine conversion, in the two-component lantibiotic, lactacin 3147. J Biol Chem 274:37544–37550

Sahl HG, Jack RW, Bierbaum G (1995) Biosynthesis and biological activities of lantibiotics with unique post-translational modifications. Eur J Biochem 230:827–853

Saito H, Watanabe T, Osasa S, Tado O (1979) Susceptibility of normal and tumor cells to mycobacteriocin and mitomycin C. Hiroshima J Med Sci 28(3):141–146

Salvucci E, Hebert EM, Sesma F, Saavedra L (2010) Combined effect of synthetic enterocin CRL35 with cell wall, membrane-acting antibiotics and muranolytic enzymes against *Listeria* cells. Lett Appl Microbiol 51:191–195

Sand SL, Haug TM, Nissen-Meyer J, Sand O (2007) The bacterial peptide pheromone plantaricin A permeabilizes cancerous, but not normal, rat pituitary cells and differentiates between the outer and inner membrane leaflet. J Membr Biol 216(2–3):61–71

Sathe SJ, Nawani NN, Dhakephalkar PK, Kapadnis BP (2007) Antifungal lactic acid bacteria with potential to prolong shelf-life of fresh vegetables. J Appl Microbiol 103:2622–2628

Schirru S, Todorov SD, Favaro L, Mangia NP, Basaglia M, Casella S, Comunian R, Franco BDGM, Deiana P (2012) Sardinian goat's milk as source of bacteriocinogenic potential protective cultures. Food Control 25(1):309–320

Schirru S, Favaro L, Mangia NP, Basaglia M, Casella S, Comunian R, Fancello F, Franco BDGM, Oliveira RPS, Todorov SD (2014) Comparison of bacteriocins production from *Enterococcus faecium* strains in cheese whey and optimised commercial MRS medium. Ann Microbiol 64(1):321–331

Schnürer J, Magnusson J (2005) Antifungal lactic acid bacteria as biopreservatives. Trends Food Sci Technol 16:70–78

Schoeman H, Vivier MA, Du Toit M, Dicks LMT, Pretorius IS (1999) The development of bactericidal yeast strains by expressing the *Pediococcus acidilactici* pediocin gene (pedA) in Saccharomyces cerevisiae. Yeast 15:647–656

Sharma N, Handa S, Gupta A (2013) A comprehensive study of different traditional fermented foods/beverages of 21 himachal pradesh to evaluate their nutrition impact on health and rich biodiversity of fermenting microorganisms. Int J Res Appl Nat Social Sci 1(3):19–28

Shiby VK, Mishra HN (2013) Fermented milks and milk products as functional foods – a review. Crit Rev Food Sci Nutr 53(5):482–496

Simova E, Beshkova D, Angelov A, Hristozova T, Frengova G, Spasov Z (2002) Lactic acid bacteria and yeasts in kefir grains and kefir made from them. J Ind Microbiol Biotechnol 28:1–6

Smaoui S, Elleuch L, Bejar W, Karray-Rebai I, Ayadi I, Jaouadi B, Mathieu F, Chouayekh H, Bejar S, Mellouli L (2010) Inhibition of fungi and Gram-negative bacteria by bacteriocin BacTN635 produced by *Lactobacillus plantarum* sp. TN635. Appl Biochem Biotechnol 162: 1132–1146

Sobrino-López A, Martín-Belloso O (2008) Use of nisin and other bacteriocins for preservation of dairy products. Int Dairy J 18(4):329–343

Stecchini ML, Aquilli V, Sarai I (1995) Behaviour of *Listeria monocytogenes* in mozzarella cheese in presence of *Lactococcus lactis*. Int J Food Microbiol 25:301–310

Stiles ME (1996) Biopreservation by lactic acid bacteria. Antonie van Leeuwenhoek 70:331–345

Stiles EM, Holzapfel WH (1997) Lactic acid bacteria of foods and their current taxonomy. Int J Food Microbiol 36(1):1–2

Stiles J, Penkar S, Plockova N, Chumchalova J, Bullerman LB (2002) Antifungal activity of sodium acetate and *Lactobacillus rhamnosus*. J Food Prot 65:1188–1191

Stoyanova LG, Ustyugova EA, Sultimova TD, Bilanenko EN, Fedorova GB, Khatrukha GS, Netrusov AI (2010) New antifungal bacteriocin-synthesizing strains of *Lactococcus lactis* ssp. *lactis* as the perspective biopreservatives for protection of raw smoked sausages. Am J Agric Biol Sci 5:477–485

Ström K, Sjögren J, Broberg A, Schnürer J (2002) *Lactobacillus plantarum* MiLAB 393 produces the antifungal cyclic dipeptides cyclo (L-Phe-L-Pro) and cyclo (L-Phe-trans-4-OH-L-Pro) and phenyllactic acid. Appl Environ Microbiol 68:4322–4327

Suzuki I, Nomura M, Morichi T (1991) Isolation of lactic acid bacteria which suppress mold growth and show antifungal action. Milchwissenschaft 46:635–639

Taira W, Funatsu Y, Satomi M, Takano T, Abe H (2007) Changes in extractive components and microbial proliferation during fermentation of fish sauce from underutilized fish species and quality of final products. Fish Sci 73:913–923

Talon R, Leroy S, Lebert I (2007) Microbial ecosystems of traditional fermented meat products: the importance of indigenous starters. Meat Sci 77:55–62

Tamime AY (2007) Fermented milk. Wiley-Blackwell, New York

Tamime AY, Robson RK (1999) Yogurt: science and technology, 2nd edn. CRC, Boca Raton, FL

Tamime AY, Muir DD, Wszolek M (1999) Kefir, koumiss and kishk. Dairy Ind Int 64:32–33
Todorov SD (2008) Bacteriocin production by *Lactobacillus plantarum* AMA-K isolated from Amasi, a Zimbabwean fermented milk product and study of adsorption of bacteriocin AMA-K to *Listeria* spp. Braz J Microbiol 38(1):178–187
Todorov SD (2010) Diversity of bacteriocinogenic lactic acid bacteria isolated from boza, a cereal-based fermented beverage from Bulgaria. Food Control 21:1011–1021
Todorov SD, Dicks LMT (2006a) Parameters affecting the adsorption of plantaricin 423, a bacteriocin produced by *Lactobacillus plantarum* 423 isolated from sorghum beer. Biotechnol J 1(4):405–409
Todorov SD, Dicks LMT (2006b) Screening for bacteriocin producer lactic acid bacteria from boza, a traditional cereal beverage from Bulgaria. Characterization of produced bacteriocins. Process Biochem 41(1):11–19
Todorov SD, Dicks LMT (2009) Bacteriocin production by *Pediococcus pentosaceus* isolated from marula (*Scerocarya birrea*). Int J Food Microbiol 132(2–3):117–126
Todorov SD, Holzapfel WH (2015) Traditional cereal fermented foods as sources of functional (bacteriocinogenic and probiotic) microorganisms. In: Holzapfel WH (ed) Advances in fermented foods and beverages: improving quality, technologies and health benefits. Woodhead, London, pp 123–153
Todorov SD, Wachsman MB, Knoetze H, Meincken M, Dicks LMT (2005) An antibacterial and antiviral peptide produced by *Enterococcus mundtii* ST4V isolated from soy beans. Int J Antimicrob Agents 25(6):508–513
Todorov SD, Danova ST, Van Reenen CA, Meincken M, Dinkova G, Ivanova IV, Dicks LMT (2006) Characterization of bacteriocin HV219, produced by *Lactococcus lactis* subsp. *lactis* HV219 isolated from human vaginal secretions. J Basic Microbiol 46(3):226–238
Todorov SD, Botes M, Danova ST, Dicks LMT (2007a) Probiotic properties of *Lactococcus lactis* subsp. *lactis* HV219, isolated from human vaginal secretions. J Appl Microbiol 103(3): 629–639
Todorov SD, Koep KSC, Van Reenen CA, Hoffman LC, Slinde E, Dicks LMT (2007b) Production of salami from beef, horse, mutton, blesbok (*Damaliscus dorcas phillipsi*) and springbok (*Antidorcas marsupialis*) with bacteriocinogenic strains of *Lactobacillus plantarum* and *Lactobacillus curvatus*. Meat Sci 77(3):405–412
Todorov SD, van Reenen CA, Dicks LMT (2007c) Pre-treatment of growth medium with Amberlite® XAD-1180 produces higher levels of bacteriocin plantaricin 423. Cent Eur J Biol 2(4):588–596
Todorov SD, Botes M, Guigas C, Schillinger U, Wiid I, Wachsman MB, Holzapfel WH, Dicks LMT (2008) Boza, a natural source of probiotic lactic acid bacteria. J Appl Microbiol 104(2): 465–477
Todorov SD, von Mollendorff JW, Moelich E, Muller N, Witthuhn RC, Dicks LMT (2009) Evaluation of potential probiotic properties of *Enterococcus mundtii*, its survival in boza and in situ bacteriocin production. Food Technol Biotechnol 47(2):178–191
Todorov SD, Ho P, Vaz-Velho M, Dicks LMT (2010a) Characterization of bacteriocins produced by two strains of *Lactobacillus plantarum* isolated from *Beloura* and *Chouriço*, traditional pork products from Portugal. Meat Sci 84(3):334–343
Todorov SD, Wachsman M, Tomé E, Dousset X, Destro MT, Dicks LMT, Franco BDGM, Vaz-Velho M, Drider D (2010b) Characterisation of an antiviral pediocin-like bacteriocin produced by *Enterococcus faecium*. Food Microbiol 27(7):869–879
Todorov SD, Vaz-Velho M, Franco BDGM, Holzapfel WH (2013) Partial characterization of bacteriocins produced by three strains of *Lactobacillus sakei*, isolated from *salpicao*, a fermented meat product from North-West of Portugal. Food Control 30(1):111–121
Todorov SD, Franco BDGM, Wiid IJ (2014) *In vitro* study of beneficial properties and safety of lactic acid bacteria isolated from Portuguese fermented meat products. Benef Microbes 5(3): 351–366
Torres AM, Menz I, Alewood PF, Bansal P, Lahnstein J, Gallagher CH, Kuchel PW (2002) D-amino acid residue in the C-type natriuretic peptide from the venom of the mammal, *Ornithorhynchus anatinus*, the Australian platypus. FEBS Lett 524:172–176

Twomey D, Ross RP, Ryan M, Meaney B, Hill C (2002) Lantibiotics produced by lactic acid bacteria: structure, function and applications. Antonie Van Leeuwenhoek 82:165–185

Van Reenen CA, Dicks LMT, Chikindas ML (1998) Isolation, purification and partial characterization of plantaricin 423, a bacteriocin produced by *Lactobacillus plantarum*. J Appl Microbiol 84:1131–1137

Van Reenen CA, Chikindas ML, Van Zyl WH, Dicks LMT (2003) Characterization and heterologous expression of a class IIa bacteriocin, plantaricin 423 from *Lactobacillus plantarum* 423, in *Saccharomyces cerevisiae*. Int J Food Microbiol 81:29–40

van Wyk J, Witthuhn RC, Britz TJ (2011) Optimisation of Vitamin B12 and folate production by *Propionibacterium freudenreichii* strains in kefir. Int Dairy J 21:69–74

Vaz-Velho M, Todorov S, Ribeiro J, Gibbs P (2005) Evaluation of *Carnobacterium divergens* V41 and supernatant V41 abilities for reducing *Listeria innocua* 2030c numbers during processing and storage of cold-smoked salmon trout. Food Control 16(6):541–549

Verellen TLJ, Bruggeman G, Van Reenen CA, Dicks LMT, Vandamme EJ (1998) Fermentation optimisation of plantaricin 423, a bacteriocin produced by *Lactobacillus plantarum* 423. J Ferment Bioeng 86:174–179

Von Mollendorff JW, Todorov SD, Dicks LMT (2006) Comparison of bacteriocins produced by lactic acid bacteria isolated from boza, a cereal-based fermented beverage from the Balkan Peninsula. Curr Microbiol 53(3):209–216

Von Mollendorff JW, Todorov SD, Dicks LMT (2009) Optimization of growth medium for production of bacteriocins produced by *Lactobacillus plantarum* JW3BZ and JW6BZ, and *Lactobacillus fermentum* JW11BZ and JW15BZ isolated from boza. Trakia J Sci 7(1):22–33

Voulgari K, Hatzikamari M, Delepoglou A, Georgakopoulos P, Litopoulou-Tzanetaki E, Tzanetakis N (2010) Antifungal activity of non-starter lactic acid bacteria isolates from dairy products. Food Control 21:136–142

Wachsman MB, Farias ME, Takeda E, Sesma F, De Ruiz Holgado AP, de Torres RA, Coto CE (1999) Antiviral activity of enterocin CRL35 against herpes virus. Int J Antimicrob Agents 12: 293–299

Wachsman MB, Castilla V, De Ruiz Holgado AP, de Torres RA, Sesma F, Coto CE (2003) Enterocin CRL35 inhibits late stages of HSV-1 and HSV-2 replication *in vitro*. Antivir Res 58: 17–24

Wang LL, Johnsonm EA (1997) Control of *Listeria monocytogenes* by monoglycerides in food. J Food Prot 60:131–138

Wang J, Chen X, Liu W, Yang M, Zhang H (2008) Identification of *Lactobacillus* from koumiss by conventional and molecular methods. Eur Food Res Technol 227:1555–1561

Wescombe PA, Upton M, Dierksen KP, Ragland NL, Sivabalan S, Wirawan RE, Inglis MA, Moore CJ, Walker GV, Chilcott CN, Jenkinson HF, Tagg JR (2006) Production of the lantibiotic salivaricin A and its variants by oral streptococci and use of a specific induction assay to detect their presence in human saliva. Appl Environ Microbiol 72:1459–1466

Wiedemann I, Breukink E, van Kraaij C, Kuipers OP, Bierbaum G, de Kruijff B, Sahl HG (2001) Specific binding of nisin to the peptidoglycan precursor lipid II combines pore formation and inhibition of cell wall biosynthesis for potent antibiotic activity. J Biol Chem 276(3): 1772–1779

Wiseman DW, Marth EH (1981) Growth and aflatoxin production by Aspergillus parasiticus when in the presence of *Streptococcus lactis*. Mycopathologia 73:49–56

Wouters JTM, Ayad EHE, Hugenholtz J, Smit G (2002) Microbes from raw milk for fermented dairy products. Int Dairy J 12:91–109

Yang Z, Suomalainen T, Mäyrä-Mäkinen A, Huttunen E (1997) Antimicrobial activity of 2-pyrrolidone-5-carboxylic acid produced by lactic acid bacteria. J Food Prot 60:786–790

Zaoutis T, Attia M, Gross R, Klein J (2004) The role of group C and group G streptococci in acute pharyngitis in children. Clin Microbiol Infect 10:37–40

Health Properties of Traditional Fermented Mongolian Milk Foods

Jie Dong, Yong Zhang, and Heping Zhang

Contents

Abstract The historical data at the beginning of the thirteenth century reflected that when the Mongolian nationality took initial shape, the people had grasped dairy fermentation technology skillfully. In the Ming and Qing dynasties, these dairy products once served as articles of tribute to the imperial court. The famous medical scientists, including Sun Simiao and Li Shizhen, concluded the medical effects of fermented dairy products based on the experience of northern nomadic nationalities. On the basis of long-term medical practice and research, the Mongolian medicine has created "koumiss therapy" for treatment of diseases. *L. casei*

J. Dong • Y. Zhang
Key Laboratory of Dairy Biotechnology and Engineering, Ministry of Education, Inner Mongolia Agricultural University, Hohhot, Inner Mongolia 010018, P. R. China

H. Zhang (✉)
Key Laboratory of Dairy Biotechnology and Engineering, Ministry of Education, Inner Mongolia Agricultural University, Hohhot, Inner Mongolia 010018, P. R. China

College of Food Science and Engineering, Inner Mongolia Agricultural University, 306 Zhaowuda Road, Hohhot, Inner Mongolia 010018, P. R. China
e-mail: hepingdd@vip.sina.com

M.-T. Liong (ed.), *Beneficial Microorganisms in Food and Nutraceuticals*, Microbiology Monographs 27, DOI 10.1007/978-3-319-23177-8_2

Zhang, isolated from koumiss in Xilin Gol grassland, is an effective probiotic for Chinese people. A series of in vitro and in vivo studies have shown its favorable probiotic functionality.

1 The History of Mongolian Traditional Fermented Dairy Products

The Mongolian nationality was originated from Shiwei. Shiwei people lived in Hulunbeier grassland, the east and west of DaXingAnLing, and on both banks of Ergun River and Heilongjiang River at early stage. They lived on hunting rather than a pastoral lifestyle. After the Tang dynasty, the Shiwei people changed from hunting to nomadism (Figs. 1, 2, 3, 4, 5, 6, and 7).

It was recorded in the *Biographical Sketches of Black Tartars* that Mongolian people "took horse milk, cow and sheep cheeses as food" (Wang 1983a). Zheng Sixiao (1241–1318) also mentioned that Mongolian people "hunted animals and ate meat-based food with few cereals. They liked to drink milk, eat cow and horse cheeses. They were very fat and sturdy" (Zheng 1991). The unfermented horse milk was too tart to drink, while cheese was generally referred to as a fermented dairy product in ancient times. Therefore, at the beginning of the twelfth to thirteenth century, Mongolian people already utilized the milks of cow, sheep, and horse to produce fermented dairy products. It was reported both by a French missionary, Guillaume de Rubru-quis (approx. 1215–1270) (Dawson 1983), and in Chap. 69, "The Magic of Tatars," of *The Travels of Marco Polo* (Feng and Dang 1999) that the animal milk was processed and dried, so that the resultant milk curd became hard. They were stored in a bag as food for the winter. When there was a lack of milk in winter, Mongolian people would put such fermented milk curd, which was called grut, in a leather bladder. Water was added to dissolve the milk cake before

Fig. 1 The fossils of animals in Mandela mountain rocks (about 6000 years ago)

Fig. 2 Herdsman and their leather bladders in a painting of Yin mountain rock (about 4000 years ago)

consumption; and the Mongolian people drank such water instead of milk and fresh water (Fig. 8).

Such milk cake referred to a fermented dairy product, and dairy fermentation technology is regarded as a very important food processing technology of Mongolian people. It was recorded in the *Biographical Sketches of Black Tartars*: "If a flagon containing cheese is tilted down, it will not function later" (Wang 1983b). Here, cheese was used as a ferment, that is to say, if a container with ferment was dumped, the milk fermentation would not occur any more, extended in meaning as no function later.

The Mongolian people not only fermented cow milk but also fermented horse milk into koumiss, with better flavor. Peng Daya (?–1245) recorded that Mongolians collected horse milk into leather containers in the *Biographical Sketches of Black Tartars*: "fermented for several nights, drank after it became a little sour, called Kumis." Xu Ting also witnessed the method of common nomads to produce koumiss; he especially investigated the fermentation technology of black koumiss drank by Mongolian aristocrats. "Upon my first arrival in the Golden Horde, the tartar host offered me Kumis to drink. It was clear and sweet, distinct from those common white and opaque ones with sour and unsavoury taste. It was named black Kumis. 'Perhaps clear resembled black.' I asked him. He answered: 'Actually it was fermented for seven to eight days. The longer it was fermented, the clearer it

ARCHAEOLOGY

Ancient cheese found with mummies

The oldest known pieces of cheese have turned up in the tombs of an early Bronze Age cemetery in Xinjiang, China.

Andrej Shevchenko at the Max Planck Institute of Molecular and Cell Biology and Genetics in Dresden, Germany, Changsui Wang at the University of Chinese Academy of Sciences in Beijing, and their colleagues analysed 3,800-year-old lumps found at the neck and chest of mummies (pictured) in the cemetery and identified them as a 'kefir' cheese.

This type of cheese is made by curdling ruminant milk with a symbiotic culture of bacteria, including *Lactobacillus kefiranofaciens*, and yeast. Evidence of a kefir dairy — which makes lactose-free products — in this region explains why large-scale ruminant herding and milking spread in a population known to have been lactose intolerant, the authors say. The origin of cheese making dates back some 4,000 years earlier, but evidence for this has relied on analysis of milk fat in pottery shards.

J. Arch. Sci. http://doi.org/rpq (2014)

Fig. 3 Ancient cheese discovered in Xinjiang (about 3600 years ago)

became. Once clear, the taste would not be unsavoury. I only drank it once, and never saw such black Kumis again in any other place'" (Wang 1983c).

The historical data at the beginning of the thirteenth century reflected that when the Mongolian nationality took initial shape, the people had grasped dairy fermentation technology skillfully. In 1260, Kublai Khan (1215–1294) took the throne in Kaiping. He encroached upon the Central Plains later and established the Yuan dynasty. This period had promoted the greatest development of Mongolian dairy products, which exerted huge influence on the development of dairy products in China (Fig. 9).

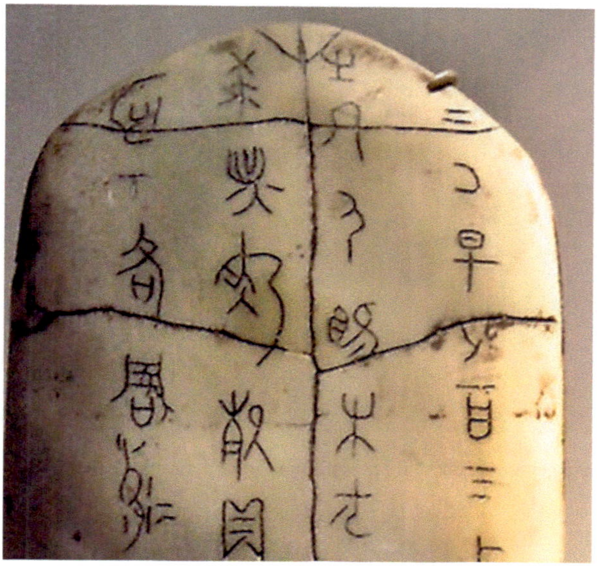

Fig. 4 Oracle bone inscriptions (about fourteenth to seventeenth century BC)

Fig. 5 Fresco of Mogao Caves showing sketches of butter production, Tang dynasty (618–907)

Chinese traditional society always emphasized the "rule of virtue." There were numerous etiquettes in a huge system; however, "Rites included five classics, sacrificial rite ranked first," so sacrificial rite was the most important one (Liao

Fig. 6 Dairy management agency in ancient time, Song dynasty (960–1127)

Fig. 7 Genghis Khan (1162–1227)

and Feng 2008). In the 29th year of Zhiyuan during the period of Emperor Shizu, the government of the Yuan dynasty started to order additional "cheese and koumiss" in the monthly sacrificial offerings in the name of "New Recommendations" (Xu 1530). These fermented dairy products were by no means of supporting role. "Kumis was highly valued in all great sacrifices. Every time before the sacrifice, Dongma Officer of Taipu Temple would be ordered imperially to offer Kumis in a drinker's leather bladder as gift in a grand way" (Song 1976a). Fermented dairy products were also used widely in the heaven worshipping ceremony of the Mongolian people. The *History of Yuan* mentioned the heaven worshipping of the rulers of the Yuan dynasty using horse milk for many times. At that time, a special clan, Kipchak people, fermented koumiss specifically for the Great Yuan Imperial Family. It was recorded in the *Achievements Stele of Jurong*

Fig. 8 Marco Polo (1254–)

Fig. 9 The figure of a feast bestowed by Kublai Khan, Marco Polo, and Li Livres du Graunt Caam, in French prose, fifteenth-century manuscript, preserved in Bodleian Library at Oxford University, MS. Bodl.264, fol.239r

King: "Emperor Shizu invaded westwards to Dali and conquered Song in the south. Such people were strong and brave, and were officially in charge of the livestock husbandry. They contributed koumiss to the Emperor as food. Black koumiss was superior, and black was called qara in Mongolian. Thus, the alias of such people was qara qumis" (Yu 1912).

As influenced by the Mongolian people, more fermented dairy products appeared in the feasts of all walks of life in the Yuan dynasty than ever before. The emperor drank with courtiers in royal feasts of the Yuan dynasty, and the consumption of fermented dairy products was considerably high (Yuan 2005). These dairy products were also consumed widely by common people. In the 21st year of Zhiyuan (1286), Lu Shirong (?–1285) focused on the expansion of financial revenue, and one of the policies was collecting transaction tax from the furs, bones and muscles, curds and cheese, etc. produced by the Mongolian people (Song 1976b). Therefore, a huge trading volume of dairy products between the Mongolian grassland and the Central Plains district was seen because of that.

Cheng Wen was a famous litterateur at the end of the Yuan dynasty, born in Wuyuan, Huizhou. He once wrote two poems, namely, *Milk Cake* and *Cow's Milk Butter*, which did not only describe the production technique but also outline the characteristics of fermented dairy products. More importantly, he expressed that the people in the Central Plains, the regions south of the Yangtze River, liked to eat dairy products, so they felt "very happy for getting such products" and that "they had really satisfied their craving for good foods" (Cheng 1984). Cheng Wen was not the only one. A poet in the Yuan dynasty, Yuan Jue (1266–1327), also depicted dairy products in his poems in the *Anthology of Qingrong Jushi* and recorded the offerings sacrificed to his dead younger brother, as well as dry cheese, butter, and other fermented dairy products in daily diet. Dairy products had not surprisingly become normal food in the life of residents in the Yuan dynasty.

In the Ming and Qing dynasties, people regarded Mongolian dairy products as delicious food. Rich businessmen went all out to seek "horse milk" to entertain the respected guests in feasts (Xie 2001). Gu Yanwu (1613–1682) recorded in his *Miscellaneous Poems from Datong to Xikou*: "In the old place of Fengzhou, now called Bansheng, large grain of salt was used in dishes and tea cakes were made with fermented milk" (Gu 1997). This poem could reflect that Fengzhou district was the trading area for tea from the Central Plains and milk from the Mongolian grassland. Liang Zhangju (1775–1849), who once assumed the Commissioner in Gansu, missed the vintage wine ("Mongolian brewed wine") in the Hexi area, as seen in his works in his later years (Liang 1935). It may be seen from analyzing the description of Liang Zhangju that he regarded the distilled liquor of fermented cow milk as "putao wine" by mistake.

Mongolian fermented dairy products once served as articles of tribute to the imperial court. Wu Zhenyu (1792–1870) recorded in his *Collection of Yangjizhai Works:* "The vassal state, Mongolia, offered sheep and Kumis as tribute upon festivals in old days. They were counted per nine, and sometimes to 12 times of nine" (Wu 2002). When Zhang Peilun (1848–1903) was exiled for penal servitude beyond the Great Wall, he would purchase more than ten packs of milk cake every time when the Mongolian carriers of articles of tribute passed by. He gave these milk cakes to Li Hongzhang (1823–1901), Chen Baochen (1848–1935), and other people and told them to contribute the excellent cheese beyond the Great Wall to their excellency (Zhang 2002).

2 The Health-Promoting Functions of Naturally Fermented Dairy Products Described by Traditional Mongolian Medicine

Mongolian people process the milk of livestock such as horse, cow, sheep, and camel into many kinds of delicious dairy products that are rich in nutrition. These products are also regarded as sacred, pure, and indispensable in the daily life of the Mongolian ethnic group. The traditional Mongolian dairy foods including dried milk cake and yogurt are good sources of high-quality protein. Additionally, they also contain abundant iron, zinc, calcium, and phosphorus (Yan et al. 1997). Mongolian people have taken cognizance of the nutritional and diet therapeutic values of dairy products in living practice, and they particularly recognize the health-related functions of fermented horse milk.

Fermented horse milk, also called koumiss (kumis, kumys, or coomys), is a traditional fermented horse milk beverage in Central Asia. Koumiss is made from fresh horse milk. It is an acidic and low alcohol-containing milk beverage, which is produced by joint autofermentation of lactic acid bacteria, yeasts, and other microorganisms (Ishii and Konagaya 2002; Jain and Prajapati 1998; Keogh and Barbara 1976; Kosikowski 1977; Lozovich 1995; Mann 2003; Ozer 2000; Qiaobenrichuren et al. 1998; Robinson 2002). Up to now, the production and consumption of koumiss are still popular among the nomadic tribes of Mongolians and Kazaksatan living in regions like Eastern Europe, Central Asia, Southeast Russia, Mongolia, and Inner Mongolia and Xinjiang of China (Ishii and Konagaya 2002; Qiaobenrichuren et al. 1998).

As early as in the Tang dynasty, a famous medical scientist, Sun Simiao (581–682), summarized the curative effect of horse milk and koumiss according to the experience of northern nomadic people: "Horse milk is pungent-warm, innocuous, and can quench thirst. Horse, cow and goat milk cheeses are sweet and sour, slightly cold in nature, innocuous, lung-nourishing, and favorable to the large intestine. *Huangdi Neijing* states that consuming sweet cheese or vinegar may lead to hematic abdominal mass and hematuria. Hua Tuo proposed the use of horse, cow and goat milk cheeses to expel common house centipedes, in case they accidentally enter the ear" (Sun 1987). The record in the *New Materia Medica* may be regarded as supplementary supporting evidence, "Both horse milk and donkey milk are cold in nature and favorable, can quench thirst and cure fever, while eating cheese of horse milk can cope with extreme coldness. Horse milk is unavailable in the regions south of the Yangtze River, so Tao Hongjing never mentioned it. Shepherds claim that horse milk cheese is a warm food, which can help to digest meat. Different food components may be regulatory to each other, no matter whether they are warm or cold in nature" (Su 1997). Li Shizhen (1518–1593) from the Ming dynasty also mentioned horse milk in his *Compendium of Materia Medica*, "sweet and cold odoured, innocuous—mainly quenches thirst, cures fever" (Li 1987). Undoubtedly, the records of the health-related functions of fermented dairy and horse milk products in these medical books are mostly originated from the knowledge and experience of the Mongolian and other northern nomadic people.

In traditional Mongolian medicine, koumiss has always been used as a medicine to treat some diseases. *The Secret History of the Mongols* recorded that once Genghis Khan (1162–1227) was injured and became unconscious in a tangled warfare with the enemy and Zhelemie ignored his own safety and sneaked into the enemy's camp to look for koumiss for quenching the thirst and curing the wound of Genghis Khan (Aerdazhabu 2005). It is mentioned in a nutrition book in the Yuan dynasty, *Principles of Correct Diet* (*Yinshan Zhengyao*), that "Horse milk is sweet, can quench thirst and can cure fever. It may be divided into three classes, namely, Shengjian, Huangheer, and Chuangwu. Among them, Shengjian is the best" (Hu et al. 2009). Marco Polo (1254–1324) also said, "Koumiss is a beverage of Mongolian and nomadic people in Asia, which may be preserved for a long time. According to the legend, it is nourishing, and it can cure phthisis. Yet its taste is not liked by all" (Marco and Feng 1999). A famous Mongolian medical expert in the nineteenth century, Zhanbra Dorje (1792–1855), recorded in the *Mongolian Medicine Canonical* (*Mengyao Zhengdian*), that "Horse milk is sour, salty, and can cure lung and arthropathic disorders" (Zhanbra and Liubaiyila 2006). It was recorded in Mongolian pharmacopeia that "The taste of koumiss is sour, sweet and acerbic. The sour taste can stimulate appetite, help digestion, clear dampness, and promote the circulation of qi; the sweet taste can strengthen the body, clear the esophagus, repair injuries, set broken bones, detoxicate, and enhance the functions of the five sense organs; acerbic taste can cure blood heat, remove blood stasis, eliminate obesity, remove necrotic but promote granulation tissue, and moisturize skin" (Zhang and Na 1993). It was recorded in *A Supplement to Compendium of Materia Medica* in the Qing dynasty that "Chigo—produced by putting horse milk into a leather bag, tying the bag with a rope, lifting and pressing it with hand for about an hour, and leaving it at a hot place for one night. Drinking Chigo keeps people warm and nourished. Daily consumption of Chigo may serve a rejuvenating function. It is hot in nature, can compensate general debility and is suitable for people who want to become stronger" (Zhao 1871). A famous Tibetan doctor, Dimaer Danzengpengcuo, recorded in his book, *Jing Zhu Materia Medica* (booked in 1835), that "The taste of milk and its fermented products is sweet. They are nourishing, can moisturize skin and improve skin tone, develop strength, and boost the vital essence. Horse milk and donkey milk are pungent, sour, salty. They are lung-nourishing and can eliminate limb coldness" (Dimaer 1986).

Koumiss is often used to treat phthisis in the former USSR, Mongolia, and Inner Mongolia of China; the therapeutic effect is notable (Hou and Lei 1989; Kosikowski 1977; Zha 1999). Currently, "Koumiss Treatment Center" has been established in these regions specializing in treating cardiovascular diseases, digestive system diseases, nervous system diseases, tuberculosis, anemia, diabetes, and other chronic wasting diseases (La and Ce 1990; Lozovich 1995; Wu 1986). Modern medicine also proves the efficacy of koumiss in reducing blood fat, decreasing blood pressure, suppressing the growth of tubercle bacillus, curing constipation and chloasma, etc. (Hou and Lei 1989; Liu and Maliyahan 2000; Sun and Peng 2003; Wu and Ma 1998; Zhamusu et al. 1994).

On the basis of long-term medical practice and medical research, the Mongolian medicine has created "koumiss therapy" (Hasisurong et al. 2003; Hu and Dagula 1996; Liu and Gao 2003; Wu 1986). Wu Zhamusu summarized the medical experiences of the predecessors together with his own medical practice and wrote the book *Koumiss Therapy*, in which the obvious auxiliary therapeutic actions of koumiss for hypertension, hyperlipidemia, coronary artery hardening, anemia, phthisis, chronic digestive tract infection, diabetes and neurological diseases, etc. are documented (Wu 1986). The Mongolian Medicine Research Institute in Xilinguole League, Inner Mongolia, has been using the traditional fermented koumiss to treat gastrointestinal tract diseases, cardiovascular diseases, phthisis, asthma, etc. for many years with high efficacy. Meanwhile, the Institute has accumulated a lot of clinical experience. In summer when koumiss is made, a patient generally drinks several liters of koumiss every day. The Inner Mongolia Chinese-Mongolian Hospital and the Mongolian Medicine Research Institute in Xilinguole League, Inner Mongolia, have conducted many clinical trials to confirm the medical effects of koumiss. The results show that koumiss (1) can reduce blood fat and extend prothrombin time. After drinking koumiss, the whole blood viscosity, plasma viscosity, packed cell volume, blood sedimentation, plasma fibrinogen, etc. of the patient may fall to different degrees, so that the adhesion and aggregation of platelets on the vascular wall are prevented. Thus, it functions well to prevent from and treat myocardial infarction or cerebral thrombosis; and long-term consumption of koumiss can prevent from hypertension, coronary heart diseases, and cerebrovascular aging; (2) has certain therapeutic action on emphysema, phthisis, and tracheitis; (3) can suppress and kill enteric pathogens and may be used for therapy of chronic gastrointestinal tract diseases; (4) can increase immunity of the body and exert antiaging effect; (5) has certain positive effect regarding neurological problems like neural headache and neurasthenia; (6) can promote hematopoiesis function and improve on hypotension; (7) can regulate insulin secretion and sugar metabolism, thus improving diabetes; (8) also effectively acts against other diseases, for example, scurvy, hemorrhoid, amenorrhea, constipation, hemorrhoid, diarrhea, etc. (Zhaoeqimude 1994).

The curative effect of the naturally fermented Mongolian koumiss might come from some special components in horse milk, while most important of all are the lactic acid bacteria and their metabolites generated in the fermentation process. Horse milk is very effective for treating chronic hepatitis and gastric ulcer, of which the high contents of phospholipid and V_A are thought to be the active ingredients. Drinking horse milk can increase the quantities of human erythrocyte and lymphocyte and restore the normal packed cell volume. The human health-promoting effect of horse milk may be viewed from different perspectives. The n-3 series fatty acids in horse milk may promote the synthesis of prostaglandins; and it has a high intrinsic content of lactoferrin and lysozyme. In addition, the bioactive peptides generated after β-casein hydrolysis may also contribute to the therapeutic action, but little research is available in this respect. Recent researches have shown that horse milk and koumiss contain antihypertensive peptides (Doreaum and Martin 2003; Park et al. 2006). In traditional horse breeding areas, koumiss is commonly used for postoperative physical recovery of patients. This is because, in addition to the original milk components, microbial metabolic products like polypeptides, bacteriocins, and vitamins are generated upon the horse milk fermentation process. These components play an important role in patients' physical recovery.

Besides being used as food, in some European countries, horse milk is applied in cosmetic products like face cream, soap, and moisturizing cream. Examples of commercially available cosmetic products include Gyda's shampoo from Norway, designed specifically for skin allergy sufferers, and the Kumylac lotion made in Germany. However, it is unclear about the functional uniqueness of horse milk-based cosmetics, as compared to those containing other animal milks (Park et al. 2006; Zollmann 1985).

The composition of lactic acid bacteria varies greatly in the naturally fermented koumisses. The quality and quantity of koumiss microorganisms are influenced by factors like the local environment, climate, fermentation temperature and time, and the production method. In the fermentation and maturation processes of koumiss, lactic acid bacteria and yeasts are the dominant microbial species. The fermentation by both lactic acid bacteria and yeasts gives the unique flavor to koumiss, and at the same time these microorganisms, especially lactic acid bacteria, and their metabolites are closely linked to the medical effects of koumiss.

Because of the interesting health effect and functions of koumiss, it has been of interest to isolate the probiotic bacteria in the products that give rise to the functions (Fuller 1992). The Key Laboratory of Dairy Biotechnology and Engineering, the Ministry of Education of China, at the Inner Mongolia Agricultural University, had performed a series of experiments to screen for probiotic lactobacilli from koumiss collected from Inner Mongolia and Mongolia. Two lactobacillus strains with desirable properties have been identified, which are *L. casei* Zhang and *L. acidophilus* MG 2-1, respectively. These strains are acid and bile salt resistant and antioxidative; and they can also regulate blood lipid level and enhance immunity (Menghebilige et al. 2004, 2005a, b; Wang et al. 2005; Wu et al. 2005; Xu et al. 2005, 2006; Yun et al. 2006; Zhang et al. 2006a, b) (Figs. 10, 11, 12, 13, 14, 15, 16, 17, and 18).

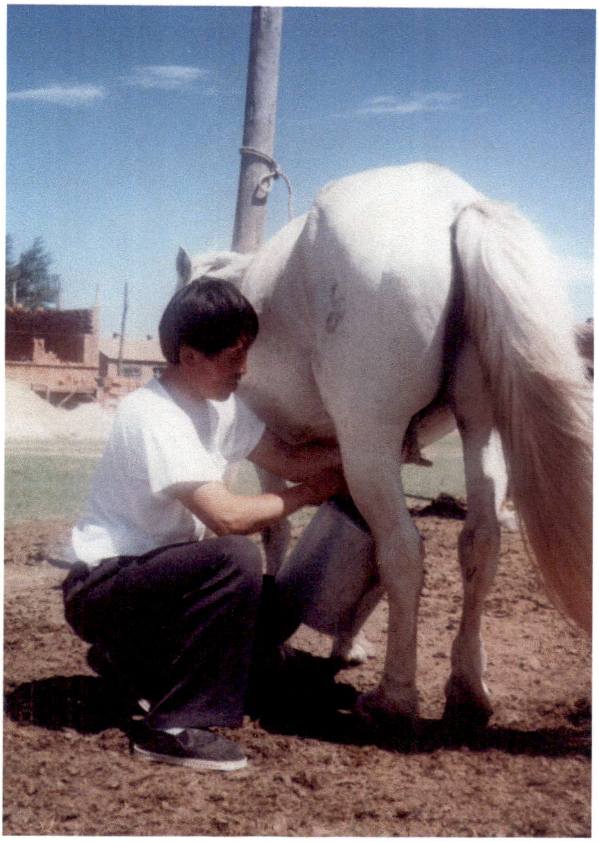

Fig. 10 Mare milking in Inner Mongolia

Fig. 11 Mare milking in Mongolia

Fig. 12 Koumiss fermented in a wooden cask in Inner Mongolia

3 Probiotics from Traditional Fermented Mongolian Foods: *Lactobacillus casei* Zhang

Koumiss is a widely consumed drink by Mongolians in the past centuries. Koumiss was considered as a traditional Mongolian adjunctive drug to cure pulmonary tuberculosis. This has led to research focused on fermented strains of commercial interest. *L. casei* Zhang (LCZ), isolated from koumiss in the Xilin Gol grassland, has been found to be an effective probiotic for the Chinese. A series of in vitro and in vivo studies have shown favorable probiotic functionality of LCZ. The present paper will summarize research findings on LCZ.

Fig. 13 Koumiss fermented in a leather bag in Xinjiang

Fig. 14 Koumiss fermented in an urn in Inner Mongolia

3.1 Screening Properties In Vitro

L. casei Zhang was originally isolated from 240 strains of *Lactobacillus* from koumiss in China. LCZ demonstrated favorable high tolerance to the destructive low pH of the stomach juice and bile salt in intestinal juice. The survival rate of LCZ was 73.5 % while subjected to pH 2.0 human artificial stomach juices for 2 h

Fig. 15 Starter and churning rod for koumiss fermentation

Fig. 16 A wooden cask for mare milk in Xinjiang

Fig. 17 Koumiss ready for drinking. Photo taken in Xinjiang

Fig. 18 Photo of a koumiss sanatorium established in Mongolia

and followed by pH 8.0 intestinal juice for 24 h. The tolerance of LCZ has proved to be similar to commercial probiotic strains such as *L. acidophilus* NCFM, *L. rhamnosus* GG, *L. casei* Shirota, and *B. animalis* Bb12 (Guo et al. 2009). In addition, gastrointestinal tolerance of LCZ in fermented bovine milk, soya milk, and mare milk was investigated (Wang et al. 2009, 2012; Zhou et al. 2009). This strain exhibited less stress in protective food materials and differentially expressed proteins observed in different materials.

3.2 Antimicrobial Activity

As one important mechanism to exhibit probiotic properties, antimicrobial activity is indispensable which leads to potential useful application (Servin 2004). Probiotic can adhere to mucosa and exhibit site competition effect to enhance local mucosal defenses in the intestine. Mice were challenged with *E. coli* O157 or *E. coli* K88 for evaluating the antimicrobial activity of LCZ. The results showed that the survival of infected mice was greatly improved by LCZ treatment (33.4 % vs. 94.4 %) and the intestinal *E. coli* counts significantly decreased (Zhang et al. 2007). It is suggested that LCZ could directly inhibit the colonization and rapid proliferation of *E. coli* via flora competition.

3.3 Lipid-Reducing Effect

Obesity has been considered as a disease by The American Medical Association. Fat is responsible for energy expenditure and is therefore important in health. But too much fat maintained in circulation is a risk factor linked to metabolic diseases (Tchkonia et al. 2013). Lipid-reducing effect has been found in *Lactobacillus acidophilus*, *Lactobacillus casei*, and *Lactobacillus plantarum* (Ooi and Liong 2010). *L. casei* Zhang had been reported to lower cholesterol, triglycerides, and low-density lipoprotein cholesterol (LDL-C) in hypercholesterolemia rats (Zhong et al. 2012). Further microarray investigation revealed that LCZ could accelerate the progression of β-oxidation of fatty acids, coenzyme A synthesis, and dehydrogenation by regulating expression of genes such as *Acsl1*, *Hadh*, *Acaa2*, *Acads*, and *gcdH* (Zhong et al. 2012). Importantly, the biosynthesis of bile acids by *Acaa2* upregulation was consistent with the discovery that probiotics potently enhanced the intestinal bile acid secretion. Additionally, *Nr3c1* (a glucocorticoid receptor) upregulation might be correlated with the prevention of high-fat-induced chronic inflammation. This study documented a global gene expression in regulating lipid metabolism by LCZ.

Oxidative stress is always accompanied by hyperlipidemic and oxidative damage, which are considered as accelerating risk factors to cardiovascular diseases (Yang et al. 2008). It has been well established that *Lactobacillus* strains possess antioxidative effects (Amaretti et al. 2013). LCZ has been shown to increase the total antioxidative status in the serum and liver of hyperlipidemic rats (Zhang et al. 2010b). Moreover, antioxidative SOD and GSH-Px activities were markedly improved, and lipid peroxidation was favorably inhibited after LCZ administration. In addition, biomarkers for fatty liver damage, the liver GOT and GPT, also decreased by LCZ administration. The antioxidative mechanisms of LCZ remain to be explored.

3.4 Liver Damage

In spite of high-fat-induced fatty liver, alcoholic fatty liver disease is also characterized by excessive alcohol consumption. LCZ had been reported to improve the liver damage of drunken mice by regulating activities of ALP and P450 enzymes (CYP3A1 and CYP3A2) (Zhang, unpublished).

Moreover, liver can be damaged by lipopolysaccharide (LPS) from Gram-negative bacteria through impaired intestinal permeability (Cani et al. 2008). LPS and induced endotoxemia are two of the major risk factors contributing to hepatocyte apoptosis. LCZ have been demonstrated to exhibit increased hepatoprotective effects to protect rats from LPS/D-GalN challenge through lowering TNF-α level and TLR4 expression (Wang et al. 2013a).

3.5 Immunomodulatory Effect and Anticancer Activity

Immunomodulatory effect is thought to be an important function for probiotic bacteria. It has been clearly demonstrated that gut microbiota has a considerable effect on the shaping of the immune system and the development of immune responses (Round and Mazmanian 2009). LCZ had proved to adhere to human intestinal Caco-2 cells with high efficiency (Zhang et al. 2006c). Once the living probiotic arrived in the intestine, bacteria adhering to intestinal cells may influence their binding activity with other substances. An in vitro study indicated that LCZ could interact with macrophage cells from intestinal epithelium and stimulate the production of nitric oxide and immune responses (Wang et al. 2013b). Moreover, LCZ significantly suppressed the exaggerated immune responses of macrophage induced by foreign antigens (Wang et al. 2013b). Later evidence from in vivo experiments suggested that LCZ is not only capable of inducing gut mucosal responses by secreting more SIgA but also of regulating systematic immune responses through IFN-γ and IL-12 secretion in mice (Ya et al. 2008). To sum up, LCZ provides important stimuli to maintain host immunity homeostasis.

Cancer has become a major public health concern all over the world. Immune deficiency is known to facilitate the growth of uncontrolled cancer cells and generate a variety of tumors throughout the body. LCZ could greatly inhibit the growth of tumor and accelerate apoptosis of cancer cells through regulating beneficial immune responses in tumor-bearing rats (Ya et al. 2010; Ya and Zhang 2010).

3.6 Regulation of Blood Glucose Level

It has recently become apparent that gut microbiota alterations are associated with blood glucose changes in both humans and mice (Hartstra et al. 2015). Microbiota disturbance-induced chronic inflammation is thought to be a leading role in the pathogenesis of metabolic syndrome and type 2 diabetes (Everard and Cani 2013). Two animal studies have been conducted with LCZ in the treatment and prevention of hyperglycemia.

In a previous study, LCZ was able to prevent the emergence of hyperinsulinemia in high-fructose-fed rats involving gut microbiota-driven improvement of GLP-2 (Zhang et al. 2014a). This result suggests that LCZ administration contributes to amelioration of gut permeability since GLP-2 is an important biomarker of gut permeability. In addition, LCZ could also alleviate symptom of rats with impaired glucose tolerance. Further investigation reveals that LCZ is a modulator of *B. fragilis*-vitamin K2-osteocalcin axis through a PPAR-gamma signaling.

The function of LCZ was found to exhibit high identity to a probiotic, *Lactobacillus acidophilus* (LA) ATCC4356. This is because a functional feature of stimulation of Cl^- absorption was found both in LA and LCZ (Zhang et al. 2014b; Raheja et al. 2010). LCZ could modify the gut microbiota and tissue Cl^- movement of

2-week high-fat-sucrose-fed rats (Zhang et al. 2014b). Hence, we further treated the mice with low dose of STZ to mimic the progression of T2DM. As a result, overnight fasting and postprandial 2-h blood glucose levels were markedly reduced in LCZ-pretreated group.

3.7 Altered Gut Microbiota in Human

Next-generation sequencing technology allows a more precise assessment of bacterial community in gut microbiota. By using 16S rRNA 454 pyrosequencing, Zhang et al. evaluated the efficacy of LCZ in 24 healthy volunteers belonging to three different ages (Zhang et al. 2014c; Kwok et al. 2014). They were subjected to LCZ treatment for 4 weeks and subsequently without LCZ for 2 weeks. The analysis of gut microbiota showed that LCZ altered the composition of intestinal microbiota such as *Bifidobacterium* and *Prevotella*. Furthermore, LCZ remained in high number in the gut of the subjects without LCZ administration for 2 weeks, suggesting the colonization of LCZ in human gastrointestinal tract. Interestingly, the composition of adult microbiota has become more similar to that of the young population after consumption of LCZ.

3.8 Research on Genomics and Proteomics

Important advances in the fields of genomics and proteomics have emerged in recent years. There is a growing interest in the molecular mechanism of the probiotic effects. Sequencing of the genome of LCZ was undertaken in 2008 and the GenBank accession number is CP001084. LCZ genome consists of a 2,861,848-bp circular chromosome and a 36-kb plasmid with a 46.5 % GC ratio in chromosome (Zhang et al. 2010a). Moreover, the LCZ genome is characterized by a number of phosphotransferase system (PTS)-related genes, mucus-binding-related genes, and a bacteriocin biosynthetic system. On the other hand, a reference map of protein expression in LCZ during different growing stages was constructed by proteomic analysis (Wu et al. 2009). Microbial colonization of living probiotics in the gastrointestinal tract is limited to the low pH of gastric juice and bile salt tolerance. Further to this, the proteomic changes of LCZ subjected to mimic gastric juice and gastrointestinal transit environment were conducted. Differentially expressed proteins were mainly belonging to stress response proteins and central and intermediary metabolic proteins (Wu et al. 2009). These genomic and proteomic studies have paved the way for future utilization and fast industrial screening strains.

Overall, we have shown the functional characteristics of LCZ in various respects. Although probiotics clearly play a significant role in regulating host gut microbiota, probiotic has its own specificity, and the mechanism remains unclear.

LCZ is a relatively novel probiotic strain that possesses multiple desirable properties exerted by unique mechanisms compared to other well-studied probiotics. Therefore, it merits further evaluation of LCZ, as well as its beneficial effects, functions, and mechanisms in different disease models.

References

Aerdazhabu (2005) New translation and annotation for the *secret history of the Mongols* (translation and annotation). Inner Mongolia University Publishing House, Hohhot, 265p

Amaretti A, di Nunzio M, Pompei A, Raimondi S, Rossi M, Bordoni A (2013) Antioxidant properties of potentially probiotic bacteria: *in vitro* and *in vivo* activities. Appl Microbiol Biotechnol 97(2):809–817

Cani PD, Delzenne NM, Amar J, Burcelin R (2008) Role of gut microflora in the development of obesity and insulin resistance following high-fat diet feeding. Pathol Biol (Paris) 56:305–309

Cheng W (1984) Erudite poetry. Bibliography and Document Publishing House (Book 1), Beijing, 117p (Yuan)

Dawson (1983) The record of an envoy to the Mongolia (translated by Lv P and annotated by Zhou L). China Social Science Press, Beijing, China, pp 117 [UK]

Dimaer D (1986) Jing Zhu Materia Medica. Shanghai Scientific and Technical Publishers, Shanghai, pp 12–13

Doreaum, Martin RW (2003) Horse. In: Encyclopedia of dairy science, Vol 2. Academic, London, pp 630–636

Everard A, Cani PD (2013) Diabetes, obesity and gut microbiota. Best Pract Res Clin Gastroenterol 27(1):73–83

Feng C, Dang B (1999) Travels of Marco Polo (translated and annotated by authors). Hebei People's Publishing House, Shijiazhuang, 242p

Fuller R (1992) Probiotics. In: The scientific basis. Chapman and Hall, New York, 398p

Gu Y (1997) Miscellaneous poems in Xikou, Datong. Zhuo Erkan. Poems of the adherent poets. In: Collection of the books unincluded in the imperial collection of four, Vol 5. Collections. Beijing Publishing House, Beijing, pp 21–532 (Qing)

Guo Z, Wang J, Yan L, Chen W, Liu X, Zhang H (2009) In vitro comparison of probiotic properties of *Lactobacillus casei* Zhang, a potential new probiotic, with selected probiotic strains. LWT - Food Sci Technol 42:1640–1646

Hartstra AV, Bouter KE, Bäckhed F, Nieuwdorp M (2015) Insights into the role of the microbiome in obesity and type 2 diabetes. Diabetes Care 38(1):159–165

Hasisurong, Amuguleng, Manglai (2003) Koumiss and its medical value. China J Chin Mater Med 28(1):11–14

Hou W, Lei T (1989) Test of horse milk fat to resist mycobacterium tuberculosis. J Anim Sci Vet Med 2:1–4

Hu C, Dagula (1996) Mongolian koumiss therapy. J Med Pharm Chin Minor 2(1):32

Hu S, Shang Y, Sun L, Lin H (eds) (2009) Notes of the principles of correct diet (annotated by Lin H). Central University for Nationalities Publishing House, Beijing, 213p (Yuan)

Ishii S, Konagaya Y (2002) Beneficial role of kumiss intake of Mongolian nomads. J Jpn Soc Nutr Food Sci 55(5):281–285

Jain AK, Prajapati JB (1998) Associative growth in dairy starter cultures—a review. J Dairy Foods Home Sci 17(1):1–11

Keogh, Barbara P (1976) Microorganisms in dairy products—friends and foes. Pag Bull 6 (4):34–41

Kosikowski FV (1977) Cheese and fermented milk foods, 2nd edn. Edwards Brothers, Ann Arbor, MI, pp 40–46

Kwok LY, Wang L, Zhang J, Guo Z, Zhang H (2014) A pilot study on the effect of *Lactobacillus casei* Zhang on intestinal microbiota parameters in Chinese subjects of different age. Benef Microbes 5(3):295–304

La B, Ce N (1990) Mongolian Airag. Inner Mongolia People's Publishing House, Hohhot, pp 137–138

Li S (1987) Compendium of Materia Medica. Complete library of the four branches of literature, Vol 774. Shanghai Ancient Books Publishing House, Shanghai, 445p (Ming)

Liang Z (1935) Langji Santan, Vol 5. Shanghai Jinbu Book Store, Shanghai, 194p (Qing)

Liao X, Feng F (2008) Political function and influence of national ritual in ancient China. Pursuit 2:56–58

Liu H, Gao K (2003) Nutrition value and medical effect of horse milk and koumiss (Fermented horse milk). Food Nutr China 4:45–46

Liu G, Maliyahan (2000) Koumiss therapy for constipation. China's Naturopathy 8(6):27–28

Lozovich S (1995) Medical uses of whole and fermented mare milk in Russia. Cultur Dairy Prod J 30(1):18–19

Mann E (2003) Kefir and koumiss. Dairy Ind Int 10:36–37

Marco Polo, Feng C (1999) The travels of Marco Polo (narration from Marco Polo and translated by Feng C). Shanghai Bookstore Publishing House, Shanghai, pp 155–156

Menghebilige WR, Wang L, Yang X, Xu J, Dong Y, Sun Z, Zhang H (2004) Research on the separation of lactobacillus from koumiss and its biological properties in different districts. China Dairy Ind 32(11):6–11

Menghebilige ZH, Chen Y, Guan H, Zhou D (2005a) Research on the impact of *Lactobacillus acidophilus* MG2-1 on serum lipid metabolism of rats. Acta Microbiol Sin 45(6):52–54

Menghebilige ZY, Zhang H, Chen Y, Wang L, Xu J, Zhou D (2005b) Research on the antioxidative function of Lactobacillus MG2-1 strain in koumiss. China Dairy Ind 33(9):22–23

Ooi LG, Liong MT (2010) Cholesterol-lowering effects of probiotics and prebiotics: a review of *in vivo* and *in vitro* findings. Int J Mol Sci 11:2499–2522

Ozer B (2000) Fermented milks-products of eastern European and Asia. In: Robinson RK, Batt CA, Patel PD (eds) Encyclopedia of food microbiology. Academic Press, London, pp 803–804

Park YW, Zhang H, Zhang B, Zhang L (2006) Mare milk. In: Park YW, Haenlein GFW (eds) Handbook of milk of non-bovine mammals. Blackwell Publishing, Hoboken, NJ, pp 275–329

Qiaobenrichuren, Li S, Menghebilige, Edungerile (1998) The techniques and methods of Inner Mongolia people to make traditional dairy products. Inner Mongolia Agricultural University, Japan International Cooperation Agency, Hohhot, pp 20–29

Raheja G, Singh V, Ma K, Boumendjel R, Borthakur A, Gill RK, Saksena S, Alrefai WA, Ramaswamy K, Dudeja PK (2010) *Lactobacillus acidophilus* stimulates the expression of SLC26A3 via a transcriptional mechanism. Am J Physiol Gastrointest Liver Physiol 298 (3):395–401

Robinson RK (2002) Dairy microbiology handbook, 3rd edn. Wiley, New York, pp 417–419

Round JL, Mazmanian SK (2009) The gut microbiota shapes intestinal immune responses during health and disease. Nat Rev Immunol 9(5):313–23

Servin AL (2004) Antagonistic activities of lactobacilli and bifidobacteria against microbial pathogens. FEMS Microbiol Rev 28(4):405–440

Song L (1976a) The history of Yuan. In: Records of sacrificial offering 3, Ancestral temple 1, Vol 74, Annal 25. Zhonghua Book Company, Beijing, 1841p (Ming)

Song L (1976b) The history of Yuan. In:·Collected biographies, 92 Treacherous court official, Vol 250. Zhonghua Book Company, Beijing, 4566p (Ming)

Su J (1997) New Materia Medica. In: Revised continuance of complete library in the four branches of literature, Vol 989. Shanghai Ancient Books Publishing House, Shanghai, 661p (Tang)

Sun S (1987) Essential recipes for emergent use worth a thousand gold. In: Complete library of the four branches of literature, Vol 735. Shanghai Ancient Books Publishing House, Shanghai, 817p (Tang)

Sun H, Peng G (2003) In vitro test and observation of horse milk fat to suppress the growth of *Mycobacterium tuberculosis*. J Pract Med Tech 10(10):112

Tchkonia T, Thomou T, Zhu Y, Karagiannides I, Pothoulakis C, Jensen MD, Kirkland JL (2013) Mechanisms and metabolic implications of regional differences among fat depots. Cell Metab 17:644–656

Wang G (1983a) Annotation of the biographical sketches of Black Tartars. Posthumous papers of Wang Guowei (13). Photographic print of Shanghai Ancient Bookstore according to the 1940 Edition of the Commercial Press, Shanghai, p 6a

Wang G (1983b) Annotation of the biographical sketches of Black Tartars. Posthumous Papers of Wang Guowei (13). Photographic print of Shanghai Ancient Bookstore according to the 1940 Edition of the Commercial Press, Shanghai, p 15a-b

Wang G (1983c) Annotation of the biographical sketches of Black Tartars. Posthumous papers of Wang Guowei (13). Photographic print of Shanghai Ancient Bookstore according to the 1940 Edition of the Commercial Press, Shanghai, p 19b

Wang L, Xu J, Yun Y, Menghebilige WR, Zhang H (2005) Research on the potential probiotic properties of lactobacillus in Mongolia traditional fermented koumiss. China Dairy Ind 33 (4):4–10

Wang J, Guo Z, Zhang Q, Yan L, Chen W, Liu XM, Zhang HP (2009) Fermentation characteristics and transit tolerance of probiotic *Lactobacillus casei* Zhang in soymilk and bovine milk during storage. J Dairy Sci 92(6):2468–2476

Wang J, Zhong Z, Zhang W, Bao Q, Wei A, Meng H, Zhang H (2012) Comparative analysis of the gene expression profile of probiotic *Lactobacillus casei* Zhang with and without fermented milk as a vehicle during transit in a simulated gastrointestinal tract. Res Microbiol 163 (5):357–365

Wang Y, Li Y, Xie J, Zhang Y, Wang J, Sun X, Zhang H (2013a) Protective effects of probiotic *Lactobacillus casei* Zhang against endotoxin- and d-galactosamine-induced liver injury in rats via anti-oxidative and anti-inflammatory capacities. Int Immunopharmacol 15(1):30–37

Wang Y, Xie J, Wang N, Li Y, Sun X, Zhang Y, Zhang H (2013b) *Lactobacillus casei* Zhang modulate cytokine and Toll-like receptor expression and beneficially regulate Poly I:C-induced immune responses in RAW264.7 macrophages. Microbiol Immunol 57(1):54–62

Wu Z (1986) Koumiss therapy. Inner Mongolia People's Publishing House, Hohhot

Wu Z (2002) Records of Yangjizhai. Revised continuance of complete library in the four branches of literature, Vol 24. Zi Radical. Eclectics Category. Shanghai Ancient Books Publishing House, Shanghai, 475p (Qing)

Wu J, Ma F (1998) Efficacy observation of 44 cases in the therapy of chegee (koumiss) cream reverse mold facial mask for chloasma. Chin J Dermatovenereol 12(2):110

Wu R, Zhang H, Menghebilige (2005) 16S rDNA gene order and cluster analysis of *L. casei* Zhang and ZL12-1 in koumiss. China Dairy Ind 33(6):4–9

Wu R, Wang W, Yu D, Zhang W, Li Y, Sun Z, Wu J, Meng H, Zhang H (2009) Proteomics analysis of *Lactobacillus casei* Zhang, a new probiotic bacterium isolated from traditional home-made koumiss in Inner Mongolia of China. Mol Cell Proteomics 8(10):2321–38

Xie Z (2001) Wuzazu records, Vol 11, Material Part 3. Shanghai Bookstore Publishing House, Shanghai, 218p (Ming)

Xu Y (1530) Etiquettes in Ming Dynasty. In: New recommendations, Vol 4. Collected by Waseda University, Japan, continued block-printed edition in the 9th year of Jiajing, p 9b (Ming)

Xu J, Shao Y, Menghebilige ZH (2005) Research on the fermentation characteristic of *Lactobacillus acidophilus* in a traditional Mongolian fermented dairy product–koumiss. J Dairy Sci Technol 5:202–206

Xu J, Yun Y, Zhang W, Shao Y, Menghebilige ZH (2006) Research on the fermentation characteristic of *Lactobacillus casei* in koumiss. China Dairy Ind 34(7):23–27

Ya T, Zhang H (2010) Anti-tumor mechanisms of probiotic *L. casei* Zhang. J Mol Diagn Ther 2 (6):394–399

Ya T, Zhang Q, Chu F, Merritt J, Bilige M, Sun T, Du R, Zhang H (2008) Immunological evaluation of *Lactobacillus casei* Zhang: a newly isolated strain from koumiss in Inner Mongolia, China. BMC Immunol 19(9):68

Ya T, Du R, Zhang H (2010) Anti-tumor effect and its mechanism of *Lactobacillus casei* Zhang on H22 tumor-bearing mice. Cancer Res Prev Treatm 37(4):463–465

Yan S, Ga E, Zhang F, Shuang Q, He Y (1997) Analysis and evaluation of nutrition value of the traditional dairy products in Xilinguole League, Inner Mongolia. J Inner Mongol Coll Agric Anim Husb 18(4):31–36

Yang RL, Shi YH, Hao G, Li W, Le GW (2008) Increasing oxidative stress with progressive hyperlipidemia in human: relation between malondialdehyde and atherogenic index. J Clin Biochem Nutr 43:154–158

Yu J (1912) Historical data recorded in Daoyuan. In: Achievements Stele of Jurong King, Vol 23. Cungu Book Store, Chengdu, p 9b (Yuan)

Yuan J (2005) Textual research on imperial feasts in the Yuan Dynasty. Research on the history of Mongolia, 8th Collection. Inner Mongolia University Publishing House, Hohhot, pp 116–132

Yun Y, Wang L, Zhang H, Chen Y, Menghebilige (2006) The impact of feeding *Lactobacillus casei* Zhang on lipid metabolism in rats. Microbiol China 33(3):60–64

Zha H (1999) Clinical observation of 56 cases in koumiss therapy for phthisis. Chin J Ethnomed Ethnopharm 1:13–14

Zhamusu, Liu Y, Jin Z (1994) Clinical observation of 50 cases for the lipid-lowering and anticoagulation of koumiss. Inner Mongol J Tradit Chin Med 139(4):16

Zhanbra D, Liubaiyila (2006) Mongolian medicine canonical (edited by Zhanbra and processed by Liubaiyila). National Publishing House, Beijing, 349p (Qing)

Zhang P (2002) Jianyu collections. In: Revised continuance of complete library in the four branches of literature, Vol 4 (Letters). Ji Radical. Collections Amendments. Shanghai Ancient Books Publishing House, Shanghai, 505p (Qing)

Zhang H, Na R (1993) Mongolian koumiss. China Dairy Ind 21(4):165–168

Zhang H, Menghebilige, Wang J, Sun T, Xu J, Wang L, Yun Y, Wu R (2006a) Research on potential probiotic properties of *L. casei* Zhang separated from traditional fermented koumiss of Inner Mongolia. China Dairy Ind 34(4):4–10

Zhang H, Zhang Q, Menghe B, Amuer J (2006b) Effect of oral administration of *L. casei* Zhang on T-lymphocyte subclass, serum IgG and intestinal mucous SIgA of mouse. China Dairy Ind 34:4–8

Zhang Q, Sun T, Menghebilige, Ren G, Amuerjirigala, Zhang H (2006c) Research on the impact of lactic acid bacteria preparation on Newcastle disease HI antibody valence. Chin J Anim Quar 23(2):22–24

Zhang H, Zhang Q, Ren G, Bao Q (2007) The antagonism of *Lactobacillus casei* Zhang to pathogenic *Escherichia coli* in mice and the influence on the microbial population in gut. Microbiol China 5:63–68

Zhang W, Yu D, Sun Z, Wu R, Chen X, Chen W, Meng H, Hu S, Zhang H (2010a) Complete genome sequence of *Lactobacillus casei* Zhang, a new probiotic strain isolated from traditional homemade koumiss in Inner Mongolia, China. J Bacteriol 192(19):5268–9

Zhang Y, Du R, Wang L, Zhang H (2010b) The antioxidative effects of probiotic *Lactobacillus casei* Zhang on the hyperlipidemic rats. Eur Food Res Technol 231(1):151–158

Zhang J, Wang L, Guo Z, Sun Z, Gesudu Q, Kwok L, Menghebilige, Zhang H (2014a) 454 pyrosequencing reveals changes in the faecal microbiota of adults consuming *Lactobacillus casei* Zhang. FEMS Microbiol Ecol 88(3):612–22

Zhang Y, Guo X, Guo J, He Q, Li H, Song Y, Zhang H (2014b) *Lactobacillus casei* reduces susceptibility to type 2 diabetes via microbiota-mediated body chloride ion influx. Sci Rep 4:5654

Zhang Y, Wang L, Zhang J, Li Y, He Q, Li H, Guo X, Guo J, Zhang H (2014c) Probiotic *Lactobacillus casei* Zhang ameliorates high-fructose-induced impaired glucose tolerance in hyperinsulinemia rats. Eur J Nutr 53:221–232

Zhao X (1871) A supplement to compendium of Materia Medica, Vol 9. Block-Printed Edition of Jixintang in the 10th Year of Tongzhi, Qing Dynasty, 64p (Qing)
Zhaoeqimude (1994) Nutrition and medical effect of traditional Mongolian dairy product Chegee (Koumiss). In: Proceedings of the second symposium for processing of animal products on both sides of the Taiwan Straits, pp 116–119
Zheng S (1991) Sketches of righteous cause (collated by Chen Fukang). In: The works of Zheng Sixiao. Shanghai Ancient Books Publishing House, Shanghai, 187p (Song)
Zhong Z, Zhang W, Du R, Meng H, Zhang H (2012) Effect of *Lactobacillus casei* Zhang on global gene expression in the liver of hypercholesterolemic rats. Eur J Lipid Sci Technol 114:244–252
Zhou Q, Wang J, Guo Z, Yan L, Zhang Q, Chen W, Liu X, Zhang H (2009) Fermentation characteristics and transit tolerance of probiotic *Lactobacillus casei* Zhang in mare milk during storage. Int J Dairy Technol 62(2):249–254
Zollmann H (1985) Mare's milk. Electoral Stud Farm, Waldbrunn-Muelben, p 8

Microencapsulation of Probiotic Bacteria

Anthony N. Mutukumira, Jolyn Ang, and Sung Je Lee

Contents

Abstract The global popularity of functional foods containing probiotics has generated increased interest in developing protective materials for the bacteria during food processing, storage and consumption. To confer beneficial effects to the host, a probiotic product should contain at least 10^6 CFU/g or ml of the product. Probiotic microorganisms are sensitive to food processing environments and to conditions in the gastrointestinal tract. Microencapsulation technology can be used to protect the probiotic bacteria against adverse conditions. This chapter discusses the potential of using various microencapsulation techniques to protect probiotic bacteria.

A.N. Mutukumira (✉) • J. Ang • S.J. Lee
School of Food and Nutrition, College of Health, Massey University, North Shore Mail Centre, Private Bag 102904, Auckland 0745, New Zealand
e-mail: a.n.mutukumira@massey.ac.nz; tony.mutukumira@gmail.com

M.-T. Liong (ed.), *Beneficial Microorganisms in Food and Nutraceuticals*,
Microbiology Monographs 27, DOI 10.1007/978-3-319-23177-8_3

1 Introduction

The interest in using live probiotics for health maintenance and disease prevention through their incorporation into foods has been continuously increasing in the food industry (Ouwehand et al. 2002; Picot and Lacroix 2003). *Bifidobacterium* and *Lactobacillus* species have been the focus of probiotic interest since the presence of a large population of these bacteria in the gastrointestinal (GI) tract is indicative of a healthy microbiota (Crittenden et al. 2001; Picot and Lacroix 2003). However, some studies have shown poor viability of probiotic bacteria in functional foods when exposed to intrinsic and extrinsic factors such as moisture, oxygen, light, relative humidity, product composition and heat during food processing and storage (Kailasapathy 2002; Ding and Shah 2009; Peighambardoust et al. 2011). A significant proportion of probiotics is destroyed by high stomach and bile acids during their passage through the GI tract (Sultana et al. 2000; Krasaekoopt et al. 2003). Since probiotics are required at specific target sites in the host, it is therefore essential that probiotics are able to withstand the host's natural barriers against the ingested bacteria and remain viable (Kailasapathy and Chin 2000; Kailasapathy 2002; Picot and Lacroix 2003, 2004). In order for probiotics to confer beneficial effects on human health by improving the balance of intestinal microflora, high levels of viable cells are recommended in probiotic foods for efficacy (Gardiner et al. 2002). According to the FAO/WHO (2002), the level of probiotic bacteria required in a food product at consumption should be at least $>10^6$ cells/g or ml.

Microencapsulation is a potential technology that protects sensitive lactic cultures against harsh conditions during their transit through the GI tract of the host (Kailasapathy 2002; Mortazavian et al. 2007; Solanki et al. 2013). Encapsulating live cells within a shell material can minimize cell injury and enhance cell viability. Different encapsulation methods such as spray-drying, spray freezing, extrusion, emulsion and fluidized bed coating have been successfully used to encapsulate probiotics using various materials to protect them and extend their shelf life.

2 Probiotic Bacteria

Probiotic bacteria can be defined as live microorganisms that are beneficial to human health by positively influencing the intestinal microbial balance (Crittenden et al. 2001; Ananta et al. 2005). Previous studies have demonstrated numerous beneficial effects of probiotics in human health, including the prevention of diarrhoea caused by certain pathogenic bacteria and viruses (Andersson et al. 2001; Isolauri 2001). The term "probiotic" includes a large range of microorganisms mainly bacteria, but yeasts are also included (Ouwehand et al. 2002; Mortazavian et al. 2007). With the ability to remain alive until they reach the colon and provide beneficial effects to the host, selected lactic acid bacteria, non-lactic acid bacteria and yeasts can all be considered as probiotics.

2.1 Types of Probiotic Microorganisms

Probiotics can be classified into three types, namely, lactic acid bacteria (LAB), non-lactic acid bacteria and yeasts. LABs are mainly gram-positive and usually live in anaerobic environments, but they can also grow under aerobic conditions. Bifidobacteria can grow at a pH range of 4.5–8.5, but the most important characteristic of this group of LAB is that they are strictly anaerobic. Other LABs such as *Lactococcus lactis*, non-lactic acid bacteria (e.g. *Escherichia coli* Nissle 1917 and *Enterococcus faecium*) and some yeasts (e.g. *Saccharomyces cerevisiae*, *Saccharomyces boulardii*) are considered as probiotics (Burgain et al. 2011). However, only strains classified as LAB are considered important in food and nutrition (Mortazavian et al. 2007). The characteristics and optimum growth conditions of probiotic cultures vary with different microorganisms.

2.2 Health Beneficial Roles of Probiotics

There is strong evidence that probiotics have the potential to exert beneficial effects to human health (Isolauri 2001; Isolauri et al. 2001; Kalliomäki et al. 2001; Ouwehand et al. 2002). The effects of probiotics are strain specific, and therefore, it is important to specify the genus and species of probiotic bacteria when claiming health benefits (Saarela et al. 2000). The probiotic health benefits may be due to various factors, such as the production of bacteriocins, competition for nutrients with pathogens and enhanced immune system (Isolauri 2001; Peighambardoust et al. 2011). However, the exert mechanisms of how probiotics confer health benefits to human health are still not well studied.

Probiotics play significant therapeutic roles in human nutrition (O'Riordan et al. 2001). The main therapeutic and health benefits of most probiotics include enhancing the immune system against intestinal infections; prevention of diarrheal diseases, colon cancer, hypercholesterolaemia and upper gastrointestinal tract diseases; and stabilization of gut mucosal barrier (Kailasapathy and Chin 2000; Isolauri 2001; Isolauri et al. 2001; Kalliomäki et al. 2001; Desmond et al. 2002; Ouwehand et al. 2002; Krasaekoopt et al. 2003). Probiotic bacteria not only compete with the growth of pathogens and suppress "unhealthy fermentations" in the human intestine, but also confer several beneficial effects on the host by improving its intestinal microbial balance (Kailasapathy and Chin 2000).

3 Microencapsulation Techniques

Microencapsulation is defined as a technology of packaging solids, liquids or gaseous materials in miniature, sealed capsules that can release their contents at controlled rates under specific conditions (Anal and Stevens 2005). Encapsulation can be used for many applications in the food industry, including stabilization of materials, control of oxidative reaction and degradation, sustained and/or controlled release, masking flavours, colours or odours, extension of shelf life and protection of components against nutritional loss. A variety of different techniques are available that can be used for encapsulation including spray-drying, spray-congealing, fluidized bed coating, extrusion and coacervation (Anal and Singh 2007).

Microencapsulation techniques can be classified into encapsulation process and drying process (Solanki et al. 2013). In the encapsulation process, probiotics are commonly encapsulated by a method involving cell immobilization through their physical entrapment in polymeric networks, such as extrusion and emulsion. Immobilization of LAB can offer many advantages for biomass and metabolite production compared with free-cell systems, such as high cell concentration, reuse of biocatalysts, improved resistance to contamination and bacteriophage attack, enhancement of plasmid stability and resistance to washing out during continuous cultures as well as physical and chemical protection of cells (Champagne et al. 1994; Doleyres and Lacroix 2005). For food applications, cell entrapment in a food-grade porous gel matrix has been widely used.

In the dairy industry, immobilized cells can be utilized for continuous yoghurt fermentation and optimized continuous biomass production through re-harvest but are not directly incorporated in foods (Heidebach et al. 2009a). This could be because these microcapsules are not optimized towards the requirements for capsules intended to pass through the human GI tract. In the drying process for microencapsulation of microorganisms, spray-drying has been widely investigated, particularly for the preservation of LAB (Santivarangkna et al. 2007). The ideal cell culture for distribution is in a dry form, sufficiently active to produce rapid growth when added to foods and yet dormant that it can be stored for extended periods without losing its activity. This section describes the preparation of microcapsules entrapping probiotics by using cell immobilization techniques, such as emulsion and extrusion, and spray-drying.

3.1 Cell Immobilization for Microencapsulation

3.1.1 Emulsion Technique

The emulsion technique allows the encapsulation of probiotic living cells by using hydrocolloids (e.g. alginate, carrageenan and pectin) as encapsulating materials. The principle of this technique is based on the addition of a cell-polymer suspension (e.g. alginate) into a vegetable oil (e.g. soybean oil, corn oil) to form a water-in-oil emulsion (Burgain et al. 2011). The emulsion containing small droplets is formed by agitating the mixture usually with a magnetic stirrer to form tiny gel particles within the oil phase. The solidified beads are then formed by adding a cross-linking agent (e.g. calcium chloride) to the solution while stirring. The calcium alginate gel beads formed can be further introduced into a second polymer solution to create a coating layer that provides added protection to the encapsulated cells (Burgain et al. 2011). An advantage of the emulsion technique for microencapsulation of probiotics is that the process is relatively easy to scale up and gives high survival rates of microbial cells (Burgain et al. 2011). However, the main disadvantage of this method is that it provides relatively large sizes of beads (Burgain et al. 2011) ranging from 200 to 1000 μm in diameter that can influence the texture and mouthfeel when added into food products (Capela et al. 2007).

The size of capsules or beads entrapping cells varies and is affected by various factors, such as the type, concentration and viscosity of wall material polymers used, and the shear stress and pressure applied. In most instances, hydrocolloids, such as alginate, κ-carrageenan, gellan gum and xanthan gum, have been used as matrix materials for the immobilization of probiotics using the emulsion technique. The high viscosities of the aqueous solutions of these polymers even at low concentrations hinder the formation of small droplets in the continuous oil phase, resulting in the formation of large gel beads. Some methods that could be employed in reducing the bead size is to use homogenization using mechanical devices such as high shear mixer and high pressure valve homogenizer that generate intensive disruptive forces. In a study by Capela et al. (2007), the effect of homogenization on the size of calcium alginate beads was demonstrated using different types of homogenizers (e.g. ultra-turrax, Silverson mixer and Avestin piston homogenizer). It was shown that homogenization induced the formation of small beads from 3 % (w/v) alginate solution with average diameters between 30 and 200 μm. However, the disadvantage of these processes involving homogenization can be its negative impact on the encapsulation efficiency and viability of probiotics as the study showed a significant loss in the survival of probiotics (35.6–94.5 %) due to the shearing stress. This was also affected by the type of homogenizers used. Individual species of probiotic microorganisms may vary in their sensitivity to external stresses, such as those encountered with high shear force and pressure during homogenization. It is therefore essential to ensure that the method and conditions selected for the microencapsulation process of probiotic microorganisms is gentle to sensitive probiotic microorganisms.

Novel encapsulation techniques based on emulsification and enzymatic cross-linking or cold gelation were also introduced for the delivery of heat-sensitive bioactives including probiotics using proteins as encapsulating matrix agents. Milk protein, such as sodium caseinate, was used for encapsulation of probiotic micro-organisms using similar principle and technique as described above, but gelation of caseins was induced by using enzymes (transglutaminase or rennet) during the emulsifying process (Heidebach et al. 2009a, b). The studies demonstrated small capsules lower than 200 μm in diameter being formed by gentle agitation using a magnetic stirrer without the addition of extra emulsifiers or application of high shearing during the emulsifying process. This could be attributed to good emulsifying properties of milk proteins and the low viscosity of the milk concentrate regardless of having a high solid concentration of 35 % (Heidebach et al. 2009a). Cold gelation provides an alternative gel matrix development method based on adding cations to a preheated protein suspension (Maltais et al. 2005). With this method, a heating step is required where whey proteins are denatured and polymerized into soluble aggregates. This is followed by a cooling step and subsequent salt addition which results in the formation of a network via Ca^{2+}-mediated interactions of soluble aggregates (Roff and Foegeding 1996). The formation of cold-set gels opens interesting opportunities for whey proteins as carriers of heat-sensitive bioactive compounds (Chen and Subirade 2007).

3.1.2 Extrusion Technique

Extrusion can be used to encapsulate probiotic living cells and uses hydrocolloids (alginate and carrageenan) as encapsulating materials because of their ability to form gels under mild conditions. Extrusion is a simple and cheap method that uses a gentle operation which does not damage probiotic cells and gives high probiotic viability. With this method, the solution containing cells is projected through a nozzle at high pressure into a hardening solution (e.g. calcium chloride), and beads with a few millimetres in diameter are produced (Burgain et al. 2011). Extrusion leads to a relatively narrow particle size distribution which is determined by the geometry of the nozzle (Heidebach et al. 2009a). The inner diameter of the nozzle or openings influences the size of capsules formed (De Vos et al. 2010).

Large-scale droplet production can be achieved by multiple-nozzle systems, spinning disc atomizer or jet-cutter techniques (De Vos et al. 2010; Kailasapathy 2002). The advantage of extrusion technology is that it is in most cases a true encapsulation procedure and not just an immobilization technology (De Vos et al. 2010). Extrusion technology has many advantages for encapsulation of microbes. This technology does not involve the use of deleterious solvents or oils and can be conducted under both aerobic and anaerobic conditions. This is especially advantageous with anaerobic microorganisms by placing an extrusion device in a sterile cabinet where oxygen is substituted for nitrogen (De Vos et al. 2010).

3.1.3 Encapsulating Materials for Cell Immobilization

Entrapment of LAB in calcium alginate beads has been commonly used for cell immobilization (Rowley et al. 1999). Alginate is a naturally derived polysaccharide extracted from brown sea weeds which is composed of α-L-guluronic and β-D-mannuronic acids (Koo et al. 2001). The simple, mild aqueous-based gel formation of alginate is completed in the presence of divalent cations such as Ca^{2+} (Koo et al. 2001). Calcium alginate beads entrapping cells are made by extruding a solution of alginate mixed with cells into a calcium chloride solution. Calcium chloride solution containing cells can also be dropped as tiny droplets into alginate solution while being stirred and then instantly wrapped with calcium alginate gel membrane. The characteristics of hydrogel membranes such as thickness, pore size, surface charge and mechanical strength can be fabricated by varying some processing conditions and ingredient formulations. These include reaction time for gelation, concentration of alginate and calcium ions and different types of gel-forming polymers (Nigam et al. 1988).

Encapsulation of bacteria in alginate gel can improve survival rates of bacteria by one log compared to free cells when stored in skim milk for 24 h (Rokka and Rantamaki 2010). Alginate gel cross-linked with calcium ions is preferred for encapsulating probiotics due to its simplicity, non-toxicity, biocompatibility and low cost (Krasaekoopt et al. 2003). However, some disadvantages may be attributed to its sensitivity to strong acidic environment which is not resistant to decomposition in the stomach conditions (Mortazavian et al. 2007). Calcium alginate capsules are also chemically unstable upon contact with cation chelating agents, such as phosphate, citrate and lactate, which can cause disruption or dissolution of alginate gel matrix (Castroa et al. 2009). In addition, the microparticles formed with alginate are porous which is a disadvantage for good protection of cells. Since alginate is anionic, cationic polymer coatings, such as polylysine, polyvinylamine and chitosan, have been used to increase the stability of alginate capsules or to minimize the loss of encapsulated material. However, these can be compensated by mixing alginate with starch that has been commonly used to improve the effectiveness of probiotic encapsulation (Sultana et al. 2000; Hansen et al. 2002; Krasaekoopt et al. 2003).

Gellan gum is a microbial polysaccharide derived from *Pseudomonas elodea*. It is composed of a repeating unit of tetrasaccharide consisting of glucose, glucuronic acid, glucose and rhamnose. Xanthan gum is also a microbial synthesized polysaccharide. A mixture of xanthan gum and gellan gum has been used to encapsulate probiotic cells, and contrary to alginate, the mixture presents high resistance towards acid conditions (Sultana et al. 2000).

κ-carrageenan is a natural polysaccharide extracted from marine macroalgae (e.g. red seaweeds). Elevated temperatures (60–80 °C) are required to dissolve the polymer at concentrations ranging from 2 % to 5 %. Cooling the mixture to room temperature can result in gelation, and the addition of potassium ions stabilizes the microparticles (Krasaekoopt et al. 2003). However, the produced gels tend to be brittle and are not able to withstand potential stresses. Locust bean gum, at a ratio of carrageenan to locust bean gum of 2:1, increases the strength of gels through specific interaction of galactomannan chains with carrageenan (Krasaekoopt et al. 2003).

Chitosan is a positively charged linear polysaccharide formed by the deacetylation of chitin extracted from crustacean shell. It is water soluble below pH 6 and forms a gel by ionotropic gelation. Chitosan, a polycation with amine groups, can be cross-linked by anions or polyanions, such as polyphosphate and polyaldehydrocarbonic acid. Chitosan has not shown good efficiency for increasing cell viability but exhibited inhibitory effects on different types of LAB (Groboillot et al. 1993). To overcome viability problems with chitosan, the polymer is used as a coat for alginate beads to deliver the viable non-LAB cells to the colon.

Gelatin is a protein that is useful for making a thermo-reversible gel for probiotic encapsulation. Due to its amphoteric nature, it cooperates well with anionic polysaccharides like gellan gum. These hydrocolloids are miscible at pH > 6 since they both carry net negative charges and repel each other. However, when the pH is adjusted below the isoelectric point of gelatin, the net charge on the protein becomes positive, and this causes a strong interaction with the negatively charged gellan gum (Anal and Singh 2007; Krasaekoopt et al. 2003).

Milk proteins are natural vehicles for probiotic cells owing to their structural and physicochemical properties. Milk proteins are also widely available, inexpensive, natural and generally regarded as safe (GRAS) raw materials with high nutritional value and good sensory properties. They have many different structural properties and functionalities which make them highly suitable as vehicles for delivering various bioactives. Milk proteins such as caseins have excellent gelation properties (Heidebach et al. 2009a, b) that can be caused by micelle aggregation, based on their isoelectric precipitation and the proteolytic cleavage of κ-casein's hydrophilic "hairy layer" from casein micelles. This process is effective for the encapsulation of probiotic bacteria. Milk proteins also have excellent buffering capacity which provides good shielding for probiotic microorganisms against the harsh environment in the stomach.

3.2 Spray-Drying of Probiotics

Drying of bacterial cultures without losing their activity has been investigated to convert them into a dry state and facilitate storage and transportation (Anal and Singh 2007). Large-scale production of freeze-dried cultures is an expensive process with low yields. However, spray-drying represents a good low-cost alternative yielding higher production rates (Meng et al. 2008). Spray-drying of freshly prepared probiotic cell concentrates in various protein solutions with and without carbohydrates (maltodextrin, oligosaccharides, hydrocolloids) is widely used for the entrapment and drying of probiotic microorganisms in a single step (Corcoran et al. 2004; Desmond et al. 2001; Gardiner et al. 2000),whereas water-based dispersions are usually applied to spray-drying probiotics in water-soluble polymer matrices (De Vos et al. 2010).

Spray-drying is defined as the removal of water by vaporization from a solution of a non-volatile solid (Santivarangkna et al. 2007). The spray-drying process involves the injection of spray-drying medium at high velocities into the direction of the flow of hot air (typically, 150–200 °C). The atomized droplets have a very large surface area in the form of millions of micrometre-sized droplets (10–200 μm), which result in a very short drying time when exposed to hot air in the drying chamber (Santivarangkna et al. 2007). Water-based dispersions are usually applied in spray-drying. Therefore, the matrix should have very high solubility in water. The microencapsulation can be achieved with biopolymers of various sources; however, typical wall materials for microencapsulation by spray-drying are low molecular weight carbohydrates, milk or soy proteins, gelatin and hydrocolloids like gum arabic.

Ananta et al. (2005) reported the application of spray-drying in the production of skim milk-based preparations (20 % w/v) containing *L. rhamnosus* GG (ATCC 53103). Using a range of outlet temperatures between 70 and 100 °C, bacteria survival rate of >60 % was achieved at an outlet temperature of 80 °C (Ananta et al. 2005). Similarly, another study by Corcoran et al. (2004) showed good survival rates of 25–41 % with ~10^9 CFU/g of *L. rhamnosus* E800 when subjected to spray-drying conditions at inlet temperature of 170 °C and outlet temperature of 85–90 °C. Similarly, Gardiner et al. (2000) evaluated the potential of 20 % skim milk in producing powders with *L. paracasei* NFBC 338. After subjecting the slurry to spray-drying conditions at outlet temperatures between 80 and 85 °C, 65 % of the cells retained viability. The probiotic viability was further assessed for over a period of two months where the maximum survival rates for *L. paracasei* NFBC 338 of 92 % of the initial cells remained viable. Carbohydrates are also used as encapsulating wall materials including gum arabic and starches because they tend to form spherical microparticles during the drying process (De Vos et al. 2010).

Spray-drying is rapid and relatively low cost. This technique is also highly reproducible and also suitable for industrial applications. However, the major hurdle of this technique is that it is an immobilization technology than an encapsulation technology which implies that some of the encapsulating core materials may be exposed on the surface of spray-dried microcapsules. This is especially problematic when considering the encapsulation of probiotics, where the bacteria may leak into the product when some hydration occurs.

Bifidobacteria are sensitive to high inlet temperatures (O'Riordan et al. 2001). Thus, it is necessary to investigate the sensitivity of probiotics before spray-drying. Protectants, such as trehalose, can be used to improve the survival of probiotics and reduce the deleterious effect of bile salts present in the acidic environment of stomach (Burgain et al. 2011).

The performance of a variety of probiotics during spray-drying and, in general, the survival rate of probiotic cultures depend on several factors, such as type of probiotic strains, inlet and outlet temperatures and drying medium. The tolerance of different bacterial species varies with spray-drying conditions. However, improved viability could be achieved by maintaining spray-dried powders at preferred moisture of ~3.5 % (Teixeira et al. 1995a).

3.2.1 Effect of Spray-Drying Conditions on Survival of Bacteria

The survival of probiotic cells during spray-drying can be affected by various factors relating to process parameters (inlet and outlet temperatures, drying time), product parameters (type of carrier medium and concentrations) and pretreatments of cells (bacterial strain) (Desmond et al. 2001; O'Riordan et al. 2001; Lian et al. 2002).

Inlet and Outlet Temperatures

In order to enhance the efficiency of microencapsulation process with a suitable encapsulating wall material, optimal spray-drying conditions must be used. The main factors in spray-drying that need to be optimized are feed temperature, air inlet temperature and air outlet temperature. Feed temperature modifies the viscosity and fluidity of feed solutions, thus resulting in alteration in their capacity to be homogeneously sprayed (Fang and Bhandari 2012). When the feed temperature is increased, the size of droplets being sprayed in the dryer decreases due to a decrease in the feed viscosity, but high temperatures can cause degradation of some heat-sensitive ingredients to be encapsulated (Gharsallaoui et al. 2007; Medina-Torres et al. 2013). Air inlet temperature is directly related to the drying rate of droplets and the final water content of particles. High inlet temperatures increase the rate of droplet drying, thus facilitating rapid drying inside the drying chamber and leading to shorter residence times for the particles being dried (Bhandari et al. 2008). The inlet and outlet temperatures used during spry-drying vary, depending on the type and purpose of products being spray-dried. In food applications, high inlet (160–300 °C) and low outlet air temperatures (60–100 °C) are used to achieve the high thermal efficiency of the drier (Bhandari et al. 2008).

An increase in air inlet temperature normally decreases cell viability, but the bacterial cell survival is highly correlated to the outlet temperature (Peighambardoust et al. 2011). This means that the higher inlet temperature does not have a direct correlation to the inactivation of bacterial cells and has only a slight effect. This is due to the fact that the extent of cell inactivation is largely dependent on the drying temperature–time combinations. During spray-drying, the temperature of spray-dried particles increases but does not reach the inlet air temperature because of an evaporative cooling effect that occurs owing to the instant removal of moisture, and the subsequent exposure time of dried particle to the high temperature is very short (Fichtali and Namal Senanayake 2010). Therefore, it is important to maintain an optimum drying time (residence time) required for the removal of moisture without causing an increase in temperature of the dried particles for the survival of encapsulated bacterial cells during spray-drying (Santivarangkna et al. 2007).

In addition to the inlet air temperature, the outlet air temperature is another major drying parameter affecting the viability of spray-dried starter cultures (Gardiner et al. 2000, 2002). However, it is largely influenced by the inlet air temperature, air-flow rate, product-feed rate and the atomized droplet sizes. These factors highlight the importance of optimization of process parameters, in particular, inlet and outlet temperatures. Due to the difficulty in setting these variables and in turn the stabilization of outlet air temperature, there is often a great variation in the viability of the dried cultures (Ananta et al. 2005; Desmond et al. 2002).

Gardiner et al. (2000) reported that the survival rates for *L. paracasei* NFBC 338 were affected by outlet temperatures. It was 97 % at 70–75 °C, while the survival rate was close to 0 % when the outlet temperature was increased to 120 °C. The survival rates were better than for *L. salivarius* UCC 118 that had only 11 % even at the lowest outlet temperatures of 60–65 °C. These findings might be attributed to the greater thermal tolerance of strain NFBC 338. The survival rate of NFBC 338 during spray-drying was also considerably higher than the survival rate previously obtained for *L. acidophilus* or *L. curvatus* spray-dried under similar conditions. In many situations, the lower outlet air temperature correlates with higher cell viability (Santivarangkna et al. 2007). A low outlet air temperature is desirable to maintain high stability during storage. However, if the outlet air temperatures are too low, it may cause high residual moisture contents that exceed the required level for prolonged powder storage life and stability (4 %) (Gardiner et al. 2000).

Carrier Medium and Concentration

The use of different carriers has an impact on the viability of spray-dried cultures. In a study by Lian et al. (2002), the effect of different carriers (10 %, w/w) on the survival of various *Bifidobacterium* strains was demonstrated. The survival rate of *B. infantis* CCRC 14633 after spray-drying at inlet and outlet air temperatures of 100 and 50 °C, respectively, was 15.99 % for skim milk, 2.15 % for gum arabic, 1.30 % for gelatin and 0.92 % for soluble starch, while the survival of *B. longum* B6 was 83 % for reconstituted skim milk powder, 41 % for gum arabic, 64 % for gelatin and 29 % for soluble starch. Skim milk has potential for effective spray-drying of probiotic cultures (Corcoran et al. 2004; Desmond et al. 2002; Ananta et al. 2005) as skim milk proteins can prevent cellular injury by stabilizing cell membrane constituents (Ananta et al. 2005).

Soluble solid concentrations of the liquid feed can vary from 10 to 50 %, depending on the properties of feed (e.g. viscosity and heat sensitivity), the type of atomizer and the final product requirements. Higher feed concentrations improve the commercial viability of the process through thermal efficiency, but it also affects the survival of bacteria after spray-drying. Lian et al. (2002) also reported the effect of different concentrations of carriers on the survival rate of bifidobacteria. When the concentration of gum arabic, gelatin or soluble starch was increased from 10 to 20 % (w/w) or more, the survival rate was significantly lowered. For *B. infantis* CCRC 14633, the survival was 0.65, 0.52 and 0.09 %, respectively, after spray-drying with gum arabic, gelatin and soluble starch. For *B. longum* B6, it was 6.51, 2.07 and 1.56 %, respectively. However, the effect of skim milk concentration was not reported in their study.

A 10 % carrier concentration is generally considered ideal for increasing viability in spray-dried cultures (Morgan et al. 2006). Increasing feed concentration of carriers from 10 to 20 % or more can cause a reduction in the viability of spray-dried cultures. Lower viabilities at high feed concentrations may be caused by higher solid content that can result in larger particles that require longer drying times, thus subjecting the entrapped microorganisms to more heat damage (Santivarangkna et al. 2007). However, in the case of reconstituted skim milk, the total solid content of 20 % has been frequently used and considered optimum for retaining high residual viability of different strains of lactic acid bacteria (Desmond et al. 2001; Gardiner et al. 2000). In fact, the storage stability of dried powder was reduced as the amount of skim milk solids in the carrier was decreased (Ananta et al. 2005).

Combinations of different carriers can be used to improve the survival of spray-dried probiotics. For instance, a combination of skim milk and gum arabic was shown to result in good survival rates of *B. lactis* BB12 after spray-drying and also during storage in vacuum at 30 °C, compared to the control sample prepared from skim milk without adding gum arabic (Chavez and Ledeboer 2007). Desmond et al. (2002) also used gum arabic to protect probiotic cultures of *L. paracasei* NFBC 338 during spray-drying, storage and gastric transit. It was demonstrated that a mixture of reconstituted skim milk (RSM, 10 % w/v) and gum arabic (10 % w/v) rendered tenfold greater survival than the control prepared with RSM (20 % w/v). Gum arabic has emulsifying properties and exhibits high solubility and low viscosity in aqueous solution compared to other hydrocolloid gums, thus facilitating the spray-drying process.

Stress Response Factors

To improve the viability of probiotics, several approaches have been attempted, including stress adaptation technique and selection of more resistant strains from various sources (Krasaekoopt et al. 2003). Adaptive cellular response could be induced, prior to dehydration, with the exposure of microorganisms to sublethal or gradually increasing doses of stress. This enhances the resistance of bacterial cells to stressful conditions and enables them to survive during dehydration. The bacteria respond to changes in their surroundings by a metabolic programming which leads to a cellular state of enhanced resistance (Meng et al. 2008).

Most bacteria exhibit stress sensing systems as defensive mechanisms against various stresses allowing them to survive under severe conditions. The induction of these defence systems influences the tolerance against harsh conditions such as heat or osmotic stress during drying. Adaptation to heat could be induced by heat-shock treatments by placing cells in sublethal high temperatures (50 °C for 30 min) (Teixeira et al. 1995a) before spray-drying. However, heat shock has little significance to stationary cells compared to cells from the exponential phase (Teixeira et al. 1995a). Meanwhile, the exposure of cells to other nonhomologous sublethal agents such as salt can render tolerance to heat and spray-drying even at high outlet temperatures of 100–105 °C (Santivarangkna et al. 2007; Desmond et al. 2001). Starved cells show multi-resistances against stresses particularly to heat and oxidative tolerances with increasing duration of starvation (Santivarangkna et al. 2007).

Protective Substances

The inclusion of protective agents to starter cultures is a common means to protect cells during drying and storage. The excipients added can be in the form of compounds used as suspending media or carriers, such as skim milk, whey, gum acacia and gelatin (Santivarangkna et al. 2007). The effectiveness of a given protectant varies largely with each type of culture. The most extensively investigated compound is trehalose. This may be due to the phenomenon called anhydrobiosis, where organisms in nature can survive for a long and extreme dehydration period by accumulating a large amount of disaccharides, especially trehalose. The presence of trehalose has the ability to raise the glass transition temperature (T_g) of the dry matrix (Santivarangkna et al. 2007).

Prebiotics are non-digestible carbohydrates (e.g. lactulose, inulin and some oligosaccharides) that benefit the host by selectively stimulating the growth and activity of beneficial bacteria in the colon (Burgain et al. 2011). It was shown that the partial substitution of solid content of skim milk powder with Raftilose (P95) and/or polydextrose enhanced the survival of *L. rhamnosus* GG (ATCC 53103) during spray-drying, but the storage stability of the bacteria was decreased during long-term storage (Ananta et al. 2005). This might be because some oligosaccharides present in these prebiotic substances are inadequate in replacing water molecules in the dehydrated skim milk. Thus, the maintenance of structural and functional integrity of bacterial cell membrane is not as effective as in the presence of skim milk alone (Ananta et al. 2005).

4 Post-drying Conditions

Several intrinsic and extrinsic factors affect the stability of probiotics during storage. Storage conditions, such as temperature (storage), moisture content of powders, water activity, relative humidity, powder composition, oxygen content, exposure to light and storage materials, have significant impact on the survival of probiotics in dried powders. Temperature is, however, one single most important factor affecting the stability of probiotics during storage. Stability of spray-dried samples decreases during storage, and higher microbial survival rates are maintained at low storage temperatures (Corcoran et al. 2004; Desmond et al. 2002). It is therefore essential to use correct storage conditions to maintain viable populations of spray-dried probiotic bacteria (Meng et al. 2008).

Moisture content and water activity are directly affected by the efficiency of the drying process and the quality of packaging materials used. The glass transition temperature (T_g) is another factor that impacts on the survival of probiotics in dried powders as it is a thermodynamic property of materials which is altered by the presence of water. Mass transfer rates are slower in a glassy state matrix. Storage of dried cultures at temperatures lower than their T_g increases their stability as it (T_g) retards the mobility of molecules and reaction rates (Chavez and Ledeboer 2007).

Relative humidity of the storage environment has a significant effect on the survival of dried probiotic cultures (Ying et al. 2010). High relative humidity can cause the caking phenomenon in dried powders. This phenomenon which is associated with the transition of powders from a glassy state to a rubbery state is one of the most undesirable conditions for the survival of probiotics. Therefore, it is essential that the relative humidity is kept to the critical equilibrium value that corresponds to the glass/rubber transition.

Lipid oxidation of cell membrane fatty acids is a possible cause for cell death during storage (Ananta et al. 2005). The onset of membrane lipid oxidation during storage has a detrimental effect to cells. Addition of antioxidant materials such as ascorbic acid and monosodium glutamate can protect cells during storage at 4 °C (Peighambardoust et al. 2011). However, the addition of such antioxidant materials is not recommended at storage temperature of about 20 °C as this can lead to high death rate of the culture due to pro-oxidant activity of ascorbic acids at higher temperature (Santivarangkna et al. 2007).

Proper packaging for storage of the cultures is important. Packaging under vacuum or nitrogen replacement is suitable for storing anaerobic probiotics such as bifidobacteria (Peighambardoust et al. 2011). Storage of cultures under vacuum is better than storage under nitrogen or air. Since vacuum packaging also removes air humidity, packaging of dried probiotics under vacuum is recommended (Chavez and Ledeboer 2007). The package should prevent the transmission of oxygen, moisture and light which are detrimental to the dehydrated cultures. Spray-dried *S. thermophilus* and *B. longum* can survive better in laminated pouches, followed by glass bottles and polyethylene terephthalate (PET) bottles (Wang et al. 2004). Skimmed milk powders containing cells can also be stored in polythene bags and kept in aluminium-coated paper bags (Simpson et al. 2005; Chavez and Ledeboer 2007).

5 Applications of Encapsulated Probiotics in Food Products

Microencapsulation is important for the survival of probiotics during storage and also its passage through the digestive tract depending on the type of microencapsulation system used. Addition of microcapsules to food matrices should not affect the sensory properties of food products when the size of the capsules is kept below 100 μm (Heidebach et al. 2009b). Encapsulated probiotic bacteria can be used in fermented dairy products, such as liquid fermented milks (yoghurt, cultured cream), Cheddar cheese and frozen dairy desserts, and for biomass production. The use of encapsulated probiotic cells, particularly in cheese, is common. Cheddar cheese has particular advantage of being a good carrier of encapsulated probiotics because of its high pH (5.5), good buffering capacity and relatively high fat content which can protect probiotic bacteria (Burgain et al. 2011). Spray-dried powder of *L. paracasei*

NFBC 338 was successfully used in probiotic Cheddar cheese manufacture, retaining good viability levels (up to 10^8 CFU/g) for 6–8 months during ripening without affecting quality (Gardiner et al. 2002).

Due to the low pH of yoghurt (~ pH 4.5), the viability of unprotected probiotics is often affected, although the use of acid-tolerant strains may be possible. Using encapsulated probiotic bacteria would be better for their survival in liquid yoghurt without making major modifications of the traditional fermentation process. Gellan–xanthan gum can be used to increase probiotic tolerance in acidic environments. The incorporation of encapsulated bifidobacteria into stirred yoghurt can lead to the defect of grainy texture, which affects the sensory quality and consumer acceptance of the product (Adhikari et al. 2003).

Microencapsulation technology has created opportunities to introduce probiotic microorganisms into products such as frozen desserts with high acidity and high osmotic pressure and containing incorporated air introduced during the freezing step. In ice cream, high viable cells of probiotics can be further protected by adding resistant starch (Homayouni et al. 2008).

The incorporation of probiotic cells encapsulated by spray-coating technology has been successfully used in chocolate. Probiotic viability in the small intestine was three times higher when incorporated in chocolate than in dairy products. Encapsulation of cells into chocolate acts as an excellent protectant against environmental stress conditions. The lipid fraction of cocoa butter protects bifidobacteria (Lahtinen et al. 2007). Encapsulated probiotics can be also protected against bacteriophages and harsh environments such as freezing and gastric solutions (Krasaekoopt et al. 2003).

6 Conclusions and Future Trends

While there has been significant progress in the development of encapsulation technology of probiotics, there is still a lot scope for more research. There are very few encapsulation materials that fulfil all the requirements to protect probiotics and deliver the microorganisms to the site of action. Ideal properties of microencapsulation materials include protecting live cells from sublethal damages, being easily digestible in the target site of GI tract, and providing protection to cells during handling and processing. Further, the materials should not affect the sensory profile of products. Therefore, the development of new encapsulation materials and techniques is desirable. Although there is an abundance of information on in vitro studies of probiotics, there is very little published data on in vivo investigations. Information on the optimum cell density in the microcapsules is important to avoid quorum sensing (cell-to-cell interactions) which express different genes. Consequences of morphological changes of encapsulated probiotics have not been reported. There is also renewed interest in induced stress resistance of probiotics caused by adverse conditions such as high temperature and acids. Whether the induced stress resistance is transient or permanent is unknown.

References

Adhikari K, Mustapha A, Grün IU (2003) Survival and metabolic activity of microencapsulated *Bifidobacterium longum* in stirred yogurt. J Food Sci 68(1):275–280

Anal AK, Singh H (2007) Recent advances in microencapsulation of probiotics for industrial applications and targeted delivery. Trends Food Sci Technol 18(5):240–251

Anal AK, Stevens WF (2005) Chitosan-alginate multilayer beads for controlled release of ampicillin. Int J Pharm 290(1–2):45–54

Ananta E, Volkert M, Knorr D (2005) Cellular injuries and storage stability of spray-dried Lactobacillus rhamnosus GG. Int Dairy J 15(4):399–409

Andersson H, Asp NG, Bruce A, Roos S, Wadstrom T, Wold AE (2001) Health effects of probiotics and prebiotics: a literature review on human studies. Scand J Nutr 45:58–75

Bhandari B, Kamlesh C, Patel KC, Chen XD (2008) Spray drying of food materials – process and product characteristics. In: Chen XD, Mujumdar AS (eds) Drying technologies in food processing. Wiley, New York, pp 113–159

Burgain J, Gaiani C, Linder M, Scher J (2011) Encapsulation of probiotic living cells: from laboratory scale to industrial applications. J Food Eng 104(4):467–483

Capela P, Hay TKC, Shah NP (2007) Effect of homogenisation on bead size and survival of encapsulated probiotic bacteria. Food Res Int 40(10):1261–1269

Castroa GR, Chenb J, Panilaitisc B, Kapland DL (2009) Emulsan–alginate beads for protein adsorption. J Biomater Sci 20(4):411–426

Champagne CP, Lacroix C, Sodini-Gallot I (1994) Immobilized cell technologies for the dairy industry. Crit Rev Biotechnol 14(2):109–134

Chavez BE, Ledeboer AM (2007) Drying of probiotics: optimization of formulation and process to enhance storage survival. Dry Technol 25(7):119–1201

Chen L, Subirade M (2007) Effect of preparation conditions on the nutrient release properties of alginate–whey protein granular microspheres. Eur J Pharm Biopharm 65(3):354–362. doi:10.1016/j.ejpb.2006.10.012

Corcoran BM, Ross RP, Fitzgerald GF, Stanton C (2004) Comparative survival of probiotic lactobacilli spray-dried in the presence of prebiotic substances. J Appl Microbiol 96 (5):1024–1039

Crittenden R, Laitila A, Forssell P, Matto J, Saarela M, Mattila-Sandholm T, Myllarinen P (2001) Adhesion of bifidobacteria to granular starch and its implications in probiotic technologies. Appl Environ Microbiol 67(8):3469–3475

De Vos P, Faas MM, Spasojevic M, Sikkema J (2010) Encapsulation for preservation of functionality and targeted delivery of bioactive food components. Int Dairy J 20(4):292–302

Desmond C, Stanton C, Fitzgerald G, Collins K, Ross R (2001) Environmental adaptation of probiotic lactobacilli towards improvement of performance during spray drying. Int Dairy J 11 (10):801–808

Desmond C, Ross RP, O'Callaghan E, Fitzgerald G, Stanton C (2002) Improved survival of *Lactobacillus paracasei* NFBC 338 in spray-dried powders containing gum acacia. J Appl Microbiol 93(6):1003–1011

Ding WK, Shah NP (2009) An improved method of microencapsulation of probiotic bacteria for their stability in acidic and bile conditions during storage. J Food Sci 74(2):M53–M61

Doleyres Y, Lacroix C (2005) Technologies with free and immobilised cells for probiotic bifidobacteria production and protection. Int Dairy J 15(10):973–988

Fang Z, Bhandari B (2012) Spray drying, freeze drying and related processes for food ingredient and nutraceutical encapsulation. In: Garti N, McClements DJ (eds) Encapsulation technologies and delivery systems for food ingredients and nutraceuticals, Chap. 4. Woodhead, London, pp 73–109

FAO/WHO (2002) 'Guidelines for evaluation of probiotics in food'. In: Working group report on drafting guidelines for the evaluation of probiotics in food, London, Ontario, Canada, 30 Apr and 1 May 2002. Food and Agricultural Organisation of the United Nations/World Health Organisation, pp 1–11

Fichtali J, Namal Senanayake SPJ (2010) Development and commercialization of microalgae-based functional lipids. In: Smith J, Charter EA (eds) Functional food product development, Chap. 10. Wiley, New York, pp 206–228

Gardiner GE, O'Sullivan E, Kelly J, Auty MAE, Fitzgerald GF, Collins JK, Ross RP, Stanton C (2000) Comparative survival rates of human-derived probiotic *Lactobacillus paracasei* and *L. salivarius* strains during heat treatment and spray drying. Appl Environ Microbiol 66 (6):2605–2612

Gardiner GE, Bouchier P, O'Sullivan E, Kelly J, Kevin Collins J, Fitzgerald G, Ross RP, Stanton C (2002) A spray-dried culture for probiotic Cheddar cheese manufacture. Int Dairy J 12 (9):749–756

Gharsallaoui A, Roudaut G, Chambin O, Voilley A, Saurel R (2007) Applications of spray-drying in microencapsulation of food ingredients: an overview. Food Res Int 40(9):1107–1121

Groboillot AF, Champagne CP, Darling GD, Poncelet D, Neufeld RJ (1993) Membrane formation by interfacial cross-linking of chitosan for microencapsulation of *Lactococcus lactis*. Biotechnol Bioeng 42(10):1157–1163

Hansen LT, Allan-Wojtas PM, Jin YL, Paulson AT (2002) Survival of Ca-alginate microencapsulated *Bifidobacterium* spp. in milk and simulated gastrointestinal conditions. Food Microbiol 19(1):35–45

Heidebach T, Först P, Kulozik U (2009a) Microencapsulation of probiotic cells by means of rennet-gelation of milk proteins. Food Hydrocoll 23(7):1670–1677

Heidebach T, Först P, Kulozik U (2009b) Transglutaminase-induced caseinate gelation for the microencapsulation of probiotic cells. Int Dairy J 19(2):77–84

Homayouni A, Azizi A, Ehsani MR, Yarmand MS, Razavi SH (2008) Effect of microencapsulation and resistant starch on the probiotic survival and sensory properties of synbiotic ice cream. Food Chem 111(1):50–55

Isolauri E (2001) Probiotics in human disease. Am J Clin Nutr 73(Suppl):1142S–1146S

Isolauri E, Sutas Y, Kankaanpaa P, Arvilommi H, Salminen S (2001) Probiotics: effects on immunity. Am J Clin Nutr 73(Suppl):444S–450S

Kailasapathy K (2002) Microencapsulation of probiotic bacteria: technology and potential applications. Curr Iss Intest Microbiol 3(2):39–48

Kailasapathy K, Chin J (2000) Survival and therapeutic potential of probiotic organisms with reference to *Lactobacillus acidophilus* and *Bifidobacterium* spp. Immunol Cell Biol 78 (1):80–88

Kalliomäki M, Salminen S, Arvilommi H, Kero P, Koskinen P, Isolauri E (2001) Probiotics in primary prevention of atopic disease: a randomised placebo-controlled trial. Lancet 357:1076–1079

Koo SM, Cho YH, Huh CS, Baek YJ, Park J (2001) Improvement of the stability of *Lactobacillus casei* YIT 9018 by microencapsulation using alginate and chitosan. J Microbiol Biotechnol 11 (3):376–383

Krasaekoopt W, Bhandari B, Deeth H (2003) Evaluation of encapsulation techniques of probiotics for yoghurt. Int Dairy J 13(1):3–13

Lahtinen SJ, Ouwehand AC, Salminen SJ, Forssell P, Myllärinen P (2007) Effect of starch- and lipid-based encapsulation on the culturability of two Bifidobacterium longum strains. Lett Appl Microbiol 44(5):500–505

Lian WC, Hsiao HC, Chou CC (2002) Survival of bifidobacteria after spray-drying. Int J Food Microbiol 74(1–2):79–86

Maltais A, Remondetto GE, Gonzalez R, Subirade M (2005) Formation of soy protein isolate cold-set gels: protein and salt effects. J Food Sci 70(1):C67–C73

Medina-Torres L, Garcia-Cruz EE, Calderas FJ, González Laredo RF, Sánchez-Olivares G, Gallegos-Infante JA, Rocha-Guzmán NE, Rodríguez-Ramírez J (2013) Microencapsulation

by spray drying of gallic acid with nopal mucilage (Opuntia ficus indica). LWT-Food Sci Technol 50(2):642–650
Meng XC, Stanton C, Fitzgerald GF, Daly C, Ross RP (2008) Anhydrobiotics: the challenges of drying probiotic cultures. Food Chem 106(4):1406–1416
Morgan CA, Herman N, White PA, Vesey G (2006) Preservation of micro-organisms by drying: a review. J Microbiol Methods 66(2):183–193
Mortazavian A, Razavi SH, Ehsan MR, Sohrabvandi S (2007) Principles and methods of microencapsulation of probiotic microorganisms. Iran J Biotechnol 5(1):1–18
Nigam SC, Tsao IF, Sakoda A, Wang HY (1988) Techniques for preparing hydrogel membrane capsules. Biotechnol Techn 2(4):271–276
O'Riordan K, Andrews D, Buckle K, Conway P (2001) Evaluation of microencapsulation of a *Bifidobacterium* strain with starch as an approach to prolonging viability during storage. J Appl Microbiol 91(6):1059–1066
Ouwehand AC, Salminen S, Isolauri E (2002) Probiotics: an overview of beneficial effects. Antonie van Leeuwenhoek 82:279–289
Peighambardoust SH, Golshan Tafti A, Hesari J (2011) Application of spray drying for preservation of lactic acid starter cultures: a review. Trends Food Sci Technol 22:215–224
Picot A, Lacroix C (2003) Effects of micronization on viability and thermotolerance of probiotic freeze-dried cultures. Int Dairy J 13(6):455–462
Picot A, Lacroix C (2004) Encapsulation of bifidobacteria in whey protein-based microcapsules and survival in simulated gastrointestinal conditions and in yoghurt. Int Dairy J 14(6):505–515
Roff CF, Foegeding EA (1996) Dicationic-induced gelation of pre-denatured whey protein isolate. Food Hydrocoll 10(2):193–198
Rokka S, Rantamaki P (2010) Protecting probiotic bacteria by microencapsulation: challenges for industrial applications. Eur Food Res Technol 231(1):1–12
Rowley JA, Madlambayan G, Mooney DJ (1999) Alginate hydrogels as synthetic extracellular matrix materials. Biomaterials 20(1):45–53
Saarela M, Mogensen G, Fonde F, Mättö J, Mattila-Sandholm T (2000) Probiotic bacteria: safety, functional and technological properties. J Biotechnol 84:197–215
Santivarangkna C, Kulozik U, Foerst P (2007) Alternative drying processes for the industrial preservation of lactic acid starter cultures. Biotechnol Prog 23(2):302–315
Simpson PJ, Stanton C, Fitzgerald GF, Ross RP (2005) Intrinsic tolerance of Bifidobacterium species to heat and oxygen and survival following spray drying and storage. J Appl Microbiol 99(3):493–501
Solanki HK, Pawar DD, Shah DA, Prajapati VD, Jani GK, Mulla AM, Thakar PM (2013) Development of microencapsulation delivery system for long-term preservation of probiotics as biotherapeutics agent. Biomed Res Int 2013:21
Sultana K, Godward G, Reynolds N, Arumugaswamy R, Peiris P, Kailasapathy K (2000) Encapsulation of probiotic bacteria with alginate-starch and evaluation of survival in simulated gastrointestinal conditions and in yoghurt. Int J Food Microbiol 62(1-2):47–55
Teixeira P, Castro H, Kirby R (1995a) Spray drying as a method for preparing concentrated cultures of *Lactobacillus bulgaricus*. J Appl Microbiol 78(4):456–462
Wang YC, Yu RC, Chou CC (2004) Viability of lactic acid bacteria and bifidobacteria in fermented soymilk after drying, subsequent rehydration and storage. Int J Food Microbiol 93 (2):209–217
Ying DY, Phoon MC, Sanguansri L, Weerakkody R, Burgar I, Augustin MA (2010) Microencapsulated *Lactobacillus rhamnosus* GG powders: relationship of powder physical properties to probiotic survival during storage. J Food Sci 75(9):E588

Effects of Incorporation of Lactic Acid Bacteria on Microbiological Quality and Shelf Life of Raw 'Satar'

Mohd Nizam Lani, Noraisikin Mohamad Nor, Nurul Atiqah Ramli, Zuraihan Radhuan, Mazwani Mohd Rizan, Nur Hidayah Lokman, and Zaiton Hassan

Contents

Abstract Maintaining a safe food supply has become an ever-changing endeavour as some emerging pathogens are discovered. Relying on traditional methods of thermal processing to create microbiologically safe foods is not sufficient. Research on finding other methods of controlling the growth and multiplication of pathogenic and spoilage bacteria needs to be explored. The use of crude bacteriocin produced by lactic acid bacteria may be one promising solution of controlling microbial growth in ready-to-eat (RTE) foods. The ability of lactic acid bacteria (LAB) to produce metabolites with broad-spectrum inhibitory activity that are heat stable is an important criterion for the application of LAB as preservative in food. 'Satar' was used as a model for this study because it is highly perishable and has a short

M.N. Lani (✉) • N.M. Nor • N.A. Ramli • Z. Radhuan • M.M. Rizan • N.H. Lokman
School of Food Science and Technology, Universiti Malaysia Terengganu (UMT), Terengganu, Malaysia
e-mail: nizamlani@umt.edu.my

Z. Hassan
Faculty of Science and Technology, Universiti Sains Islam Malaysia (USIM), Negeri Sembilan, Malaysia

M.-T. Liong (ed.), *Beneficial Microorganisms in Food and Nutraceuticals*, Microbiology Monographs 27, DOI 10.1007/978-3-319-23177-8_4

shelf life (<12 h) at ambient temperature and, therefore, is unable to be stored for a long period of time. This chapter briefly describes the background of 'Satar' and its relations to microbiological safety. The study focused on choosing the suitable strains of LAB, identifying the isolates phenotypically using biochemical tests and VITEK 2 Compact System. The isolates were tested on their ability to inhibit LAB microflora, ability to inhibit a broad spectrum of Gram-negative and Gram-positive bacteria and ability to exhibit the antimicrobial activity after being subjected to heating temperatures. Among nine isolates of LAB from fermented fish, supernatants of four isolates were studied extensively for their heat stability at different heating temperatures (70, 80, 90, 100 and 121 °C) and heating times (5 and 20 min). Two strains, *Lb. acidophilus* and *Lb. plantarum*, were chosen for the incorporation of their crude bacteriocin in raw 'Satar', and their characteristics and microbiological shelf life were evaluated. Incorporation of crude bacteriocin of *Lb. acidophilus* and *Lb. plantarum* at 3 % and 6 % did not significantly affect ($P > 0.05$) the water activity and pH, but significantly increased the moisture content when Satar was stored more than 20 h at ambient temperature. There was no significant difference ($P > 0.05$) for a*value and b*value of 'Satar' among all samples at 0 h of storage time, except after 3 h of storage at ambient temperature. The colour analysis of samples showed a range of colour between grey and light grey. The incorporation of 3 % and 6 % crude bacteriocin of *Lb. acidophilus* and *Lb. plantarum* in raw 'Satar' could extend the shelf life from 8 h to 20 h and 17 h, respectively. This study has proven that LAB can be used to extend the shelf life of ready-to-eat food.

1 Introduction

Lactic acid bacteria (LAB) are well-known for their antimicrobial activity that can be used as food preservatives. It is commonly known as biopreservative agent because it comes from natural resources without using chemical additives to increase the shelf life (Pratush et al. 2012). LAB play a major role for preservation for fermented product and inhibit the growth of spoilage and pathogenic bacteria (Matilla-Sandholm et al. 1999). The mechanism of inhibition of LAB against foodborne pathogens is through metabolite production, which contained various antimicrobial compounds such as organic acids (lactic acid), hydrogen peroxide, diacetyl and bacteriocins (Messens and De Vugst 2002) that are inhibitory to bacteria and fungi (Hassan and Bullerman 2008). LAB are safe and they have the status of generally recognised as safe (GRAS). The antimicrobial compounds produced by LAB against bacteria and fungi are active in range of pH 3–4.5 and heat stable at 100 °C (Lavermicocca et al. 2000).

Lactic acid bacteria from local sources in various food and fermented foods in Malaysia have been extensively studied for their antimicrobial properties (Muhialdin et al. 2012; Aween et al. 2012; Muhialdin and Hassan 2011). Most of these studies reported on the role of bacteriocin-like inhibitory substances found in these Malaysian fermented foods that are able to maintain the inhibition activity against nonpathogenic and pathogenic food-associated and human pathogenic bacteria (Liasi et al. 2009). Bacteriocins are proteins or complexed proteins biologically active with antimicrobial action against other bacteria, principally closely related species (Parada et al. 2007). Bacteriocins are normally produced by LAB that are able to inhibit their own LAB species. In food matrices, the bacteriocin activity may be affected by changes in solubility and the charge of the bacteriocins, binding of bacteriocins to food components, inactivation by proteases or changes in the cell envelope of the target microorganism as a response to environmental factors (Cleveland et al. 2001).

'Satar' is a Malaysian food that is famous in the East Coast of Peninsular Malaysia, especially in Terengganu and Kelantan. 'Satar' contains boneless fish and a mixture of spices, wrapped in banana leaf and grilled over a flaming charcoal of fire. Many studies have recently been published on the microbiological quality of 'Satar', which indicate this product is highly perishable and prone to microbial contamination that affects the organoleptic quality of this product (Lani et al. 2014a, b; Ramli et al. 2011, 2014).

Therefore, research on finding ways to prolong the shelf life of 'Satar' is important in order to help small and medium entrepreneurs of 'Satar' to sustain the production of 'Satar'. The addition of LAB cell supernatant in food is one way to preserve food and may prolong the shelf life. However, this study is the first report on application of LAB to extend shelf life of raw 'Satar'. The raw 'Satar' was used in this study because this will exclude the effect of grilling. Thus, the results obtained from this study solely depend on direct application of LAB on raw 'Satar'.

This study creates a new knowledge on understanding the role of LAB for the preservation of 'Satar'. The use of different LAB as a potential starter culture for food preservation would extend the functionality of LAB cultures related to food preservation. This new technique of preservation can be further studied on the application of LAB and their metabolites in food products in Malaysia for extending shelf life.

2 'Satar'

Malaysia is rich with variety of foods. Some of the foods are categorised as traditional foods, foods that are influenced by the state and native people (Cayot 2007). Terengganu is a part of East Coast of Peninsular Malaysia that welcomes people with a variety and unique dishes with special taste. The elements of Terengganu cuisine are influenced by Thai, Chinese and Indian cuisine as a result of its geographical location and historical background (Tourism Terengganu 2011). The Terengganu cuisine is inexpensive and easy to get either at hotels or streets. 'Satar', 'Otak-otak', Nasi Dagang, 'Kuih Akok' and Keropok Lekor are among the famous cuisine in Terengganu.

'Satar' is also considered as a 'street food' because this ready-to-eat (RTE) food is prepared and sold on the roadside by hawkers for immediate consumption or later time without further preparation. In Malaysia, like other developing countries (Uganda, Thailand and India), street food industry generates income to households (FAO 2005; Jayasuriya 1994; Muyanja et al. 2011). These street foods are also a source of employment in Terengganu offering business opportunities to the communities majorly dominated by women (Muyanja et al. 2011). These street foods become important and essential foods for maintaining the nutritional status of the populations (Ekanem 1998).

'Satar' in Terengganu is inexpensive, has a good taste and is convenient that contained important sources of nutrition. RTE 'Satar' is usually served as a snack during tea time. The authenticity of this food and the aroma of the banana leaves and ingredients after grilling have attracted many people to try and enjoy it. 'Satar' is sold almost everywhere in Terengganu. However, from the survey, the famous areas of RTE 'Satar' are in Kuala Terengganu, Marang and Kemaman (Tourism Terengganu 2011).

'Satar' is a mixture of deboned fish and spices and wrapped in banana leaf and grilled over the charcoal fire. The processed fish normally used are yellow-striped scad (*Selaroides leptolepis*), crimson jobfish (*Nemipterus* spp., *Pentapodus* spp. and *Scolopsis* spp.) and Spanish mackerel (*Scomberomorus* spp.) (Department of Fisheries, Malaysia 2014). Table 1 shows the percentages of the basic ingredient of 'Satar'. Some of the basic ingredients used in the preparation of 'Satar' are mixture of deboned fresh fish, otoshimi, surimi, sugar, shredded coconut, onions, garlic, red chillies, ginger, fennel seed and salt. 'Satar' has high moisture content, about 66.67 %, protein 14.36 %, fat 13.67 %, ash 1.67 % and carbohydrate 3.33 % (Lani et al. 2014a, b).

Table 1 Percentage of basic ingredients of 'Satar'

Ingredients	Percentage (%)
Fresh fish	21.74
Otoshimi	26.09
Surimi	21.74
Sugar	8.70
Grated coconut	16.52
Salt and spices (onion, ginger and Fennel seed)	4.35
Total	100.00

Source: Nurul Atiqah Ramli (2013)

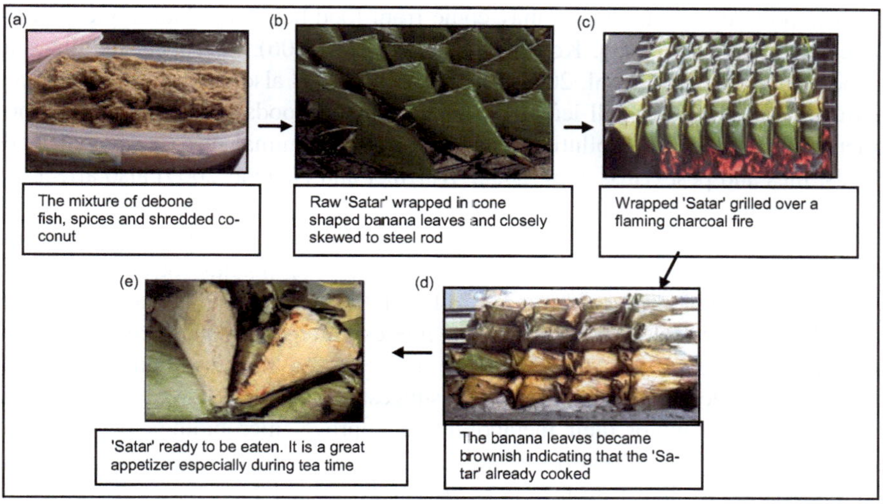

Fig. 1 The processing steps of 'Satar'

2.1 Traditional Method of Preparing 'Satar'

The traditional method of making 'Satar' is shown in Fig. 1. Firstly, the fish is deboned to get the flesh. Then, all other ingredients (onion, shallot, spices, sugar, salt and shredded coconut) are mixed thoroughly. The mixture is wrapped in banana leaf into a cone-like shape, skewered into stainless steel rods, grilled for 10 min at 80–90 °C until it is cooked as indicated by the banana leaves becoming brownish in colour and quickly removed from the griller to avoid overcooking followed by cooling to room temperature.

3 LAB as Biopreservation Strategy for Reducing Microbial Contamination in Foods

Microbial contamination may occur from farm to fork where microbes can enter at different stages of the food chain; microorganisms are highly versatile and can adapt to the environment admitting survival, growth and production of toxic compounds (Havelaar et al. 2010). Microorganisms present in raw materials and ingredients may allow contamination in foods (Cho et al. 2011; Dawnes and Ito 2001). A food also can be contaminated with different types of microorganisms coming from outside sources such as air, soil, sewage, water, feed, human, food ingredients, equipment, packages and insects (Ray and Bhunia 2008; Warriner and Namvar 2009).

The route of contamination may come from food handlers. Poor hygiene practices (Ayçiçek et al. 2004; Ko 2010; Shojaei et al. 2006), improper handling and processing (Sachindra et al. 2005; Jay et al. 2005) and also poor storage practices (Muyanja et al. 2011) will lead to contamination of foods in the street. Environmental factors such as pollution (Jayasuriya 1994), animal disease, animal feed, birds, flies and pest (Huss et al. 2004; Warriner and Namvar 2009) also affect the hygiene of the foods.

Biopreservation is the extension of shelf life and food safety by the use of natural or controlled microbiota and/or their antimicrobial compounds (Stiles 1996). The commonly used biopreservative is LAB. The increasing demand for safe foods with less chemical additives has increased the interest in replacing these compounds by natural products, which do not injure the host or the environment (Parada et al. 2007). De Martinis et al. (2001) suggested that the use of nonpathogenic microorganisms and/or their metabolites to improve microbiological safety and extend the shelf life of foods is defined as biopreservation. Antagonistic properties of LAB allied to their safe history of use in traditional food fermented products make them very attractive to be used as biopreservatives (Caplice and Fitzgerald 1999). Fermentation is one of the oldest and most economical methods of producing and preserving foods (Chavan and Kadam 1989), particularly in the tropical countries where there is high temperature and high humidity, which favour food spoilage.

3.1 Overview of Lactic Acid Bacteria

Lactic acid bacteria have been used in food fermentations for more than 4000 years. It is important to acknowledge that the widespread term 'lactic acid bacteria (LAB)' has no official status in taxonomy used to describe the group of functionally and genetically related bacteria. LAB consists of bacterial genera within the *Firmicutes* comprised of about 20 genera (Saranraj et al. 2013).

The main members of the Lactic acid bacteria are genera *Lactococcus*, *Lactobacillus*, *Streptococcus*, *Leuconostoc*, *Pediococcus*, *Carnobacterium*, *Aerococcus*, *Enterococcus*, *Oenococcus*, *Tetragenococcus*, *Vagococcus* and *Weissella*. *Lactobacillus* is the largest genus of this group, comprising around 80 recognised species. LAB are characterised as Gram-positive, usually nonmotile, non-sporulating bacteria that produce lactic acid as a major or sole product of fermentative metabolism. The LAB produce an array of antimicrobial substances (such as organic acids, diacetyl, acetoin, hydrogen peroxide, reuterin, reutericyclin, antifungal peptides and bacteriocins (Holzapfel et al. 1995; Obadina et al. 2006). LAB are nutritionally fastidious, requiring carbohydrates, amino acids, peptides, nucleic acid derivates and vitamins.

Some members of LAB produce bacteriocins and bacteriocins-like substances which may inhibit the growth of spoilage and pathogenic microorganisms (Klaenhammer 1988). Bacteriocins of LAB are considered as safe natural preservatives or biopreservatives as it is assumed that they are degraded by the proteases in gastrointestinal tract (Saranraj et al. 2013). Bacteriocins are ribosomally synthesised antimicrobial peptides that are active against other bacteria, either of the same species (narrow spectrum) or across genera (broad spectrum) (Bowdish et al. 2005; Cotter et al. 2005). Bacteriocins may be produced by both Gram-negative and Gram-positive bacteria (Savadogo et al. 2006).

Lactic acid bacteria are commonly found in the gastrointestinal tract of various endothermic animals such as mice, rats and pigs, in milk and dairy products (Sharpe 1981), in seafood products (Maugin and Novel 1994) and on some plant surface (Keddie 1959). They are generally used in the production and preservation of food products like cheese, sauerkraut and meat (Ringø and Gatesoupe 1998). LAB are either homofermentative or heterofermentative based on the amount of lactic acid formed during glucose fermentation (Frank et al. 2002).

3.2 *Application Strategies of LAB in 'Satar'*

The high demand of consumers for natural antimicrobial in foods has led to an increased research for finding alternative for chemical preservation. With the objective of using lactic acid bacteria as biopreservative agent, cell-free culture supernatants (with pH range from 3.8 to 4.0) of two bacterial strains for their efficacy to extend the shelf life of raw 'Satar' was evaluated. Principally, several possible strategies for the application of bacteriocins in the preservation of foods may be considered: (1) inoculation of the food with LAB that produce the bacteriocin in the product (production in situ), (2) addition of the purified or semi-purified bacteriocins as a food preservative and (3) use of a product previously fermented with a bacteriocin-producing strain as an ingredient in food processing (Schillnger et al. 1996).

3.2.1 Isolation and Identification of LAB from Fermented Fish

In this study, natural microflora of LAB was isolated from fermented fish (*Tilapia niloticus* added with 9 % black pepper, 9 % turmeric and 6 % chilli fermented for 15 days) using MRS agar (Oxoid), MRS agar with 0.8 % $CaCO_3$, M17 agar (Oxoid) and tomato juice agar (Oxoid) (Ismail et al. 2014). A total of 218 isolates were obtained and 89 isolates were presumptive LAB (Gram-positive, catalase and oxidase negative organism), and 27 isolates were randomly selected for their phenotypic identification using Vitek 2 Compact System of the microbial identification as described by Funke et al. (1998). Table 2 shows the results for identification of LAB isolated from fermented fish.

Previous study revealed that LAB are the dominant microorganisms in many fermented fish products produced by spontaneous fermentation (Hati et al. 2013). During fermentation process, LAB can utilise carbohydrate substrates available in

Table 2 Identification of LAB isolates using Vitek 2 Compact System

No	Code of LAB isolates	Shape	Name of LAB	Similarity index (%)
1	FGJ2*	Rod	*Lactobacillus acidophilus*	96.0
2	D1A1B	Rod	*Lactobacillus plantarum*	91.0
3	D0A4B	Rod	*Lactobacillus plantarum*	86.0
4	D0A2B	Rod	*Lactobacillus plantarum*	86.0
5	D3B3B	Rod	*Lactobacillus plantarum*	90.0
6	D9A3C	Rod	*Lactobacillus plantarum*	96.0
7	D9T1A 1	Rod	*Lactobacillus plantarum*	94.0
8	D9T1A 2	Rod	*Lactobacillus plantarum*	94.0
9	D15C1A	Rod	*Lactobacillus plantarum*	96.0
10	D15C2B	Cocci	*Pediococcus pentosaceus*	90.0
11	D9T1B	Rod	*Lactobacillus plantarum*	91.0
12	D159A1	Rod	*Lactobacillus plantarum*	97.0
13	D9A3A	Rod	*Lactobacillus plantarum*	95.0
14	D15902	Rod	*Lactobacillus plantarum*	96.0
15	CBM 1(a)*	Rod	*Lactobacillus plantarum*	96.0
16	D0A1D	Rod	*Lactobacillus plantarum*	91.0
17	CBS 2	Rod	*Lactobacillus plantarum*	89.0
18	DTM1	Cocci	*Pediococcus pentosaceus*	90.0
19	DTSI	Rod	*Lactobacillus plantarum*	97.0
20	DBJ1*	Rod	*Lactobacillus plantarum*	93.0
21	CBM 1 (b)	Rod	*Lactobacillus plantarum*	99.0
22	ECM 1	Rod	*Lactobacillus plantarum*	91.0
23	FBR 1	Rod	*Lactobacillus plantarum*	91.0
24	FBJ 1	Rod	*Lactobacillus plantarum*	93.0
25	FBS 2	Rod	*Lactobacillus plantarum*	92.0
26	FBM 2(a)	Rod	*Lactobacillus plantarum*	93.0
27	FTS 1(b)*	Rod	*Lactobacillus plantarum*	96.0

*Isolates of FGJ2, CBM 1(a), DBJ1 and FTS 1(b) were used for heat stability study

the fermentation matrix and produce organic acids, especially lactic acid that not only contributes to the taste, aroma and texture of the product but also lowers the product's pH that is the main factor to ensure the quality and safety of the product (Hati et al. 2013). Saithong et al. (2010) also stated that the combination of low pH and organic acids is the main preservation factor in fermented fish products. Generally, the pH should be below 4.5 to inhibit the growth of pathogenic and spoilage bacteria. Besides that, strains of certain LAB species display probiotic activity and are widely used in the food industry due to their potential health benefits (Hati et al. 2013).

It was found that 24 from 27 (88.9 %) identified strains of LAB were *Lb. plantarum*. This finding is aligned with Desniar et al. (2013) who also reported that *Lb. plantarum* was the dominant species from species of *Lactobacillus* during isolation of LAB from various samples of fresh and frozen fish and prawn.

Lactobacillus is widely recognised as being phylogenetically very heterogeneous databases, and this is evidenced by the broad range of %GC values exhibited within the genus of about 32–53 %. Within this genus, *Lb. plantarum* has a significantly larger genome of 3.3 Mb compared to other LAB, which is relatively uniform at 1.8–2.6 Mb (Khanh et al. 2011). As many traditional lactic acid-fermented foods all over the world contained a high number of *Lb. plantarum* and they have a reputation for being safe and wholesome, this strongly indicates that *Lb. plantarum* can be safely consumed (Danisco 2008).

3.2.2 Screening of Bacteriocin-Producing LAB Isolated from Fermented Fish

Antibacterial agents produced by LAB can be categorised as either non-proteinaceous or protein bacteriocins. The spectrum of activity of the main bacteriocins of LAB against food spoilage and for food-borne pathogenic microorganisms has been reviewed by Piard and Desmazeaud (1992). The discovery of nisin, the first bacteriocin used on a commercial scale as a food preservative, has prompted research on bacteriocin of LAB in the last two decades, searching for novel bacteriocin-producing strains from dairy, meat, plant and fermented food products in terms of their structure and genetic (Saranraj et al. 2013).

Most lactic acid bacterial species can produce bacteriocins which are active against the lactic acid flora itself (Piard and Desmazeaud 1992). All nine selected strains of LAB can inhibit other LAB indicating that all strains studied are bacteriocin-producing LAB (Table 5). The spot on lawn method was used to assess the inhibition degree of each LAB isolate against another LAB isolates as one of the characteristics of bacteriocin-producing LAB following the method was described by Cadirci and Citak (2005). In this study, nine pure lactic acid bacteria (LAB) cultures isolated from fermented fish were grown independently in de Man, Rogosa and Sharpe (MRS) broth at 30 °C for 24 h anaerobically. Then, the cultures were spotted on the surface of plates of de Man, Rogosa and Sharpe (MRS) agar (Merck) and incubated anaerobically at 30 °C for 24 h. After 24 h, 0.45 ml of indicator LAB suspensions (1.5 % v/v) was added to 30 ml MRS soft agar, and this mixture was poured onto MRS agar plate which has been inoculated with target strain of LAB for measuring the antagonistic activity. The inhibition zone after 24 h at 30 °C anaerobic incubation was measured in mm. Inhibition was scored positive if the zone was wider than 2 mm in diameter as shown in Fig. 2 (Mohankumar and Murugalatha 2011). The results are summarised in Table 3.

The results showed that all the LAB isolates had antagonistic activity against other LAB microflora as indicated by the presence of a clear zone of more than 2 mm except the antagonistic activity between *Lb. plantarum* (DBJ1) and *Lb. plantarum* (ECM1) (Mohankumar and Murugalatha 2011; Naimi and Khaled 2014).

The degree of antagonistic activity observed for the LAB isolates was different between each strain, suggesting that the antagonistic activity is influenced by the

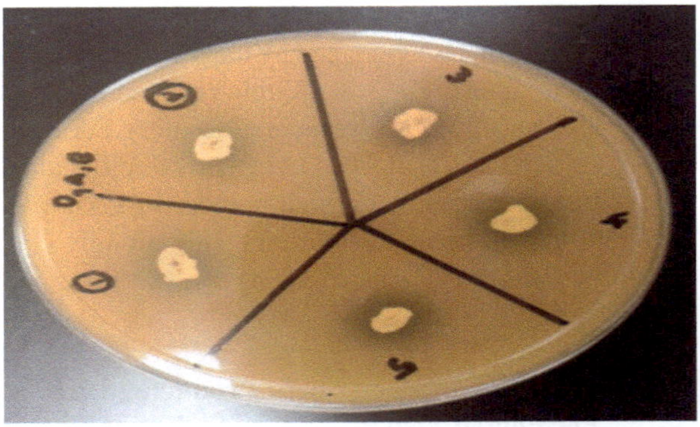

Fig. 2 Isolate *Lb. plantarum* showed antagonistic activity against other five isolates of LAB

different metabolites produced and possibly strain specific. Anas et al. (2008) described that important variations in the spectra of antagonistic activity are noted during the characterisation of the bacteriocins. It is also noted that the sensitivity of a strain depends on the genera, the species and even on the subspecies (Kalchayanand et al. 1994).

Bacteriocin-producing strains of LAB may be very important in competing with other organisms in the intestine. Bacteriocins consist of a biologically active protein moiety, have a bactericidal mode of action and attach to specific cell receptors (Soomro et al. 2002). Most of the Gram-positive bacteriocins are membrane-active compounds that increase the permeability of the cytoplasmic membrane (Jack et al. 1995). They often show a much broader spectrum of bactericidal activity than Gram-negative bacteriocins, for example, colicins.

The bacteriocins produced by LAB offer several desirable properties that make them suitable for food preservation; (1) are generally recognised as safe substances; (2) are not active and nontoxic on eukaryotic cells; (3) become inactivated by digestive proteases, having little influence on the gut microbiota; (4) are usually pH and heat tolerant; (5) have a relatively broad antimicrobial spectrum, against many food-borne pathogenic and spoilage bacteria; and (6) show a bactericidal mode of action, usually acting on the bacterial cytoplasmic membrane, no cross resistance with antibiotics, and (7) their genetic determinants are usually plasmid encoded, facilitating genetic manipulation (Galvez et al. 2008).

3.2.3 Antimicrobial Activity of LAB Isolates from Fermented Fish Against Food-Borne Pathogens

Control of both pathogenic and spoilage microbe in a variety of foods is important to guarantee food quality and safety (Anas et al. 2008). Microorganisms, such as lactic acid bacteria (LAB) produce antimicrobial compounds that can be applied as food preservatives (Stiles 1996). Many studies on LAB isolated from foods in Malaysia have concentrated on their use as organic acid producers (Ahmad and Irene 2007), production of bacteriocins, antibacterial agents (Liasi et al. 2009;

Table 3 Screening of bacteriocin-producing LAB among seven selected LAB isolates from fermented fish

LAB*	OGM2	FTR1b	FBR1	FTS1b	ECM1	DBJ1	CBM1a	FBJ1	FGJ2
OGM2	4.50 ± 0.00	4.00 ± 0.00	3.00 ± 0.00	2.75 ± 0.25	2.05 ± 0.05	4.00 ± 0.00	4.40 ± 4.45	3.90 ± 0.10	3.00 ± 0.00
FTR1b	8.90 ± 0.10	8.95 ± 0.05	8.00 ± 0.00	8.95 ± 0.05	9.25 ± 0.25	7.75 ± 0.25	8.55 ± 0.05	10.1 ± 0.45	7.55 ± 0.05
FBR1	3.05 ± 0.05	5.65 ± 0.15	4.00 ± 0.00	4.65 ± 0.15	4.45 ± 0.05	4.55 ± 0.05	3.55 ± 0.05	4.20 ± 0.20	3.45 ± 0.05
FTS1b	3.75 ± 0.25	3.90 ± 0.10	3.50 ± 0.00	4.45 ± 0.05	3.10 ± 0.10	4.45 ± 0.05	2.40 ± 0.10	7.4 ± 0.10	2.05 ± 0.05
ECM1	4.45 ± 0.05	5.25 ± 0.25	5.50 ± 0.00	5.00 ± 0.00	3.00 ± 0.00	1.50 ± 0.00	2.50 ± 0.00	4.50 ± 0.00	2.55 ± 0.05
DBJ1	6.55 ± 0.05	6.40 ± 0.10	5.50 ± 0.00	6.35 ± 0.35	8.50 ± 0.00	8.50 ± 0.00	5.90 ± 0.10	6.00 ± 0.10	8.90 ± 0.10
CBM1a	5.60 ± 0.10	6.20 ± 0.20	8.50 ± 0.00	7.40 ± 0.10	6.60 ± 0.10	6.10 ± 0.10	6.05 ± 0.05	6.10 ± 0.10	6.05 ± 0.05
FBJ1	6.55 ± 0.05	8.05 ± 0.05	7.50 ± 0.00	7.75 ± 0.25	8.05 ± 0.45	7.75 ± 0.25	8.15 ± 0.15	7.85 ± 0.15	9.00 ± 0.00
FGJ2	7.45 ± 0.05	7.05 ± 0.05	7.00 ± 0.00	6.90 ± 0.10	7.55 ± 0.05	8.45 ± 0.05	8.40 ± 0.10	8.05 ± 0.05	5.45 ± 0.05

*All isolates studied were *Lb. plantarum* except for isolate FGJ2

Data represents mean and standard deviation from triplicate independent replicates ($n = 3$) for each sample

Aween et al. 2012; Muhialdin et al. 2012; Ismail et al. 2014), antibacterial activity against multiple antibiotic-resistant (MAR) Gram-positive bacteria (Aween et al. 2012) and Gram-negative bacteria (Salleh et al. 2014) and antifungal activity (Muhialdin et al. 2012).

Well diffusion assay test was used to measure the antimicrobial activity by LAB cultures. Indicator organisms used were *Listeria monocytogenes*, *Enterobacter aerogenes*, *Klebsiella pneumoniae*, *Salmonella*, *Vibrio inaba*, *Staphylococcus epidermis*, *Proteus mirabilis* and *Escherichia coli*. Table 4 shows antimicrobial activities of selected LAB isolates from fermented fish against various pathogenic and spoilage microbes. The antimicrobial activity of LAB isolates was significantly different ($P < 0.05$) among targeted indicator organisms. It is found that all isolates of LAB from fermented fish inhibited *E. coli*, *P. mirabilis* and *V. inaba*. This result is similar to the review written by Doores (1993), where lactic acid produced by LAB is able to inhibit the growth of Gram-negative species of the families *Enterobacteriaceae* and *Pseudomonadaceae*. The antibacterial action of lactic acid is largely influenced by its ability to be in the undissociated form to penetrate the cytoplasmic membrane, resulting in reduced intracellular pH and disruption of the transmembrane proton motive force (Ray and Sadine 1992). Figures 3 and 4 showed the antimicrobial activity of selected LAB cultures *against V. inaba* and *S. epidermis*.

3.2.4 Heat Stability of Cell-Free Supernatant of LAB Isolates Against *Salmonella* and *E. coli*

LAB and their metabolites have potential as natural preservatives to control the growth of spoilage and pathogenic bacteria in foods. Heat stability was tested on the crude bacteriocin of LAB to different heating temperatures for different periods of time (based on usual processing operation of foods) against *Salmonella* and *E. coli*; these pathogens are commonly associated with ready-to-eat foods. The stability of crude supernatant from LAB isolates was determined by heating the crude supernatant in thermostatically controlled water bath for 5 and 20 min in different temperature ranges: 70, 80, 90 and 100 °C. In addition, sample was subjected to autoclaving at 121 °C for 20 min to evaluate the remaining activity of crude supernatant at sterilisation temperature. The results for the effects of heating temperature at different times of exposures on the antimicrobial activity of crude bacteriocin against *Salmonella* and *E. coli* are summarised in Tables 5 and 6, respectively. The use of 37 °C as control is important to ensure no heating is subjected to LAB cultures for comparison purposes.

As shown in Tables 5 and 6, the antagonistic activity of *Lb. plantarum* and *Lb. acidophilus* isolated from fermented fish was significantly affected ($P < 0.05$) by the combined effects of heating temperature and exposure time. There were significant interaction effects of heating temperature and exposure time of crude bacteriocin *Lb. plantarum* and *Lb. acidophilus* to inhibit the growth of *Salmonella* and *E. coli* that is commonly associated with raw 'Satar'.

Table 4 Antimicrobial activity of nine isolates of LAB isolated from fermented fish against selected pathogenic and spoilage microbes as described by the diameter of clear zones (mm) among tested lactic acid bacteria

Microbes to be challenged with	OGM2	FTR1b	FBR1	FTS1b	ECM1	DBJ1	CBM 1a	FBJ1	FGJ2*
Listeria monocytogenes	12.5 ± 0.30^{c}	15.0 ± 0.30^{b}	0.00 ± 0.00^{f}	0.00 ± 0.00^{e}	13.0 ± 0.10^{d}	15.0 ± 0.40^{b}	14.0 ± 0.20^{b}	15.0 ± 0.27^{b}	12.0 ± 0.36^{c}
Enterobacter aerogenes	13.0 ± 1.20^{bc}	13.0 ± 1.00^{c}	13.0 ± 0.35^{c}	12.0 ± 1.32^{c}	14.0 ± 0.44^{c}	0.00 ± 0.00^{f}	0.00 ± 0.00^{f}	12.0 ± 0.70^{d}	14.0 ± 0.53^{ab}
Klebsiella pneumoniae	14.0 ± 0.44^{b}	14.5 ± 0.36^{b}	14.0 ± 0.30^{b}	13.0 ± 0.56^{c}	15.5 ± 0.21^{b}	0.00 ± 0.00^{f}	0.00 ± 0.00^{f}	13.5 ± 0.87^{c}	13.0 ± 0.66^{b}
Salmonella enterica	14.0 ± 0.80^{b}	12.5 ± 0.50^{c}	12.0 ± 0.30^{d}	12.0 ± 0.50^{c}	14.0 ± 0.36^{c}	0.00 ± 0.00^{f}	0.00 ± 0.00^{f}	12.0 ± 0.10^{d}	13.0 ± 2.00^{b}
Vibrio inaba	12.5 ± 0.14^{de}	13.0 ± 1.00^{c}	14.5 ± 0.29^{b}	14.5 ± 0.10^{b}	16.0 ± 0.27^{b}	13.0 ± 0.20^{c}	12.0 ± 0.44^{d}	12.5 ± 1.00^{cd}	13.0 ± 0.17^{b}
Staphylococcus epidermis	14.0 ± 0.30^{b}	16.0 ± 0.27^{ab}	14.0 ± 0.62^{b}	15.0 ± 1.00^{b}	13.0 ± 0.44^{d}	0.00 ± 0.00^{f}	0.00 ± 0.00^{f}	11.0 ± 1.00^{d}	13.5 ± 1.05^{ab}
Proteus mirabilis	12.0 ± 0.10^{c}	11.0 ± 0.17^{d}	10.5 ± 0.70^{e}	9.0 ± 0.27^{d}	14.0 ± 0.36^{c}	12.0 ± 0.35^{d}	10.0 ± 0.44^{e}	12.0 ± 0.20^{d}	13.0 ± 0.17^{b}
Escherichia coli	12.0 ± 0.00^{a}	10.0 ± 0.00^{d}	11.0 ± 1.41^{b}	10.5 ± 0.71^{c}	10.5 ± 0.71^{c}	10.5 ± 0.71^{c}	9.75 ± 1.06^{e}	11.75 ± 0.35^{ab}	11.00 ± 0.00^{b}

*All isolates studied were *Lb. plantarum* except for isolate FGJ2

Data represent mean and standard deviation from triplicate independent replicates ($n = 3$) for each sample. Values with the same superscript letter within same column are not significantly different ($P > 0.05$)

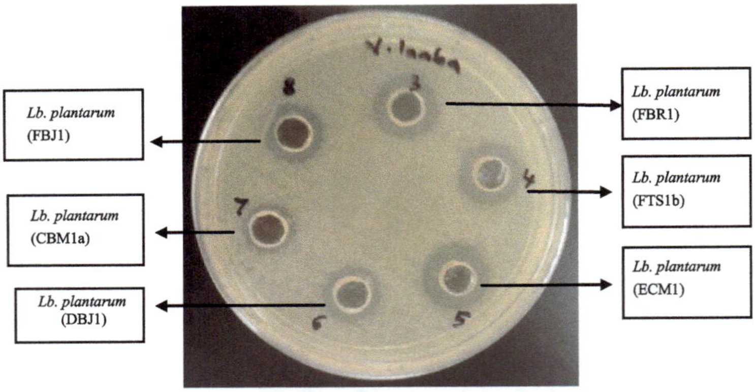

Fig. 3 Selected LAB cultures against *V. inaba*

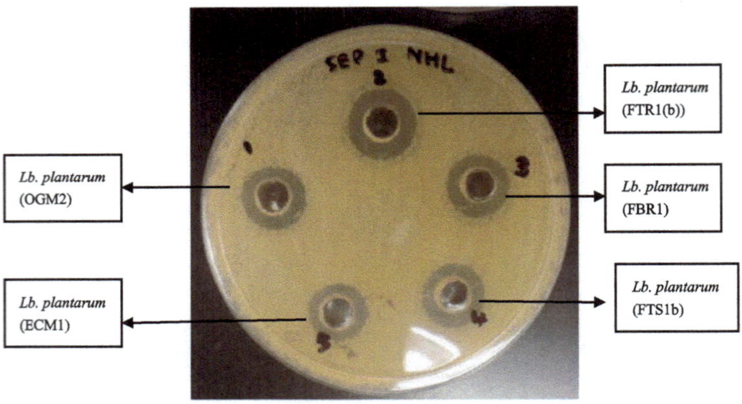

Fig. 4 Selected LAB cultures against *S. epidermis*

Heating the crude bacteriocin of *Lb. plantarum* (FTR1(b), DBJ1, CBM1(a)) and *Lb. acidophilus* (FGJ2) at 70–100 °C did not affect the inhibitory activity against *Salmonella*. However, heating the supernatant at 121 °C diminished the inhibitory activity. Similar results were reported by Naimi and Khaled (2014) who found that crude bacteriocin of *Lb. acidophilus* isolate NO.1 remained active when it was heated at 60 °C, 80 °C and 100 °C for 10 min against *Lb. acidophilus* R0052, but the activity loss after boiling at 100 °C for 30 and 60 min.

As the pH supernatant used in this study was in the range from 3.8 to 4.0, it was obvious that the heat stability of crude bacteriocin of LAB was influenced by pH. The undissociated, more hydrophobic form of the acid diffuses over the cell membrane and dissociates inside the cell, releasing H^+-ions that acidify the cytoplasm (Axelsson 1990). This has been supported by Ogunbanwo et al. (2003) who had reported that the highest antibacterial activity was exhibited in an acidic pH range of 2–6, while inactivation occurred at pH 8–12. Two bacteriocins, namely,

Table 5 Effects of heating temperatures and exposure time on the antimicrobial activity of crude bacteriocin of selected LAB isolates against *Salmonella**

			Diameter of inhibition zone (mm)				
LAB	Exposure time	Control (37 °C)	70 °C	80 °C	90 °C	100 °C	121 °C
Lb. plantarum (FTR1b)	5 min	12.25 ± 0.25	14.00 ± 0.00^{ab}	11.25 ± 0.35^{d}	12.25 ± 0.35^{cd}	14.00 ± 0.00^{ab}	0.00 ± 0.00^{e}
	20 min	12.00 ± 0.00	11.25 ± 0.35^{de}	12.25 ± 0.35^{bcd}	12.25 ± 0.35^{bcd}	13.25 ± 0.35^{ab}	0.00 ± 0.00^{f}
Lb. plantarum (DBJ1)	5 min	13.00 ± 0.00	13.25 ± 0.35^{abc}	12.50 ± 0.00^{bcd}	14.50 ± 0.71^{a}	12.25 ± 0.35^{cd}	0.00 ± 0.00^{e}
	20 min	12.00 ± 0.00	11.25 ± 0.35^{de}	11.25 ± 0.35^{de}	13.75 ± 0.35^{a}	13.00 ± 0.00^{ab}	0.00 ± 0.00^{f}
Lb. plantarum (CBM1a)	5 min	12.25 ± 0.25	13.00 ± 0.00^{abc}	12.00 ± 0.00^{cd}	13.50 ± 0.00^{abc}	12.25 ± 0.35^{cd}	0.00 ± 0.00^{e}
	20 min	12.00 ± 0.00	11.50 ± 0.71^{bc}	11.75 ± 0.35^{cde}	12.75 ± 0.35^{abc}	13.25 ± 0.35^{ab}	0.00 ± 0.00^{f}
Lb. acidophilus (FGJ2)	5 min	12.00 ± 0.00	11.00 ± 1.41^{d}	13.00 ± 0.00^{abc}	12.25 ± 0.35^{cd}	13.25 ± 0.35^{abc}	0.00 ± 0.00^{e}
	20 min	12.25 ± 0.25	13.75 ± 0.00^{a}	13.25 ± 0.35^{ab}	11.00 ± 0.00^{e}	13.50 ± 0.00^{a}	0.00 ± 0.00^{f}

*Data represent mean and standard deviation from triplicate independent replicates ($n = 3$) for each sample. Values with the same superscript letter within same column are not significantly different ($P > 0.05$)

Table 6 Effects of heating temperatures and exposure time on the antimicrobial activity of crude bacteriocin of selected LAB isolates against *E. coli**

LAB	Exposure time	Control (37 °C)	Diameter of inhibition zone (mm)				
			70 °C	80 °C	90 °C	100 °C	121 °C
Lb. plantarum (FTR1b)	5 min	12.50 ± 0.00	14.00 ± 0.00^{ab}	11.25 ± 0.35^{d}	12.25 ± 0.35^{cd}	14.00 ± 0.25^{ab}	0.00 ± 0.00^{e}
	20 min	14.25 ± 0.25	10.75 ± 0.35^{f}	12.00 ± 0.00^{cdef}	13.50 ± 0.00^{ab}	13.50 ± 0.35^{ab}	0.00 ± 0.00^{f}
Lb. plantarum (DBJ1)	5 min	12.25 ± 0.25	13.25 ± 0.35^{abc}	12.50 ± 0.00^{bcd}	14.50 ± 0.71^{a}	12.25 ± 0.35^{cd}	0.00 ± 0.00^{e}
	20 min	13.25 ± 0.25	12.50 ± 0.71^{bcd}	11.50 ± 0.00^{def}	13.00 ± 0.00^{abc}	12.25 ± 0.35^{bcde}	11.50 ± 0.71^{def}
Lb. plantarum (CBM1a)	5 min	13.00 ± 0.00	13.00 ± 0.00^{abc}	12.00 ± 0.00^{cd}	13.50 ± 0.00^{abc}	12.25 ± 0.35^{cd}	0.00 ± 0.00^{e}
	20 min	12.75 ± 0.75	12.50 ± 0.71^{bcd}	12.25 ± 0.35^{bcde}	13.00 ± 0.00^{abc}	14.00 ± 0.00^{a}	11.75 ± 0.35^{cdef}
Lb. acidophilus (FGJ2)	5 min	11.00 ± 0.00	11.00 ± 1.41^{d}	13.00 ± 0.00^{abc}	12.25 ± 0.35^{cd}	13.25 ± 0.35^{abc}	0.00 ± 0.00^{e}
	20 min	11.25 ± 0.25	10.75 ± 0.35^{f}	11.00 ± 0.35^{ab}	12.25 ± 0.35^{bcde}	11.00 ± 0.00^{ef}	0.00 ± 0.00^{g}

*Data represent mean and standard deviation from triplicate independent replicates ($n = 3$) for each sample. Values with the same superscript letter within same column are not significantly different ($P > 0.05$)

bulgarican and lactobulgarican, isolated from *Lb. bulgaricus*, were shown to have the highest activity and stability at pH 2.2 and 4.0, respectively, against a range of bacteria (Reddy et al. 1984; Abdel-Bar et al. 1987).

In general, heat stability reflects the ability of crude bacteriocin in maintaining its inhibitory activity when subjected to different heating temperatures. This is an important characteristic for a bacteriocin to be used as a food preservative because heating is one of the most important steps in food processing.

4 Effect of Incorporation of Crude Bacteriocin of LAB on Several Characteristics of Raw 'Satar' and Their Microbiological Shelf Life at Ambient Temperature

In an attempt to evaluate the effectiveness of crude bacteriocin of *Lb. plantarum* (isolate CBM1a) and *Lb. acidophilus* (isolate FGJ2) as biopreservative in 'Satar', crude bacteriocins was incorporated in raw 'Satar'. The Satar was freshly prepared in controlled environment at the Food Technology Laboratory, UMT. The fish (Selayang) was ground and was mixed with all ingredients such as sugar, salt, onion, tamarind, shredded coconut and chillies using food processor. The fish was purchased from Pulau Kambing, Kuala Terengganu. Other ingredients were purchased from the local market, Kuala Terengganu. The ingredients used are summarised in Table 7.

Then, 3 % and 6 % (v/w) crude bacteriocin of *Lb. plantarum* (CBM1a) and *Lb. acidophilus* (FGJ2) were added together with other ingredients during mixing. Satar sample without incorporation of bacteriocin was used as control. The prepared samples were put in sterile closed container and stored at ambient temperature for 0, 3, 16 and 20 h, and some physical and chemical characteristics as well as microbiological shelf life were determined.

Table 7 Ingredients of 'Satar'

Ingredients	Percentage (%)
Fresh fish	50.0
Grated coconut	25.0
Onion	10.0
Chilli	5.0
Sugar	5.0
Tamarind	2.5
Salt	2.0
MSG	0.5

Source: Modified from personal communication, Pok Nor Satar (2011)

4.1 *Effect of Incorporation of Crude Bacteriocin of LAB on Several Characteristics of Raw 'Satar'*

4.1.1 Water Activity Analysis

Water activity (a_w) is the amount of 'available water' or water available for bacterial activity. Water activity is a measure of the amount of water that is not bound to the food and therefore available for bacterial growth (McSwane et al. 2005). It is one of the important factors that promote or limit the microbial growth (Forsythe 2000). The result of water activity analysis is shown in Fig. 5.

The water activity (a_W) of 'Satar' ranged between 0.89 ± 0.006 and 0.98 ± 0.006; thus, the raw product has high water activity. Incorporation of bacteriocins to raw 'Satar' resulted in significant difference ($P < 0.05$) in water activity (a_w) between the control with 6 % *Lb. acidophilus* and 6% *Lb. plantarum*. Bacteria require higher values of a_W for growth than fungi, with Gram-negative bacteria having higher requirements than Gram-positives. According to McSwane et al. (2005), disease-causing bacteria can only grow in foods that have a water activity higher than 0.85. Most spoilage bacteria do not grow below a_W 0.91, whereas spoilage moulds can grow as low as 0.80 (Jay et al. 2005). Previous study has found that a_W was recorded between 0.97 and 0.99 in 'keropok lekor', another Malaysian traditional product with fish as main ingredient (Nor-Khaizura et al. 2009).

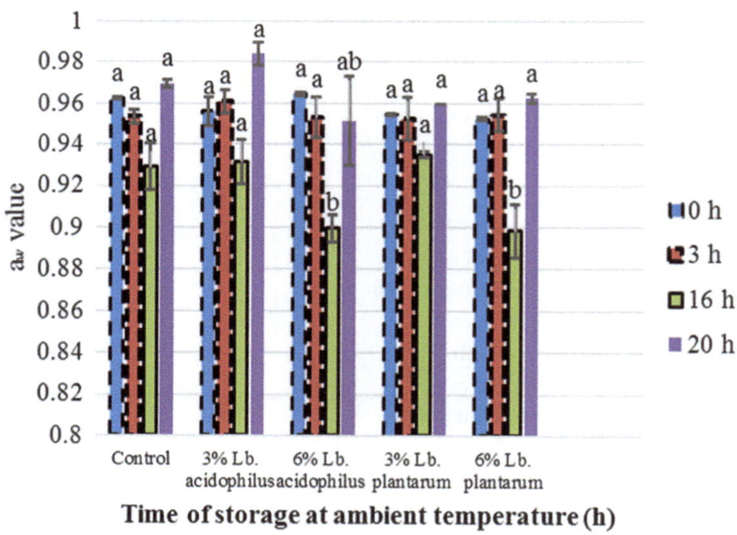

Fig. 5 Changes in water activity of raw 'Satar' incorporated with crude bacteriocin of *Lb. plantarum* and *Lb. acidophilus* stored in ambient temperature (25 ± 2 °C)

4.1.2 pH Analysis

Most foods are acidic and have pH less than 7.0. The bacteria mostly prefer a neutral environment (pH of 7.0) but are capable of growing in foods that have a pH in the range of 4.6–9.0 (McSwane et al. 2005). The initial pH values of raw 'Satar' were in the range of 6.15 ± 0.08 to 6.43 ± 0.04 (Fig. 6). A reduction in pH of about 1 pH unit was observed after 16 and 20 h storage.

From Fig. 6, prolonged storage after 16 h and 20 h had significantly reduced the pH of the samples. It was clear that this reduction of pH had benefited the samples as preservation. Similar results also were found to samples incorporated with crude bacteriocin and control samples (without crude bacteriocin). In the previous study, the pH of 'keropok lekor' samples was below 7.0 where the pH values did not change significantly ($P > 0.05$) at the different stages of 'keropok lekor' processing (Nor-Khaizura et al. 2009).

4.1.3 Moisture Content Analysis

Moisture content is an important factor in bacterial growth. Foods that are high in moisture content will favour the growth of spoilage microorganisms. Figure 7 shows the moisture content analysis of raw 'Satar' incorporated with *Lb. acidophilus* and *Lb. plantarum* supernatant stored at ambient temperature (25 ± 2 °C).

As the time of storage had increased, there was significant differences ($P < 0.05$) between samples stored at 0, 3 and 16 h compared to samples at 20 h of storage. Prolonged incubation at ambient temperature had significantly increased the moisture content of 'Satar' for all treated samples. Moisture migration is a major cause

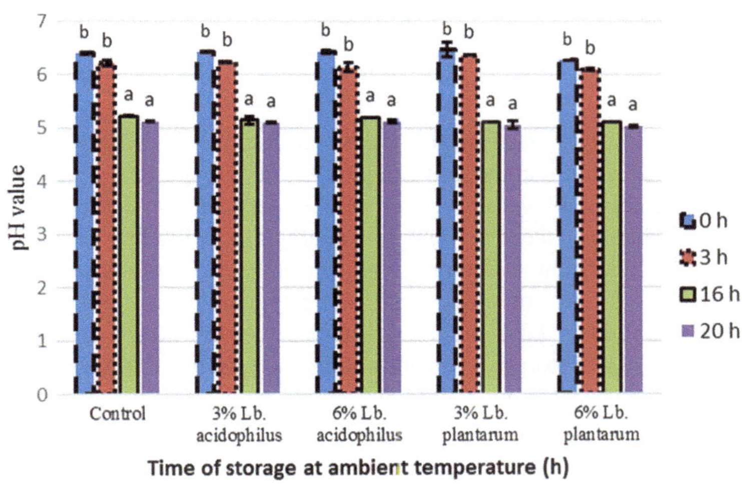

Fig. 6 Changes in pH of raw 'Satar' incorporated with crude bacteriocin of *Lb. plantarum* and *Lb. acidophilus* stored in ambient temperature (25 ± 2 °C)

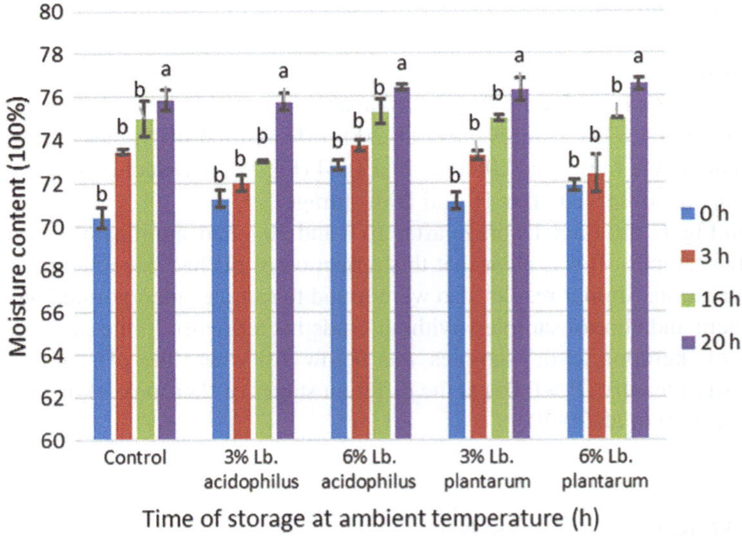

Fig. 7 Moisture content of raw 'Satar' incorporated with crude bacteriocin of *Lb. plantarum* and *Lb. acidophilus* stored at ambient temperature (25 ± 2 °C)

of deteriorative physical changes in foods, which significantly happen after 20 h storage at ambient temperature. The loss of moisture can result in loss of nutrients and increase the possibility of rancidity of food products (NZFSA 2005). The results have confirmed that the microbial growth was influenced by moisture.

4.1.4 Colour Analysis

Changes in the colour of a food will make it undesirable and can cause economical loss.

From Table 8, it was observed that there is no significant difference ($P > 0.05$) for a* and b* of colours among all samples before storage (0 h), but significant ($P < 0.05$) changes in a* were observed after 3 h storage between control sample and all other samples except for samples incorporated with 3 % *Lb. acidophilus* which has no significant difference with the control. However, no significant difference ($P > 0.05$) was observed for b* in Satar samples incorporated with 6 % *Lb. acidophilus* and 6 % *Lb. plantarum*. The colour of raw Satar remains the same during 16 h and 20 h of storage. From the colour analysis, it was observed the colour of raw 'Satar' was between grey and light grey.

Table 8 Colour analysis of raw 'Satar' incorporated with crude bacteriocin of LAB

Formulation	Control	3% *Lb. acidophilus*	6% *Lb. acidophilus*	3% *Lb. plantarum*	6% *Lb. plantarum*
0 h (control)					
L*	47.04 ± 0.716	54.075 ± 0.417	53.195 ± 0.474	53.045 ± 0.728	52.395 ± 2.454
a*	5.825 ± 1.167	5.195 ± 1.747	6.30 ± 0.58	6.91 ± 0.919	6.44 ± 0.721
b*	21.515 ± 2.25	22.81 ± 0.806	23.97 ± 0.919	24.53 ± 0.509	23.685 ± 1.478
Colour	Light grey	Light grey	Light grey	Light grey	Light grey
3 h of storage at ambient temperature					
L*	36.08 ± 0.69	55.85 ± 2.18	53.96 ± 0.41	41.4 ± 0.61	54.29 ± 1.51
a*	3.39 ± 0.63	3.63 ± 1.33	5.85 ± 0.34	7.71 ± 0.68	6.06 ± 0.37
b*	16.12 ± 0.88	20.00 ± 0.16	24.24 ± 1.12	23.28 ± 0.86	24.22 ± 0.49
Colour	Grey	Light grey	Light grey	Light grey	Light grey
16 h of storage at ambient temperature					
L*	57.91 ± 0.58	49.23 ± 0.44	61.00 ± 0.65	51.91 ± 0.26	59.72 ± 0.85
a*	2.01 ± 1.03	2.65 ± 0.693	3.46 ± 0.79	1.83 ± 0.83	2.48 ± 0.32
b*	22.1 ± 3.79	18.46 ± 0.69	21.01 ± 2.21	15.37 ± 0.52	22.21 ± 1.02
Colour	Light grey	Light grey	Light grey	Grey	Light grey
20 h of storage at ambient temperature					
L*	51.73 ± 1.64	59.73 ± 0.38	54.80 ± 2.72	53.29 ± 0.11	49.75 ± 1.01
a*	3.53 ± 0.66	3.21 ± 1.47	3.38 ± 0.58	2.29 ± 0.23	4.08 ± 0.64
b*	18.24 ± 1.68	21.09 ± 2.36	22.11 ± 2.68	15.87 ± 2.23	21.07 ± 1.07
Colour	Light grey	Light grey	Light grey	Grey	Light grey

4.2 *Microbiological Quality and Shelf Life of Raw 'Satar' After Incorporating with Crude Bacteriocin of LAB*

Specific microbial growth and changes during storage are generally influenced by some factors such as the initial microbial load at the first time the product is stored; the physicochemical properties of the food, such as moisture content, pH, presence of preservatives; the processing methods used in the production of the food; and the external environments of the food, such as the surrounding gas composition and storage temperature (Kilcast et al. 2000).

The food chain during food manufacturing can influence the combination of intrinsic and extrinsic factors that affect the quality and safety shelf life of foods (NZFSA 2005). The processes can be conveniently classified as microbiological changes, physicochemical changes and temperature-related changes (Kilcast et al. 2000).

Microbiological proliferation of spoilage and/or pathogenic microorganism can be formulated using predictive microbiology approach, as it is a reliable tool for providing an estimation of the course of the bacteria in the foods and, indirectly, providing an estimation of shelf life of the product when the cause of spoilage is known to be microbiological. The main concept behind the application of predictive microbiology is the use of specific spoilage organisms (SSO) as indicator for spoilage. The detailed principles and methodologies for the determination of shelf life are described comprehensively by Valero et al. (2012).

According to New Zealand Food Safety Authority (NZFSA 2005), the end of shelf life is defined based on the maximum numbers of microorganisms allowable to be present in the foods based on the recommended guideline by regulatory and peer review from food microbiology experts. However, in other cases, the end of shelf life also can be determined by sensory or biochemical deterioration. Microbiological guidelines for ready-to-eat (RTE) foods sampled at the point of sale act as standard for the microorganisms in the food (Gilbert et al. 2000). The guideline levels for determining the microbiological quality of ready-to-eat foods is summarised in Table 9.

4.2.1 Aerobic Plate Count of 'Satar' After Incorporating with Crude Bacteriocin of LAB

Figures 8 and 9 show the APC of raw 'Satar' incorporated with different concentrations (3 and 6 %) of crude bacteriocin *Lb. acidophilus* or *Lb. plantarum* stored at ambient temperature and different times, respectively.

The control sample (without the incorporation of bacteriocin) reached unsatisfactory level ($>10^6$ CFU/g) after 8 h stored at ambient temperature. The APC of samples 'Satar' incorporated with 3 % crude bacteriocin of *Lb. acidophilus* remained at $\log_{10}$ 5 which is within the acceptable level even after 20 h storage at ambient temperature. However, incorporation of 6 % crude bacteriocin of *Lb.*

Table 9 Guideline levels for determining the microbiological quality of ready-to-eat foods

Test	Microbiological result (CFU/g unless otherwise stated)			
	Good	Acceptable	Unsatisfactory	Potentially hazardous
Standard plate count				
Category A	$<10^4$	$<10^5$	$\geq 10^5$	N/A
Category B	$<10^6$	$<10^7$	$\geq 10^7$	N/A
Category C	N/A	N/A	N/A	N/A
Indicators				
Enterobacteriaceae	$<10^2$	10^2 to $<10^4$	$\geq 10^4$	N/A
E. coli	<3	3 to $<10^2$	$\geq 10^2$	N/A
Pathogens				
C. perfringens	$<10^2$	10^2 to $<10^3$	10^3 to $<10^4$	$\geq 10^4$
B. cereus	$<10^2$	10^2 to $<10^3$	10^3 to $<10^4$	$\geq 10^4$

Source: Modified from NSW Food Authority (2009)
Note: Category A, fully cooked for immediate sale or consumption; Category B, fully cooked with further handling or processing before consumption; Category C, uncooked fermented ingredients or fresh fruit and vegetables; N/A, Not available

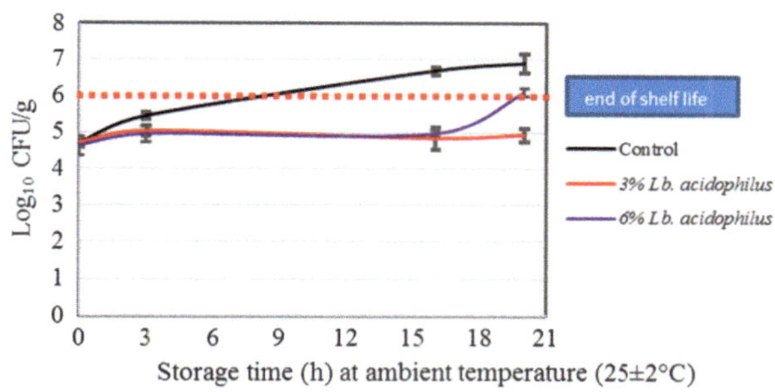

Fig. 8 Aerobic plate count of raw 'Satar' incorporated with crude bacteriocin of *Lb. acidophilus* stored at ambient temperature (25 ± 2 °C) and different storage times

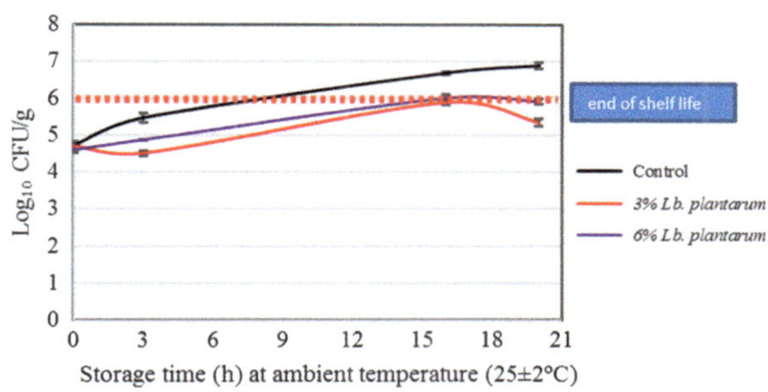

Fig. 9 Aerobic plate count of raw 'Satar' incorporated with crude bacteriocin of *Lb. plantarum* stored at ambient temperature (25 ± 2 °C) and different storage times

acidophilus was able to prevent the growth of spoilage microorganisms ($<10^6$ CFU/g) within 18 h only as shown in Fig. 8.

The results clearly demonstrated that the incorporation of 3 % and 6 % crude bacteriocin of *Lb. acidophilus* in raw 'Satar' can be used to extend the end of shelf life from 7 to 8 h (control) to about 18 h. The results demonstrated that bacteriocin produced by *Lb. acidophilus* can be used as biopreservative to extend the shelf life of raw 'Satar'. It is reported that *Lb. acidophilus* produce class II acidosis CH5 that have restricted activity towards Gram-positive bacteria (Chumchalova et al. 2004).

Similar reduction in APC was observed in raw 'Satar' samples incorporated with 3 and 6 % crude bacteriocin of *Lb. plantarum* in which APC was below the spoilage level after 12 h storage. However, studies by Todorov (2008) on the bacteriocin AMA-K produced by *Lb. plantarum*, temperatures of 4 °C and 15 °C resulted in reduction of 50 % of adsorption of this bacteriocin AMA-K to cells of *L. innocua* LMG13568 and *L. ivanovii* subsp. *ivanovii* ATCC19119.

4.2.2 *Enterobacteriaceae* Count of 'Satar' After Incorporating with Crude Bacteriocin of LAB

Enterobacteriaceae is pathogenic towards human. The family is subdivided into a number of genera that are *Escherichia*, *Salmonella*, *Shigella*, *Klebsiella*, *Enterobacter*, *Yersinia* and others based on their biochemical characteristics and pathogenicity. Generally, the *Enterobacteriaceae* count of raw 'Satar' was $<10^4$ within 16 h storage. Incorporation of crude bacteriocin from *Lb. acidophilus* and *Lb. plantarum* to raw 'Satar' at 3 and 6 % further reduced the Enterobacteriaceae count to $<10^3$ CFU/g within 16 h storage at ambient temperature (Figs. 10 and 11). It was expected that ingredients from Satar contribute to the presence of *Enterobacteriaceae* and mesophilic bacteria (APC).

In ready-to-eat foods that are fully cooked, *Enterobacteriaceae* are used as an indication of either post-processing contamination or inadequate cooking. As they can be found in raw foods, their detection may not be an indication or any processing failure (NSW Food Authority 2009).

Enterobacteriaceae are a group of bacteria that can be found in many environments. Some *Enterobacteriaceae* can be found in the intestinal tract of humans and animals, soil, vegetable matter and marine environments. The presence of Enterobacteriaceae in raw 'Satar' is expected since ingredients such as raw fish, raw grated coconut and other ingredients may contain *Enterobacteriaceae*.

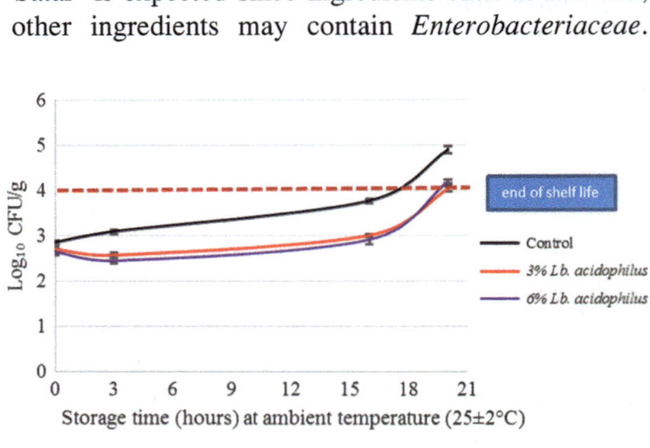

Fig. 10 *Enterobacteriaceae* count of raw 'Satar' incorporated with crude bacteriocin of *Lb. acidophilus* stored at ambient temperature (25 ± 2 °C) and different storage times

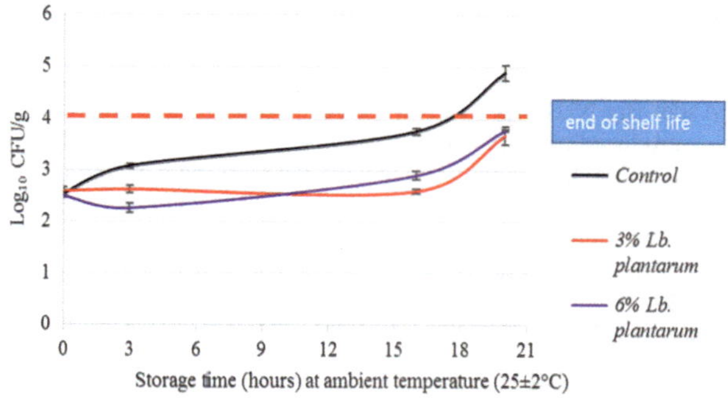

Fig. 11 *Enterobacteriaceae* count of raw 'Satar' incorporated with crude bacteriocin of *Lb. plantarum* stored at ambient temperature (25 ± 2 °C) and different storage times

However, incorporation of the crude bacteriocin from the two LAB observed can prevent their growth within 16 h storage. Additionally, the raw 'Satar' will be grilled before consumption, and this will further ensure the safety of RTE 'Satar'. However, for ready-to-eat food, the range of contamination of *Enterobacteriaceae* must not exceed 4 $\log_{10}$ CFU/g as has been described by Food Standards Australia/New Zealand (FSANZ 2001).

5 Conclusion

Among nine isolates of LAB from fermented fish, supernatants of four isolates were studied extensively for their heat stability at different heating temperatures (70, 80, 90, 100 and 121 °C) and heating times (5 and 20 min). Two strains, *Lb. acidophilus* and *Lb. plantarum*, were chosen for the incorporation of their crude bacteriocin in raw 'Satar', and their characteristics and microbiological shelf life were evaluated. Incorporation of crude bacteriocin of *Lb. acidophilus* and *Lb. plantarum* at 3 % and 6 % did not significantly affect ($P > 0.05$) the water activity and pH, but significantly increased the moisture content when 'Satar' was stored more than 20 h at ambient temperature. There was no significant difference ($P > 0.05$) for a*value and b*value of 'Satar' among all samples at 0 h of storage time, except after 3 h of storage at ambient temperature. The colour analysis of samples showed a range of colour between grey and light grey. Although crude bacteriocin was used in this study, it was possible to see the inhibitory effects on APC and *Enterobacteriaceae* and extending the shelf life or raw 'Satar' suggesting that such bacteriocin can be used in a seafood-based food with a pH of around 3.8–4.0 as well as high water activity. The challenge of improving the shelf life of food especially food with neutral pH using preservatives is not straightforward. More researches are needed in understanding the roles of LAB in preserving the foods while extending the shelf life of seafood-based foods.

Acknowledgements Financial support for this research was provided by Fundamental Research Grant Scheme (FRGS), Vot. 59269, funded by the Ministry of Education, Malaysia.

References

Abdel-Bar N, Harris ND, Rill RL (1987) Purification and properties of an antimicrobial substance produced by *Lactobacillus bulgaricus*. J Food Sci 52:411–415

Ahmad FM, Irene KP (2007) Isolation of lactic acid bacteria from Malaysian foods and assessment of the isolates for industrial potential. Biol Technol 98:1380–1385

Anas M, Eddine HJ, Mebrouk K (2008) Antimicrobial activity of *Lactobacillus* species isolated from Algerian raw goat's milk against *Staphylococcus aureus*. World J Dairy Food Sci 3(2): 39–49

Aween MM, Hassan Z, Muhialdin BJ, Eljamel YA, Al-Mabrok ASW, Lani MN (2012) Antibacterial activity of Lactobacillus acidophilus strains isolated from honey marketed in Malaysia against selected multiple antibiotic resistant (MAR) Gram-positive bacteria. J Food Sci 77(7):M364–M371

Axelsson L (1990) *Lactobacillus reuteri*, a member of the gut bacterial flora. PhD thesis, Swedish University of Agricultural Sciences, Uppsala, Sweden

Ayçiçek H, Aydoğan H, Küçükkaraaslan A, Baysallar M, Başustaoğlu AC (2004) Assessment of the bacterial contamination on hands of hospital food handlers. Food Control 15(4):253–259

Bowdish DM, Davidson DJ, Scott MG, Hancock RE (2005) Immunomodulatory activities of small host defense peptides. Antimicrob Agents Chemother 49:1727–1732

Cadirci BH, Citak S (2005) A comparison of two methods used for measuring antagonistic activity of lactic acid bacteria. Pak J Nutr 4(4):237–240

Caplice E, Fitzgerald GF (1999) Food fermentations: role of microorganisms in food production and preservation. Int J Food Microbiol 50:131–149

Cayot N (2007) Sensory quality of traditional foods. Food Chem 101(1):154–162

Chavan J, Kadam SS (1989) Nutritional improvement of cereals by sprouting. Crit Rev Food Sci Nutr 28(5):401–437

Cho J-I, Cheung C-Y, Lee S-M, Ko S-I, Kim K-H, Hwang I-S, Kim S-H, Cho S-Y, Lim C-J, Lee K-H, Kim K-S, Ha S-A (2011) Assessment of microbial contamination levels of street-vended foods in Korea. J Food Saf 31(1):41–47

Chumchalova J, Stiles J, Josephsen J, Plockova M (2004) Characterization and purification of acidocin CH5, a bacteriocin produced by Lactobacillus acidophilus CH5. J Appl Microbiol 96:1082–1089

Cleveland J, Montville TJ, Nes IF, Chikindas ML (2001) Bacteriocins: safe, natural antimicrobials for food preservation. Int J Food Microbiol 71:1–20

Cotter PD, Hill C, Ross RP (2005) Bacteriocins: developing innate immunity for food. Nat Rev Microbiol 3:777–778

Danisco AS (2008) Lactobacillus plantarum Lp-115. Danisco Publisher, New York

Dawnes PF, Ito K (2001) Compendium of methods for the microbiological examination of foods, 4th edn. American Public Health Association, Washington, DC, p 557

De Martinis ECP, Publio MRP, Santarosa PR, Freitas FZ (2001) Antilisterial activity of lactic acid bacteria isolated from vacuum-packaged Brazilian meat and meat products. Braz J Microbiol 32:32–37

Department of Fisheries, Malaysia (2014) Retrieved from official website of Department of Fisheries, http://www.dof.gov.my

Desniar RI, Suwanto A, Mubarik NR (2013) Characterization of lactic acid bacteria isolated from an Indonesian fermented fish (bekasam) and their antimicrobial activity against pathogenic bacteria. Emirates J Food Agric 489–494

Doores S (1993). In: Davidson PM, Branen AL (eds) Organic acids. Marcel Dekker, New York, pp 95–136

Ekanem EO (1998) The street food trade in Africa: safety and socio-environmental issues. Food Control 9:211–215

FAO (2005) Informal food distribution sector: in Africa (Street foods): importance and challenges. FAO/WHO Regional Conference on Food Safety for Africa, Harare, Zimbabwe

Forsythe SJ (2000) The microbiology safe food. Blackwell Science, London, pp 10–52

Frank JC, Chill D, Maida N (2002) The lactic acid bacteria: a literature survey. Crit Rev Microbiol 28(4):281–370

FSANZ (2001) Guidelines for the microbiological examination of ready-to-eat foods. Food Standard Australia New Zealand. http://foodstandards.gov.au. Retrieved 20 Oct 2009

Funke G, Monnet D, Debernandis C, Graevenitz AV, Freney J (1998) Evaluation of the VITEK 2 system for rapid identification of medically relevant gram negative rods. J Clin Microbiol 7:1948–1952

Galvez A, Lucas-Lopez R, Abriouel H (2008) Application of bacteriocins in the control of foodborne pathogenic and spoilage bacteria. Crit Rev Biotechnol 28:125–152

Gilbert RJ, Louvois JD, Donovan T, Little C, Nye K, Ribeiro CD (2000) Guidelines for the microbiological quality of some ready-to-eat foods sampled at the point of sale. Commun Dis Public Health 3(3):163–167

Hassan YI, Bullerman LB (2008) Antifungal activity of *Lactobacillus paracasei* ssp. *tolerans* isolated from a sourdough bread culture. Int J Food Microbiol 121:112–115

Hati S, Mandal SA, Prajapati JB (2013) Novel starters for value added fermented dairy products. Curr Res Nutr Food Sci 1(1):83–91

Havelaar AH, Brul S, De Jong A, De Jonge R, Zwietering MH, Ter Kuile BH (2010) Future challenges to microbial food safety. Int J Food Microbiol 139:79–94

Holzapfel WH, Geisen R, Schillinger U (1995) Biological preservation of foods with reference to protective cultures, bacteriocins and food-grade enzymes. Int J Food Microbiol 24:343–362

Huss HH, Ababouch L, Gram L (2004) Assessment and management of seafood safety and quality. In FAO Fisheries Technical Paper (444)

Ismail A, Mohd Rizan M, Lani MN, Karim NU, Alias R, Hassan Z (2014) Identification of lactic acid bacteria from fermented *Tilapia niloticus* with selected spices and their antimicrobial activity against foodborne pathogens. In: Proceedings of 32nd symposium of the Malaysian Society for Microbiology. MSM Publisher, Bangi

Jack RW, Tagg JR, Ray B (1995) Bacteriocins of gram positive bacteria. Microbiol Rev 59: 171–200

Jay JM, Loessner MJ, Golden DA (2005) Modern food microbiology, 7th edn. Springer, New York, pp 15–59

Jayasuriya DC (1994) Street food vending in Asia: some policy and legal aspects. Food Control 5(4):222–226

Kalchayanand N, Hanlin MB, Ray B (1994) Sublethal injury makes Gram negative and resistant Gram positive bacteria sensitive to the bacteriocins, pediocin AcH and nisin. Lett Appl Microbiol 15:239–243

Keddie RM (1959) The properties and classification of lactobacilli isolated from grass and silage. Appl Bacteriol 22:403–416

Khanh TT, Bee KM, Peter MS, Thi TV, Peter JC (2011) Distribution and genetic diversity of lactic acid bacteria from traditional fermented sausages. Food Res Int 338–344

Kilcast D, Subramaniam P, Research Leatherhead Food Association (2000) Introduction. In: Kilcast D, Subramaniam P (eds) The stability and shelf-life of food. Woodhead/CRC, Cambridge, England, pp 1–19

Klaenhammer TR (1988) Genetics of bacteriocins produced by lactic acid bacteria. FEMS Microbiol Rev 12:39085

Ko W-H (2010) Evaluating food safety perceptions and practices for agricultural food handler. Food Control 21(4):450–455

Lani MN, Ahmad S, Ibrahim R, Alias R, Hassan Z (2014a) Evaluation on chemical analysis and microbiological quality of partially cooked-frozen of Malaysian heritage food ('Satar'). Int J Eng Sci 3(9):71–77

Lani MN, Mohd Azmi MF, Ibrahim R, Alias R, Hassan Z (2014b) Microbiological quality of food contact surfaces at selected food premises of Malaysian heritage food ('Satar') in Terengganu, Malaysia. Int J Eng Sci 3(9):66–70

Lavermicocca P, Valerio F, Evidente A, Lazzaroni S, Corsetti A, Gobetti M (2000) Antifungal activity of phenyllactic acid against molds isolated from bakery products. Appl Environ Microbiol 69:634–640

Liasi SA, Azmi TI, Hassan MD, Shuhaimi M, Rosfarizan M, Ariff AB (2009) Antimicrobial activity and antibiotic sensitivity of three isolates of lactic acid bacteria from fermented fish product, Budu. Malay J Microbiol 5(1):33–37

Matilla-Sandholm T, Mättö J, Saarela M (1999) LAB with health claim-interactions and interference with gastrointestinal flora. Int Dairy J 9:25–35

Maugin S, Novel G (1994) Characterization of lactic acid bacteria isolated from seafood. J Appl Bacteriol 76:616–625

McSwane D, Rue NR, Linton R (2005) Essentials of food safety and sanitation, 4th edn. Pearson, New York, pp 32–145

Messens W, De Vugst L (2002) Inhibitory substances produced by lactobacilli isolated from sourdoughs - a review. Int J Food Microbiol 72:31–43

Mohankumar A, Murugalatha N (2011) Characterization and antibacterial activity of bacteriocin producing Lactobacillus isolated from raw cattle milk sample. Int J Biol 3(3):128–143

Muhialdin BJ, Hassan Z (2011) Screening of Lactic acid bacteria for antifungal activity against Aspergillus oryzae. Am J Appl Sci 8(5):447–451

Muhialdin BJ, Hassan Z, Ahmed Imdakim MM, Abdul Kahar FKS, Mustafa Aween M (2012) Malaysian isolates of lactic acid bacteria with antibacterial activity against Gram-positive and Gram-negative pathogenic bacteria. J Food Res 1(1):110–116

Muyanja C, Nayiga L, Brenda N, Nasinyama G (2011) Practices, knowledge and risk factors of street food vendors in Uganda. Food Control 22(10):1551–1558

Naimi M, Khaled MB (2014) Control of Staphylococcus aureus in meat system by in situ and ex situ bacteriocins from Lactobacillus sakei and Pediococcus spp. Int J Biol Agric Biosyst Life Sci Eng 8(2):178–183

Nor-Khaizura MAR, Zaiton H, Jamilah B, Gulam Rusul RA (2009) Microbiological quality of keropok lekor during processing. Int Food Res J 223:215–223

NSW Food Authority (2009) Microbiological quality guide for Ready-to-eat foods (Online). http://www.foodauthority.nsw.gov.au/_Documents/science/microbiological_quality_guide_for_RTE_food.pdf

Nurul Atiqah R (2013) Microbiological quality and safety of 'Satar' in Terengganu. M.Sc. thesis, School of Food Science and Technology, Universiti Malaysia Terengganu, Terengganu, pp 40–146

NZFSA (2005) A guide to calculating the shelf life of foods. New Zealand Food Safety Authority, New Zealand, Retrieved from http://www.nzfsa.govt.nz

Obadina AO, Oyewole OB, Sanni LO, Tomlins KI (2006) Bio preservative activities of Lactobacillus plantarum strains in fermenting Cassava 'fufu'. Afr J Biotechnol 5(8):620–623

Ogunbanwo ST, Sanni AI, Onilude AA (2003) Characterisation of bacteriocin produced by Lactobacillus plantarum F1 and Lactobacillus brevis OG1. Afr J Biotechnol 2(8):219–227

Parada JL, Caron CR, Medeiros ABP, Ricardo C (2007) Bacteriocins from lactic acid bacteria: purification, properties and use as biopreservatives. Braz Arch Biol Technol 50(May):521–542

Piard JC, Desmazeaud M (1992) Inhibiting factors produced by lactic acid bacteria. 2. Bacteriocins and other antibacterial substances. Lait 72:113–142

Pratush A, Anupama G, Kumar A, Vyas G (2012) Application of purified bacteriocin produced by *Lactococcus lactis* AP2 as food biopreservative in acidic foods. Ann Food Sci Technol 13: 82–87

Ramli NA, Lani MN, Ibrahim R, Alias R, Razak RA (2011) LSP12 microbiological quality of "Satar" produced in Kemaman, Terengganu. In: UMT 10th international annual symposium on sustainability science and management (UMTAS 2011), pp 279–283

Ramli NA, Lani MN, Ibrahim R, Alias R, Hassan Z (2014) Microbiological quality of Malaysian heritage food ('Satar') sold in Marang and Kemaman, Terengganu, Malaysia. Int J Eng Sci 3(11):27–32

Ray B, Sandine WE (1992) Acetic, propionic, and lactic acids of starter culture bacteria as biopreservatives. In: Ray B, Daeschel M (eds) Food preservatives of microbial origin. CRC Press, Boca Raton, FL, pp 103–136
Ray B, Bhunia A (2008) Fundamental food microbiology, 4th edn. CRC/Taylor and Francis, Boca Raton, FL, pp 53–62
Reddy GC, Shahani KM, Friend BA, Chandan RC (1984) Natural antibiotic activity of *Lactobacillus acidophilus* and *bulgaricus*, production and partial purification of *bulgaricus* cultured. J Dairy Prod 8:15–19
Ringø E, Gatesoupe FJ (1998) Lactic acid bacteria in fish: a review. Aquaculture 160:177–203
Sachindra NM, Sakhare PZ, Yashoda KP, Narasimha Rao D (2005) Microbial profile of buffalo sausage during processing and storage. Food Control 16(1):31–35
Saithong P, Panthavee W, Boonyaratanakornkit M, Sikkhamondhol C (2010) Use of a starter culture of lactic acid bacteria in plaa-som, a Thai fermented fish. J Biosci Bioeng 110:553–557
Salleh F, Lani MN, Ismail N (2014) Antimicrobial activity of cell free supernatant of lactic acid bacteria isolated from fermented Durian flesh against multiple antibiotic resistance's Salmonella associated with food poisoning cases in Malaysia. IOSR J Pharm Biol Sci 9(6):60–65
Saranraj P, Naidu MA, Sivasakthivelan P (2013) Lactic acid bacteria and its antimicrobial properties: a review. Int J Pharm Biol Arch 4(6):1124–1133
Savadogo A, Ouattara CAT, Basssole IHN, Traoer SA (2006) Bacteriocins and lactic acid bacteria—a minireview. Afr J Biotechnol 5:678–683
Schillnger U, Geisen R, Holzapfel WH (1996) Potential of antagonistic microorganisms and bacteriocins for the biological preservation of foods. Trends Food Sci Technol 7:58–64
Sharpe ME (1981) The genus Lactobacillus. In: Balows A, Schelegel HG, Starr MP, Stolp H, Truperm HG (eds) The prokaryotes. A handbook on habitat, isolation and identification of bacteria. Springer, Berlin, pp 1653–1679
Shojaei H, Shooshtaripoor J, Amiri M (2006) Efficacy of simple hand-washing in reduction of microbial hand contamination of Iranian food handlers. Food Res Int 39(5):525–529
Soomro AH, Masud T, Anwaar K (2002) Role of lactic acid bacteria (LAB) in food preservation and human health – a review. Pak J Nutr 1(1):20–24
Stiles ME (1996) Biopreservation by lactic acid bacteria. Antonie van Leeuwenhoek J 70:331–345
Todorov SD (2008) Bacteriocin production by *Lactobacillus plantarum* AMA-K isolated from Amasi, a Zimbabwean fermented milk product and study of the adsorption of bacteriocin AMA-K to Listeria sp. Braz J Microbiol 39:178–187
Tourism Terengganu. (2011). Culinary. http://tourism.terengganu.gov.my/culinary.htm. Retrieved 10 Jan 2011
Valero A, Carrasco E, Garcia-Gimeno R (2012) Principles and methodologies for the determination of shelf life in foods. In: Eissa AA (ed) Trends in vital food and control engineering. Intech, Rijeka
Warriner K, Namvar A (2009) What is the hysteria with *Listeria*? Trends Food Sci Technol 20(6–7):245–254

Leuconostoc spp. as Starters and Their Beneficial Roles in Fermented Foods

So-Yeon Shin and Nam Soo Han

Contents

Abstract *Leuconostoc* spp. are gram-positive and heterofermentative bacteria, which are capable of transforming glucose molecules into carbon dioxide, ethanol, and lactate. *Leuconostoc* spp. exist in vegetables, silage, fermented food products, and feces, among other places. These bacteria are used as a starter culture in food and beverage fermentation in order to improve the nutritional and organoleptic quality and to extend shelf life. They produce exo-polysaccharides (dextran or levan), oligosaccharides, mannitol, bacteriocins, and vitamins. In this chapter, we present an extensive discussion on *Leuconostoc* spp., especially general

S.-Y. Shin • N.S. Han (✉)
Department of Food Science and Biotechnology, Chungbuk National University, Cheongju, Chungbuk 361-763, South Korea
e-mail: namsoo@cbnu.ac.kr

M.-T. Liong (ed.), *Beneficial Microorganisms in Food and Nutraceuticals*,
Microbiology Monographs 27, DOI 10.1007/978-3-319-23177-8_5

information including morphology, taxonomy, growth characteristics, metabolism, starter uses in fermented foods, and beneficial health effects as potential probiotics.

1 General Introduction

Leuconostoc is the major bacterial genus in fermented foods found from the initial to the middle stage of fermentation (Han et al. 1990; Plengvidhya et al. 2007). As a heterofermentative bacteria, *Leuconostoc* transforms glucose molecules into carbon dioxide, ethanol, and lactate (Caplice and Fitzgerald 1999; Kuipers et al. 2000). In addition, low molecular weight organic compounds produced by *Leuconostoc* provide aroma and flavor to the fermented products (Caplice and Fitzgerald 1999). *Leuconostoc* and other LAB are generally recognized as safe (GRAS) and have been widely used all around the world to improve the preservation, sensorial characteristics, and nutritional value of various products, including milk, meat, and vegetables (Stiles and Holzapfel 1997). *Leuconostoc* spp. exist in many natural ecological niches such as green vegetation and roots from where they are easily able to propagate into various environments such as vegetables, silage (Ennahar et al. 2003), and fermented food products. In addition, *Leuconostoc* spp. have been isolated from feces and breast milk (Auge et al. 1987; Dal Bello et al. 2003).

Among LAB, *Leuconostoc* spp. are implicated in a variety of aspects of economic importance: (1) fermentation of foodstuffs such as vegetables (kimchi, sauerkraut, pickles, etc.) and meat products; (2) generation of carbon dioxide gas in cheeses (especially blue-veined cheeses) providing pores; (3) production of flavoring compounds in many dairy products; (4) in situ production of dextran in saccharose-containing products, or as high-value polymers for industrial or clinical use; and (5) potential roles in functional foods.

1.1 Morphology

Leuconostoc cells are gram-positive, asporogenous, and nonmotile. Most of the *Leuconostoc* strains display a coccoid morphology, which varies depending on the growth conditions. Cells may be present singularly or in pairs, and form short- to medium-length chains (Table 1). In the presence of sucrose, several *Leuconostoc* strains produced extracellular dextran which forms an electron-dense coat on the cell surface by means of dextransucrase (Fig. 1).

Table 1 Characteristics of genus Leuconostoc (Thunell 1995)

Gram-positive coccus
Spherical cells, often lenticular on agar
Occur usually in pairs and chains
Nonmotile
Catalase-negative
Asporogenous
Facultative anaerobes
Cytochromes absent
Arginine not hydrolyzed
Non-proteolytic
Indoles not formed
Nitrates not reduced
Non-hemolytic
Vancomycin-resistant
Generally regarded as nonpathogenic
Optimal temperature 20–30 °C
Chemo-organotrophs (require rich, complex media)

1.2 Taxonomy

Phenotypic characters have been used for a long time to isolate and characterize *Leuconostoc* strains and, sometimes, to differentiate between species or subspecies. As it is difficult to identify species or subspecies unequivocally using the classical methods, new molecular techniques have been developed which permit reliable and consistent identification. In addition to traditional methods (microscopy, plate counting, etc.), several modern techniques like Random Amplified Polymorphic Polymerase Chain Reaction (RAPD-PCR), species-specific PCR, multiplex PCR, 16S rRNA sequencing, gradient gel electrophoresis, Restriction Fragment Length Polymorphism, and cluster analysis of Temporal Temperature Gradient Electrophoresis are employed to characterize and identify *Leuconostoc* spp. (Jung et al. 2014; Greppi et al. 2015). These genomic methods have been used alone or in combination for the identification of species or subspecies. Three major genera, *Leuconostoc*, *Oenococcus*, and *Weissella*, have been distinguished (Collins et al. 1993; Dicks et al. 1995). The *Leuconostoc* genus includes *L. mesenteroides* (with the three subspecies, *L. mesenteroides*, *L. dextranicum*, and *L. cremoris*) and fourteen other species, *L. citreum*, *L. carnosum*, *L. durionis*, *L. fallax*, *L. ficulneum*, *L. pseudoficulneum*, *L. fructosum*, *L. gasicomitatum*, *L. gelidum*, *L. inhae*, *L. kimchii*, *L. lactis*, *L. halzapfelii*, and *L. pseudomesenteroides* (Euzéby 1997) (Fig. 2).

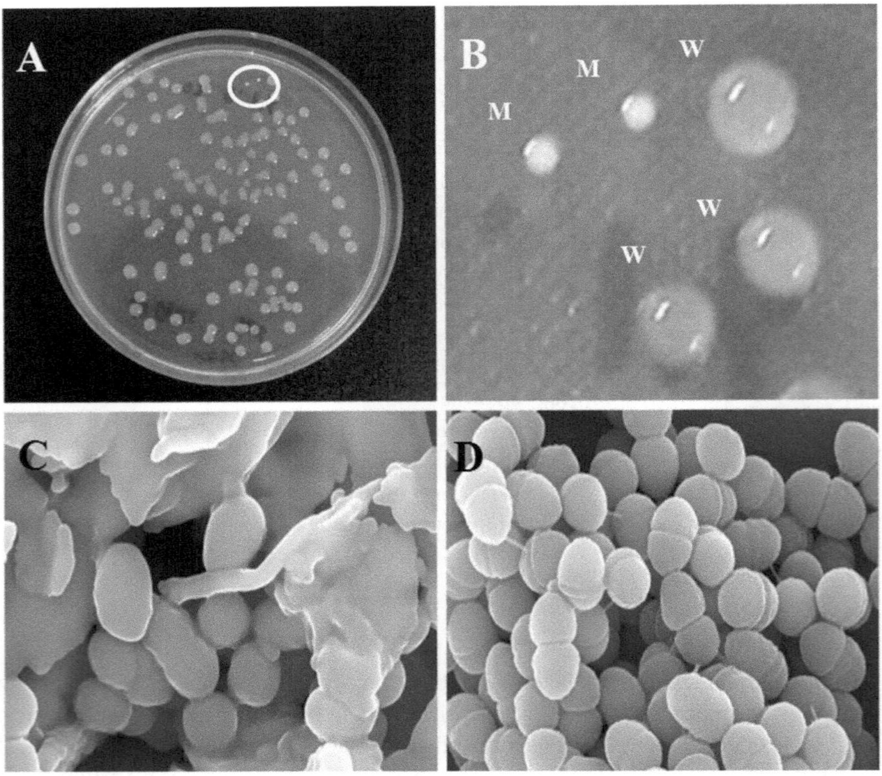

Fig. 1 Morphologies of *L. citreum* KACC 91348P (Jin et al. 2014). (**a**) Colonies on sucrose-agar medium. (**b**) *White circling* in (**a**): W, wild-type strain producing dextran; M, mutant strain producing no dextran. (**c**) Morphology of the strain producing dextran as observed by SEM. (**d**) Morphology of the strain producing no dextran as observed by SEM

1.3 Growth Characteristics

Leuconostoc spp. are facultative anaerobes (Sneath et al. 1986). All species of *Leuconostoc* require a rich medium supplemented with complex growth factors and amino acids. The cultivation of *Leuconostoc* spp. depends upon the enriched broth and selective or nonselective media used for isolating a particular genus from a mixture of microorganisms as well as to maintain the culture (Björkroth and Holzapfel 2006). Various media that meet the general nutritional requirements of *Leuconostoc* normally provide high recovery rates, without completely inhibiting other groups. The most common media satisfying the requirements are APT, Briggs, MRS, La, and BHIYE. Classical MRS medium with or without omission of citrate and meat extract is the common medium for cultivation of pure strains of *Leuconostoc*. A number of media formulae were developed for physiological studies of sugar fermentation pattern, gas production, dextran formation, citrate

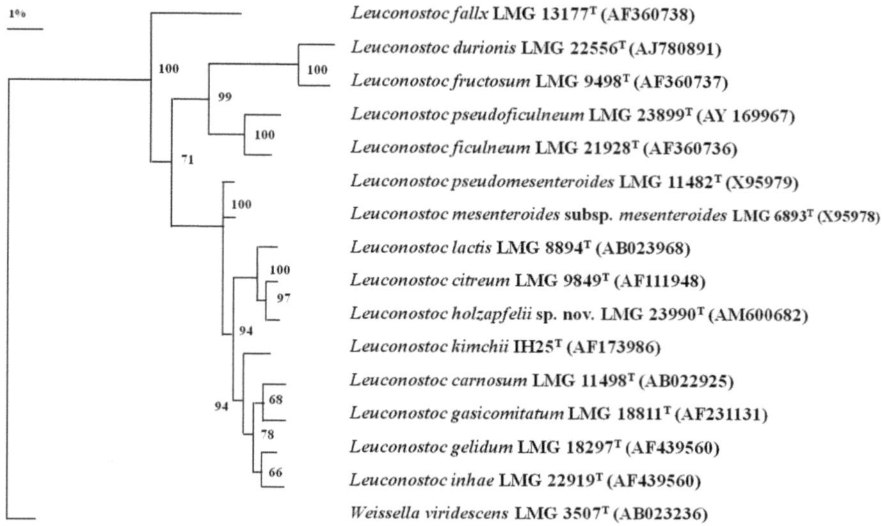

Fig. 2 Phylogenetic neighbor-joining tree based on the 16S rRNA gene sequences showing the relationships of *Leuconostoc* species. *Weissella viridescens* LMG 3507^T was used as the outgroup. *Numbers at the branching points* indicate bootstrap percentage values (>50) based on 500 tree replications

degradation, and other purposes (Björkroth and Holzapfel 2006). A chemically defined medium has been formulated, which satisfies the nutritional requirements of a representative number of *Leuconostoc* strains, and promotes quick growth comparable to that of a complex medium (Foucaud et al. 1997; Kim et al. 2012a, b). Strains of *Leuconostoc* grow at 30 °C, similar to mesophilic *Lactococcus* starter cultures, but a lower temperature is known to be more favorable. In fact, most strains grow well at 10 °C and even at 4 °C (Hamasaki et al. 2003). In contrast, some strains have been thought to be thermostable as they resist pasteurization and their presence on equipment caused contamination of milk, which is further pasteurized (Martley and Crow 1993). Optimum growth of the non-acidophilic species is achieved between pH 6 and 7, depending on the medium used. Growth of *Leuconostoc* spp. was arrested when the pH is lowered to 4.0.

1.4 Metabolism

Growth of *Leuconostoc* is dependent on fermentable carbohydrates in the medium (Garvie 1986). Glucose is converted to equimolar amounts of D-lactate, ethanol, and carbon dioxide via hexose monophosphate (6-phosphate gluconate) and pentose phosphate pathways, under micro-aerophilic condition (Schmitt et al. 1997). Key enzymes present in all *Leuconostoc* spp. are glucose 6-phosphate dehydrogenase and xylulose-5-phosphoketolase. Lactate dehydrogenase, acetaldehyde

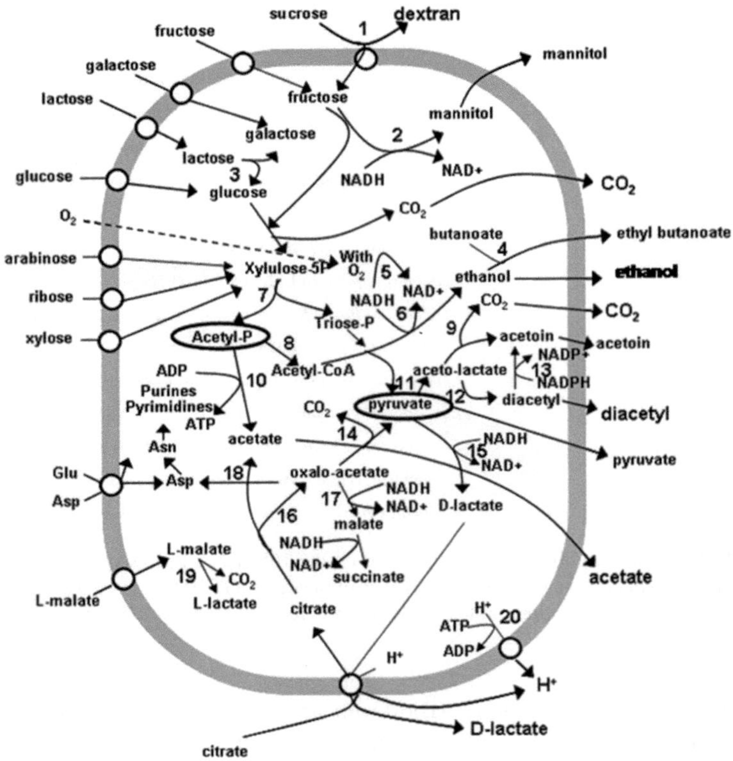

Fig. 3 General metabolism of Leuconostoc (Hemme and Foucaud-Scheunemann 2004). Major products formed are indicated in *bold. Numbers* refer to enzymes involved or steps: 1, dextransucrase; 2, mannitol-dehydrogenase; 3, β-galactosidase; 4, esterase; 5, NADH oxidase; 6, alcohol dehydrogenase; 7, phosphoketolase; 8, phosphotransacetylase, 9, α-acetolactate decarboxylase; 10, acetate kinase; 11, α-acetolactate synthase; 12, nonenzymatic formation; 13, diacetyl reductase; 14, oxaloacetate decarboxylase; 15, lactate dehydrogenase; 16, citrate lyase; 17, malate dehydrogenase; 18, formation of aspartate; 19, malolactic enzyme; 20, ATPase

dehydrogenase, and alcohol dehydrogenase regenerate NAD from reduced NAD (NADH) (Condon 1987). However, in the presence of oxygen, strains of *L. mesenteroides* use NADH oxidases and NADH peroxidases as alternative mechanisms to regenerate NAD. The general metabolism of *Leuconostoc* is presented in Fig. 3.

A number of mono- and disaccharides facilitate growth of *Leuconostoc* strains (Holt et al. 2001). *Leuconostoc* spp. incorporate carbohydrates by means of permeases that allow sugars to enter the cell without modification of the substrate (e.g., phosphorylation when the phosphoenolpyruvate (PEP)-sugar phosphotransferase system (PTS) system is involved). *Leuconostoc* spp. ferment fructose, pentose, and galactose and mannose (Cogan and Jordan 1994; von Weymarn et al. 2002). It was found that the utilization of raffinose and lactose by *Leuconostoc* varies depending on strains. Enzymes involved in this process have been characterized (Carr et al. 2002).

In common with many other properties of *Leuconostoc* strains, amino acid requirements are variable between different strains of all species. Valine and glutamic acid are the only amino acids required by all strains, and methionine is a strong growth stimulator. Riboflavin and folic acid are often required and some strains do not grow in the absence of adenine, guanine, xanthine, and uracil. *Leuconostoc* spp. do not require alanine for growth, but all strains require thiamine, panthothenic acid, and biotin (Garvie 1967). *Leuconostoc* spp. are non-proteolytic.

Citrate and malate are the organic acids most frequently fermented by *Leuconostoc* spp. These bacteria metabolize citrate to acetate, acetoin, diacetyl, 2,3-butylene glycol, and carbon dioxide (Speckman and Collins 1968). The pH of the medium has an effect on the ratio in which these compounds are formed. At neutral pH, acetate, lactate, carbon dioxide, and small amounts of formate and 2,3-butylene glycol are formed (Keenan 1968). On the contrary, acetoin and small quantities of diacetyl are produced in low pH conditions.

2 Starter Uses in Fermented Foods

2.1 *Occurrence in Foods*

Leuconostoc spp. are industrially important microbes that are used worldwide in a variety of industrial food fermentations (Table 2). The microbial flora source can either be the raw material, as is the case for milk cheeses, kimchi, and some fermented sausages, or a commercial starter culture production.

2.2 *Roles of Starters in Fermented Foods*

Historically, the production of fermented foods and beverages was obtained by spontaneous fermentation, caused by the microorganisms that existed on the raw materials. More recently, direct addition of selected starter strains to food materials has been preferred by the food industry. The advantages of direct addition included easy control over the fermentation process and standardization of the final product (Leroy and De Vuyst 2004). A starter culture can be defined as a microbial preparation of a large number of one or more microorganisms, which is introduced to a raw material enabling the production of a fermented food by accelerating and controlling its fermentation process (Caplice and Fitzgerald 1999). Industries mainly utilize commercial starter cultures which are available as frozen and freeze-dried concentrates for direct inoculation (Sandine 1996).

Table 2 *Leuconostoc* in fermented foods (Hemme and Foucaud-Scheunemann 2004)

Products	Foodstuff	Raw material	Microorganisms
Dairy	Butter and cream	Milk	LAB
	Cheese	Milk	LAB, yeasts, mold
	Fermented milks (amasi, maziwa lala, laban, filmjolk, kefir, pindidam, etc.)	Milk	LAB, yeasts
Meat	Sausages	Meat	Yeasts, molds
	Salami	Meat	LAB
Fish	Sauce foods (balacham, chinchaluk, pekasam, som-fak, etc.)	Fish, shrimp	LAB
Cereal	Beverages (beer, boza, bushera, idli, dadih, jangsu, ogi, pozol, sobia, etc.)	Malt, maize, corn, rice, millet	LAB, yeasts, molds
	Dough and starch accompaniments (bread, flour, mawe, puto, trahanas, etc.)	Maize, sorghum, tef, rice	LAB, yeasts
	Sauce foods (tsauco, etc.)	Rice, soybeans	LAB
Vegetable	Sauerkraut	Cabbage	LAB
	Kimchi	Various vegetable	LAB
	Pickles, sayur-asin	Olives, beetroot, cabbage, cucumber, carrot, sweet pepper	LAB
	Dough and starchy accompaniments (agbelima, flour, fufu, sapal, etc.)	Cassava, taro	LAB
	Cocoa		Yeast, acetic acid bacteria, LAB
	Coffee		LAB, yeasts, Enterobacteriaceae
	Juice		LAB
Fruit	Tempoyak	Durian fruit	LAB
	Kocho	*Ensete ventricosa*	LAB

Leuconostoc spp. grow poorly in milk and, therefore, must be combined with acid-producing *Lactococci* in order to act as flavor producers in mixed cheese starter cultures (Server-busson et al. 1999). During fermentation, *Leuconostoc* spp. co-metabolize lactose and citrate, leading to the production of lactate, acetate, carbon dioxide, ethanol, acetaldehyde, diacetyl, acetoin, and 2,3-butanediol (McSweeney and Sousa 2000). These compounds contribute to the organoleptic properties of unripened dairy products including buttermilk, sour cream, fresh cheese, and semi-hard cheese varieties with eyes, such as Edam and Gouda (Levata-Jovanovic and Sandine 1997). *Penicillium roqueforti* was grown in carbon dioxide which was produced using *Leuconostoc* strains (resulting in openness of the cheese) in the early stages of Roquefort production (Devoyod and Poullain 1988).

Eleven wild *Leuconostoc* strains isolated from artisanal Afuega'l Pitu cheese, a homemade acid-coagulated cheese made from raw milk in Asturias (northern Spain), were investigated to test their potential application as components of dairy starter cultures and their technological feasibilities. Metabolic activity, production of flavor compounds, resistance to NaCl, acid, nisin, and freezing, as well as genetic biodiversity were also studied (Sanchez et al. 2005).

The addition of sourdough fermented with LAB-synthesizing organic acids and oligo- and exo-polysaccharides (EPS) from sucrose enhanced texture, nutritional value, shelf life, and machinability of wheat, rye, and gluten-free bread. Comparison studies were conducted for acetate, mannitol, and oligosaccharide formation of *Weissella* and *Leuconostoc* spp. in the traditional sourdough starter *Lactobacillus sanfranciscensis* (Galle et al. 2010). *L. citreum* H012 and *W. koreensis* H020 isolated from kimchi were evaluated as starter cultures in the production of whole-wheat sourdough bread. Sourdoughs and breads with 50 % sourdough showed consistent activity to retard the growth of bread-spoilage fungi and rope-forming bacteria. Selected strains have unique fermentation characteristics and produce sourdough breads with an overall satisfactory quality (Choi et al. 2012).

Research has shown the potential benefits of using starter cultures in low-salt conditions for sauerkraut fermentation. Tolonen and others (Tolonen et al. 2002) compared cabbage inoculated with a mixed culture of *L. mesenteroides* and *Pediococcus dextrinicus* that fermented without a starter culture. Both the decline in pH and the completion of fermentation occurred more rapidly with this mixed culture. Wiander and Ryhänen (2005) found several LAB mixed starter cultures promoted a rapid decrease in pH in cabbage. Furthermore, the mixed culture containing *L. mesenteroides* and *Lb. plantarum* gave sauerkraut and sauerkraut juice the most desirable sensory characteristics. Cabbage fermented with *L. mesenteroides* consistently produced sauerkraut with a firm texture and reduced off-flavors. In addition, fermentations were rapid, with a uniform decline in pH when the starter culture was added. The application of this method to sauerkraut production showed consistent fermentation and generated nonbiodegradable chloride waste (Johanningsmeier et al. 2007).

As kimchi has become a globally popular food because of its taste and health-promoting properties, its commercial market size has gradually increased [Korean kimchi market size in 2012 was approximately USD2,300 million; Nonghyup economic research institute (http://www.nheri.re.kr/)]. The quality of the kimchi products varies depending on the raw materials even though it is processed under controlled conditions. The use of starter cultures has been suggested as an alternative for quality development and uniform kimchi production; therefore, currently, kimchi LAB including *L. mesenteroides*, *L. citreum*, and *Lb. plantarum* featured properties such as mannitol production, antimicrobial activity, and acid and bile tolerances are being used as kimchi fermentation starters (Jung et al. 2014). Besides the adaptability of kimchi starters to kimchi fermentation conditions, the production of major organic acids (lactate and acetate) and mannitol as well as other components such as minor organic acids, amino acids, flavoring compounds, vitamins, biogenic amines, and bacteriocins should be carefully considered for

Table 3 Patented starters of *Leuconostoc* genus and their uses in fermented foods

Uses	Substrate	Features	Species	Issued country	Patent No.
Kimchi	Cabbage	High dextransucrase activity	*L. citreum*	KR	1005601610000
		Helicobacter pylori growth inhibition	*L. mesenteroides*	KR	1007071020000
		Anticancer activity	*L. mesenteroides*	KR	1010414400000
		Improved storage of kimchi	*L. mesenteroides*	KR	1010559490000
		Pathogenic bacteria growth inhibition	*L. mesenteroides*	KR	1013815470000
		Anti-inflammation, intestinal survival rate enhancement	*L. mesenteroides*	KR	1013856900000
		Acid-resistant	*L. citreum*	KR	1003306740000
		Good quality	*L. mesenteroides*	KR	1005190830000
		Shelf-life extension	*L. mesenteroides*	KR	1003888660000
		Functional, organoleptic improvement	*L. mesenteroides*	KR	1005361080000
		Taste improvement	*L. citreum*	KR	1010999240000
				JP	05028458
		Mannitol production	*L. mesenteroides*	JP	WO2007074951
Functional food	*Phellinus linteus*	Functional, organoleptic improve	*L. citreum*	KR	1007183440000
	Garlic	Antioxidant activity	*L. citreum*	KR	1011592370000
	Garlic	Flavor improvement	*L. mesenteroides*	KR	1010201530000
	Ginseng	Ginsenoside bioconversion	*L. mesenteroides*	KR	1012538750000
Beverage	Soy	Bioconversion isoflavone	*L. pseudomesenteroides*	KR	1009784150000
	Whey	Stimulating the growth of *Bifidobacterium*	*L. mesenteroides*	KR	1010813760000
	Pueraria radix juice	Immune response regulation, antioxidant	*L. lactis*	KR	1014097620000
	Whey, soy, rice	Savory-flavored products	*Leuconostoc* spp.	EU	W02002085131A1
	Rice	Anti-allergy, anti-inflammation, antiobesity	*Leuconostoc* spp.	JP	04995051
	Milk	Manufacturing a sweetened lactic acid fermented food	*L. cremoris*	EU	EP0363154B1
				US	CA1330499C

Butter	Whey	Butter flavoring	*L. cremoris*	US	04670267
Polymer production		Dextran production	*L. lactis*	KR	1010911380000
		Dextran production	*L. citreum*	KR	1008146650000
		Dextran production	*L. mesenteroides*	US	US5229277A
Sugar alcohol		Mannitol production	*L. citreum*	KR	1008005300000
Probiotics		Inhibiting of harmful bacteria growth	*L. mesenteroides*	KR	1011947980000
		Exo-polysaccharide production	*Leuconostoc* spp.	KR	1008089560000
Antibiotics		Bacteriocin	*L. citreum* G7	KR	1005286410000
		Bacteriocin	*L. citreum*	KR	1013324200000
		Antifungal activity	*L. mesenteroides*	KR	1012567270000

KR Korea, *JP* Japan, *EU* European union, *US* the United States

the development of kimchi starters. In addition, a psychrotrophic *L. citreum* producing highly active dextransucrase was isolated to be used as a starter of kimchi (Eom et al. 2007, 2008).

These bacteria contribute to fermentation primarily by promoting the rapid formation of lactic acid from the available carbon source leading to acidification of the food raw material, which is a critical parameter in the preservation of fermented products. Upon acidification of the product during fermentation, the carbohydrate metabolites contribute to the flavor, texture, and the preservation. Fermentation may also increase the nutritional and functional quality of food by increasing its digestibility or by reducing its toxicity (Gueguen et al. 1997). Various patented starters of *Leuconostoc* genus and their uses in fermented foods are presented in Table 3.

2.3 Monitoring **Leuconostoc** *spp. During Fermentation*

In addition to molecular finger printing methods described in the previous part (Sect. 1.2 Taxonomy), a barcoded pyrosequencing method was used to investigate the effects of a *L. mesenteroides* strain as a starter culture for kimchi by monitoring microbial changes during fermentation (Jung et al. 2012). A competitive quantitative-PCR (CQ-PCR) method was newly developed for rapid analysis of the population dynamics of lactic acid bacteria (LAB) in kimchi (Ahn et al. 2015). In addition, a strain-specific starter monitoring system was developed by integration of chloramphenicol resistance marker gene (cat) into chromosomal DNA of *L. mesenteroides* DRC (Eom et al. 2008).

2.4 Drawbacks as Starter

Low pH is considered a growth-restrictive factor for most *Leuconostoc* spp. grown in different media. Due to the accumulation of acids in the medium, the high cell-density cultivation to overproduce the starter cells has not been achieved yet. Furthermore, the starter culture necessarily grows to become a dominant strain in the acidic condition after addition to the fermenting foods. Therefore, the ability of *Leuconostoc* to survive, grow, and metabolize actively under acidic stress conditions is very important for industrial applications. Pre-adaptation in acidic condition and addition of glutathione exhibited increase in tolerance of *Leuconostoc* (Kim et al. 2012a, b, 2014).

While various beneficial roles of *Leuconostoc* spp. are known in fermented foods, there is still risk of D-lactic acidosis; they produce D-lactic acid as major product during fermentation and accumulation of D-lactic acid in blood can be harmful to patients or infants (FAO/WHO 1974). D- and L-Lactic acids are optical isomers, which are generated in different fermented foods and metabolized into pyruvate by the enzyme lactate dehydrogenase (Jin et al. 2006). Genetic manipulation has been used to improve the adaptability and functionalities of kimchi starters. For example, *Lb. plantarum* ldhL gene encoding L-(+)-lactate dehydrogenase was cloned into the D-lactate producer *L. citreum* to increase the ratio of L-(+) lactate (Jin et al. 2009).

3 Beneficial Health Effects of *Leuconostoc*

Probiotics as a live microbial food supplements provide beneficial health factors to the consumers by improving their intestinal microbial balance (AFRC 1989). The Food and Agriculture Organization of the United Nations (FAO) and the World Health Organization (WHO) define probiotics as "live microorganisms, which when administered in adequate amount confer health benefits on the host" (FAO/WHO 2002). Probiotic strains are considered nonpathogenic and safe. The most common types of bacteria used as probiotics are LAB. *Lactobacillus acidophilus*, *Lb. casei*, *Lb. lactis*, *Lb. helveticus*, *Lb. salivarius*, *Lb. plantarum*, *Lb. bulgaricus*, *Lb. rhamnosus*, *Lb. johnsonii*, *Lb. reuteri*, *Lb. fermentum*, *Lb. delbrueckii*, *Streptococcus thermophilus*, *Enterococcus faecium*, *E. faecalis*, *Bifidobacterium bifidum*, *B. breve*, *B. longum*, and *Saccharomyces boulardii* are commonly used as bacterial probiotics (Suvarna and Boby 2005). Probiotics enhance intestinal health by limiting nutrients, inhibition of mucosal adherence of pathogens, inhibition of epithelial invasion by pathogens, the production of antimicrobial substances, and/or the stimulation of mucosal immunity (Servin and Coconnier 2003). Probiotics in humans and animals play an important role in the prevention of diarrhea and ulcer treatment and have an impact on lactose intolerance and diarrheal disease. Meanwhile, nondairy probiotic products are of significant worldwide interest due to the growing trend of vegetarianism and to an increasing incidence of lactose intolerance in many populations. China, Japan, and some African countries promote the idea of producing nondairy-based probiotics in other products such as cereals, fruits, and vegetables, due to lactose intolerance, cholesterol content, and dairy allergies (Granato et al. 2010). *Leuconostoc* genus is regarded a potential probiotic for nondairy fermented foods.

3.1 Production of Microbial Polysaccharides or Oligosaccharides

L. mesenteroides produces EPSs, which are homopolysaccharides consisting of α-D-glucans such as dextrans, mainly composed of α-1,6-linked residues with variable (strain-specific) degrees of branching, and alternates composed of α-1,3 and α-1,6 linkages. Like dextran, levan, a fructan, can be made by levan sucrase elaborated by *L. mesenteroides* (Kang et al. 2011). Dextran or levan can improve the rheology of fermented foods (viscosity and elasticity) as natural biothickeners, emulsifiers, gelling agents, and physical stabilizers to bind water and limit syneresis (Duboc and Mollet 2001). Levan has been proposed to be an antitumor agent, as a raw material suitable for cosmetics, and a component of animal food (Rhee et al. 2002).

Isomaltooligosaccharides (IMOs or branched low molecular weight dextran) are α-(1→6)-linked oligodextrans and they induced selective effect on the gut microbiota and stimulated synthesis of short chain fatty acids, indicating prebiotic sustainability in distal regions of the gut (Sarbini et al. 2013). Sanz et al. (2005) reported a similar prebiotic effect of α-(1−6) and α-(1−3)-linked oligosaccharides obtained from the reaction between sucrose and maltose, catalyzed by an alternansucrase isolated from *L. mesenteroides* NRRL B-21297.

EPSs are synthesized either extracellularly from sucrose by glycansucrase or intracellularly by glycosyltransferases from sugar nucleotide precursors (Gänzle and Schwab 2009). Specific glycosyltransferase and dextran or levansucrase enzymes are involved in the biosynthesis process (Vuyst and Degeest 1999; Monchois et al. 1999). Eight glycansucrase-encoding genes from *L. mesenteroides* were cloned (Bozonnet et al. 2002). The gene coding for an inulosucrase in *L. citreum* was also cloned, sequenced, and expressed in *E. coli*.

When *L. citreum* and the sugars (sucrose and maltose) were added in kimchi, IMOs were produced by acceptor reaction of dextransucrase (Cho et al. 2014). Same reaction was applied to milk fermentation to synthesize IMO (Seo et al. 2007).

3.2 Production of Mannitol

Mannitol is a six-carbon sugar alcohol that is synthesized by various organisms including heterofermentative LAB and has a number of applications in the food, pharmaceutical, and chemical industries. Mannitol has a lower calorie level than sugar, which it makes suitable for use as a sweetener in "light" foods. *Leuconostoc* and *Oenococcus* have been reported to produce mannitol effectively (Carvalheiro et al. 2011; Soetaert 1990). In these species, fructose can act as an external electron acceptor in a reaction involving mannitol dehydrogenase, which catalyzes the reduction of fructose to mannitol and vice versa (Ghoreishi and Shahrestani 2009). *L. citreum* KACC 91348P, isolated from kimchi, showed superior capacity of mannitol production with a higher yield showing that it could be used as an efficient mannitol producer in fructose-containing foods (Otgonbayar et al. 2011). Use of a membrane cell–recycle bioreactor has been described as a method to achieve high yields of mannitol (von Weymarn et al. 2002). The low calorie sweeteners produced by these strains possess a number of properties that contribute to potential health benefits, including having a low glycemic index, potential for use as osmotic diuretics, for weight control, and as antiplaque agents and prebiotics.

These GRAS substances could be especially useful to children, diabetic patients, and weight watchers (Patra et al. 2009; Monedero et al. 2010).

3.3 Vitamin Production

Vitamin K is involved in blood clotting, tissue calcification, and atherosclerotic plaque formation and also plays a key role in maintaining the bones and kidneys (Olson 1984). Vitamin K is present as menaquinone (vitamin K2) which is produced by some intestinal bacteria, like LAB, especially by strains of the genera *Leuconostoc*. The clinical condition of osteoporosis in infants is caused by intracranial hemorrhaging due to a vitamin K deficiency (LeBlanc et al. 2011). Strains of *L. lactis*, *L. mesenteroides* subsp. *cremoris*, and *L. mesenteroides* subsp. *dextranicum* producing significant amounts of menaquinones (vitamin K2) have been characterized, and the use of these strains would be beneficial for the prevention of vitamin K deficiency diseases (Morishita et al. 1999). In addition, the production of folate (vitamin B9) has been reported in *L. lactis* and *L. paramesenteroides* (Sybesma et al. 2003).

3.4 Production of Antimicrobial Substances

Bacteriocins are antimicrobial proteins or peptides produced by bacteria and archaea for competition against other organisms in the same environment. LAB produced various antimicrobial substances and organic acids such as lactic acid, acetic acid, and bacteriocins, during fermentation. Among them, bacteriocins safely and effectively inhibit the proliferation of pathogenic bacteria. Accordingly, bacteriocins have great potential as natural food biopreservatives (Nissen-Meyer et al. 2010). Several bacteriocin-producing *Leuconostoc* strains have been isolated and were subsequently identified as *L. gelidum* UAL 187, *L. paramesenteroides*-La7a, *L. carnosum*-Ta11a, and *L. citreum* (Hastings et al. 1994). All strains produce bacteriocins, which are active against *Listeria monocytogenes* and other LAB. Some strains isolated from different food matrices possess the capacity to produce bacteriocins such as leucocylin Q, leucocin C, and mesentericin Y105 (Masuda et al. 2011; Martinez-Murcia and Collins 1991; Hastings et al. 1994; Héchard et al. 1999; Papathanasopoulos et al. 1997; Todorov and Dicks 2004). Based on their biochemical properties, most of these bacteriocins can be classified as class II bacteriocins, which are small (<10 kDa), heat-stable membrane-active peptides that do not contain lanthionine, with an N-terminal consensus sequence (YGNGVXCaaCVaaV), and possess anti-*Listeria* activity (Klaenhammer 1993).

Besides, phenyllactic acid (PLA) is an antimicrobial compound naturally synthesized from phenylpyruvate in various fermented foods. When cultured in the presence of phenylpyruvate, growing cells of *L. mesenteroides* produced a maximum conversion yield (83 %) of D-PLA (Li et al. 2014).

3.5 Immunomodulation

As the incidence of allergic diseases has increased over decades, it has been widely accepted that decreased exposure to microbial infection in infancy due to higher hygienic standards and more frequent antibiotic treatments might contribute to a defect in the maturation of Th1-dependent immune responses (1–4). The increased generation of IgE, which is one of the typical features of allergy, has been explained by the insufficient population of Th1 cells to reciprocally inhibit Th2 cells (Gavett et al. 1995). Kang et al. (2009) reported that *L. citreum* KACC 91035-regulated serum IgE generation in allergic response in Balb/c mice to be useful in preventing the development and progression of IgE production.

3.6 Potential Probiotics

Probiotics as microbial dietary adjuvants have beneficial effect to the host by modulating mucosal and systemic immunity, as well as improving the nutritional and microbial balance in the intestinal tract. The mechanisms of LAB used as probiotics include the production of inhibitory substances against pathogens, competition for essential nutrients, adhesion sites, and tolerance for pH and bile in gastrointestinal tract.

A few studies have reported the strong potential of using *Leuconostoc* spp. as a probiotic strain. Studies for the feasibility of diarrhea control in children by feeding them fermented milk revealed that Indian Dahi containing 10^8 g/L *Lc. lactis* and *L. mesenteroides* reduced the mean duration of diarrhea by 0.3 days (Agarwal and Bhasin 2002). Kekkonen et al. (2008) showed that the use of *Leuconostoc* spp. as a probiotic strain was better at inducing cytokines than the probiotic *Lactobacillus* strain which is currently in clinical use. In aquaculture systems, it has been reported that *L. mesenteroides* isolated from snakehead fish intestine is a new potential probiotic. Moreover, *L. lactis* has been shown to have a potential as a single starter culture in Kanoon-jeen production and could reduce the microbial risk and fermentation processing time. Additionally, *Leuconostoc* spp. have potential as protective cultures for vacuum-packed meat products (Budde et al. 2003), and the bacteriocins of *Leuconostoc* can be used as protective agents in combination with another starter culture in fermented meat (Drosinos et al. 2005). *L. mesenteroides* subsp. *mesenteroides* from Algerian camel milk is a strong growth inhibitor of *Listeria* spp. Probiotic profiling and inhibition spectra against food-borne pathogens in mixed cultures were also investigated. In in vitro studies, both *L. mesenteroides* strains exhibited a significant probiotic property: showing high survival level at low pH (2–3 and 4) in the presence of bile salts and in the presence of pepsin (Benmechernene et al. 2013). In addition, *L. lactis* was isolated from the gastrointestinal tract of black porgy, *Sparus macrocephalus*, and identified by conventional biochemical characteristics and 16S rRNA gene sequence analysis. The isolated strain showed tolerance to bile contents and resistance to low pH and survived well in a trypsinase and pepsin solution. These results indicated that *L. lactis* might be an attractive candidate for probiotics in marine aquaculture (Zhang et al. 2013).

4 Conclusion

In this chapter, we introduced that *Leuconostoc* species are currently used as starter culture in different fermented foods and they are regarded potential probiotics owing to various health beneficial effects. However, given that the human gastrointestinal tract is a complex and hostile environment, it appears unlikely that common *Leuconostoc* spp. will be capable of influencing the microbial ecology of the host and of beneficially affecting gut health. It is more likely that these effects will require the introduction of superior strain having resistance to acid and bile salt after sequential screening and evaluation in in vitro and in vivo tests. Furthermore, it is essential that these probiotic strains not be developed as individual entities but rather as the active ingredients of the food products that are ultimately intended for human consumption.

References

AFRC RF (1989) Probiotics in man and animals. J Appl Bacteriol 66(5):365–378

Agarwal K, Bhasin S (2002) Feasibility studies to control acute diarrhoea in children by feeding fermented milk preparations Actimel and Indian Dahi. Eur J Clin Nutr 56:S56–S59

Ahn G-H, Moon JS, Shin S-Y, Min WK, Han NS, Seo J-H (2015) A competitive quantitative polymerase chain reaction method for characterizing the population dynamics during kimchi fermentation. J Ind Microbiol Biotechnol 42:49–55

Auge B, Donnio P, Le Deaut P, Avril J (1987) Influence of vancomycin by venous route on salivary and fecal aerobic floras. Pathol Biol 35(5):673–675

Benmechernene Z, Chentouf HF, Yahia B, Fatima G, Quintela-Baluja M, Calo-Mata P, Barros-Velázquez J (2013) Technological aptitude and applications of *Leuconostoc mesenteroides* bioactive strains isolated from algerian raw camel milk. BioMed Res Int. doi:10.1155/2013/418132

Björkroth J, Holzapfel W (2006) Genera *Leuconostoc*, *Oenococcus* and *Weissella*. In: The prokaryotes. Springer, Berlin, pp 267–319

Bozonnet S, Dols-Laffargue M, Fabre E, Pizzut S, Remaud-Simeon M, Monsan P, Willemot R-M (2002) Molecular characterization of DSR-E, an α-1, 2 linkage-synthesizing dextransucrase with two catalytic domains. J Bacteriol 184(20):5753–5761

Budde BB, Hornbaek T, Jacobsen T, Barkholt V, Koch AG (2003) *Leuconostoc carnosum* 4010 has the potential for use as a protective culture for vacuum-packed meats: culture isolation, bacteriocin identification, and meat application experiments. Int J Food Microbiol 83(2): 171–184
Caplice E, Fitzgerald GF (1999) Food fermentations: role of microorganisms in food production and preservation. Int J Food Microbiol 50(1):131–149
Carr FJ, Chill D, Maida N (2002) The lactic acid bacteria: a literature survey. Crit Rev Microbiol 28(4):281–370
Carvalheiro F, Moniz P, Duarte LC, Esteves MP, Gírio FM (2011) Mannitol production by lactic acid bacteria grown in supplemented carob syrup. J Ind Microbiol Biotechnol 38(1):221–227
Cho SK, Eom H-J, Moon JS, Lim S-B, Kim YK, Lee KW, Han NS (2014) An improved process of isomaltooligosaccharide production in kimchi involving the addition of a Leuconostoc starter and sugars. Int J Food Microbiol 170:61–64
Choi H, Kim Y-W, Hwang I, Kim J, Yoon S (2012) Evaluation of *Leuconostoc citreum* HO12 and *Weissella koreensis* HO20 isolated from kimchi as a starter culture for whole wheat sourdough. Food Chem 134(4):2208–2216
Cogan TM, Jordan KN (1994) Metabolism of *Leuconostoc* bacteria. J Dairy Sci 77(9):2704–2717
Collins M, Samelis J, Metaxopoulos J, Wallbanks S (1993) Taxonomic studies on some Leuconostoc-like organisms from fermented sausages: description of a new genus *Weissella* for the *Leuconostoc paramesenteroides* group of species. J Appl Bacteriol 75(6):595–603
Condon S (1987) Responses of lactic acid bacteria to oxygen. FEMS Microbiol Lett 46(3): 269–280
Dal Bello F, Walter J, Hammes W, Hertel C (2003) Increased complexity of the species composition of lactic acid bacteria in human feces revealed by alternative incubation condition. Microb Ecol 45(4):455–463
Devoyod J, POULLAIN F (1988) Les *Leuconostocs*. Propriétés: leur rôle en technologie laitière. Lait 68(3):249–279
Dicks L, Dellaglio F, Collins M (1995) Proposal to reclassify *Leuconostoc oenos* as *Oenococcus oeni* [corrig.] gen. nov., comb. nov. Int J Syst Bacteriol 45(2):395–397
Drosinos EH, Mataragas M, Xiraphi N, Moschonas G, Gaitis F, Metaxopoulos J (2005) Characterization of the microbial flora from a traditional Greek fermented sausage. Meat Sci 69(2): 307–317
Duboc P, Mollet B (2001) Applications of exopolysaccharides in the dairy industry. Int Dairy J 11(9):759–768
Ennahar S, Cai Y, Fujita Y (2003) Phylogenetic diversity of lactic acid bacteria associated with paddy rice silage as determined by 16S ribosomal DNA analysis. Appl Environ Microbiol 69 (1):444–451
Eom H-J, Seo DM, Han NS (2007) Selection of psychrotrophic *Leuconostoc* spp. producing highly active dextransucrase from lactate fermented vegetables. Int J Food Microbiol 117:61–67
Eom H-J, Park JM, Seo MJ, Kim MD, Han NS (2008) Monitoring of *Leuconostoc mesenteroides* DRC starter in fermented vegetable by random integration of chloramphenicol acetyltransferase gene. J Ind Microbiol Biotechnol 35(9):953–959
Euzéby JP (1997) List of bacterial names with standing in nomenclature: a folder available on the Internet. Int J Syst Bacteriol 47(2):590–592
FAO/WHO (1974) Toxicological evaluation of certain food additives with a review of general principles and of specifications. Seventeenth report of the Joint FAO-WHO Expert Committee on Food Additives. WHO Technical Report Series, vol 539. WHO, pp 1–40
FAO/WHO (2002) Guidelines for the evaluation of probiotics in food. Report of a Joint FAO/WHO Working Group on Drafting Guidelines for the Evaluation of Probiotics in Food. FAO/WHO, Rome
Foucaud C, Francois A, Richard J (1997) Development of a chemically defined medium for the growth of *Leuconostoc mesenteroides*. Appl Environ Microbiol 63(1):301–304
Galle S, Schwab C, Arendt E, Ganzle M (2010) Exopolysaccharide-forming *Weissella* strains as starter cultures for sorghum and wheat sourdoughs. J Agric Food Chem 58(9):5834–5841

Gänzle M, Schwab C (2009) Ecology of exopolysaccharide formation by lactic acid bacteria: sucrose utilization, stress tolerance, and biofilm formation. In: Ullrich M (ed) Bacterial polysaccharides: current innovations and future trends. Caister Academic, Norfolk, pp 263–278

Garvie E (1967) The growth factor and amino acid requirements of species of the genus Leuconostoc, including *Leuconostoc paramesenteroides* (sp. nov.) and *Leuconostoc oenos*. J Gen Microbiol 48(3):439–447

Garvie E (1986) Genus *Leuconostoc*. In: Garrity GM (ed) Bergey's manual of systematic bacteriology, vol 2. William and Wilkins, Baltimore, pp 1071–1075

Gavett SH, O'Hearn DJ, Li X, Huang SK, Finkelman FD, Wills-Karp M (1995) Interleukin 12 inhibits antigen-induced airway hyperresponsiveness, inflammation, and Th2 cytokine expression in mice. J Exp Med 182:1527–1536

Ghoreishi S, Shahrestani R (2009) Subcritical water extraction of mannitol from olive leaves. J Food Eng 93(4):474–481

Granato D, Branco GF, Nazzaro F, Cruz AG, Faria JA (2010) Functional foods and nondairy probiotic food development: trends, concepts, and products. Comp Rev Food Sci Food Saf 9(3):292–302

Greppi A, Ferrocino I, La Storia A, Rantsiou K, Ercolini D, Cocolin L (2015) Monitoring of the microbiota of fermented sausages by culture independent rRNA-based approaches. Int J Food Microbiol. doi:10.1016/j.ijfoodmicro.2015.01.016

Gueguen Y, Chemardin P, Labrot P, Arnaud A, Galzy P (1997) Purification and characterization of an intracellular β-glucosidase from a new strain of *Leuconostoc mesenteroides* isolated from cassava. J Appl Microbiol 82(4):469–476

Hamasaki Y, Ayaki M, Fuchu H, Sugiyama M, Morita H (2003) Behavior of psychrotrophic lactic acid bacteria isolated from spoiling cooked meat products. Appl Environ Microbiol 69(6): 3668–3671

Han H, Lim C-R, Park H-K (1990) Determination of microbial community as an indicator of kimchi fermentation. Korean J Food Sci Technol 22(1):26–32

Hastings JW, Stiles ME, von Holy A (1994) Bacteriocins of *Leuconostocs* isolated from meat. Int J Food Microbiol 24(1):75–81

Héchard Y, Berjeaud J-M, Cenatiempo Y (1999) Characterization of the mesB gene and expression of bacteriocins by *Leuconostoc mesenteroides* Y105. Curr Microbiol 39(5):265–269

Hemme D, Foucaud-Scheunemann C (2004) *Leuconostoc*, characteristics, use in dairy technology and prospects in functional foods. Int Dairy J 14(6):467–494

Holt S, Al-Sheikh H, Shin KJ (2001) Characterization of dextran-producing *Leuconostoc* strains. Lett Appl Microbiol 32(3):185–189

Jin Q, Yoon H-S, Han NS, Lee J, Han JS (2006) Effect of lactic acid bacteria on D-and L-lactic acid contents of kimchi. Food Sci Biotechnol 15(6):948–953

Jin Q, Jung JY, Kim YJ, Eom HJ, Kim SY, Kim TJ, Han NS (2009) Production of L-lactate in *Leuconostoc citreum* via heterologous expression of L-lactate dehydrogenase gene. J Biotechnol 144(2):160–164

Jin Q, Li L, Kim YJ, Han NS (2014) Construction of a dextran-free *Leuconostoc citreum* mutant by targeted destruction of the dextransucrase gene. J Appl Microbiol 117:1104–1112

Johanningsmeier S, McFeeters RF, Fleming HP, Thompson RL (2007) Effects of *Leuconostoc mesenteroides* starter culture on fermentation of cabbage with reduced salt concentrations. J Food Sci 72(5):M166–M172

Jung JY, Lee SH, Lee HJ, Seo HY, Park WS, Jeon CO (2012) Effects of *Leuconostoc mesenteroides* starter cultures on microbial communities and metabolites during kimchi fermentation. Int J Food Microbiol 153(3):378–387

Jung JY, Lee SH, Jeon CO (2014) Kimchi microflora: history, current status, and perspectives for industrial kimchi production. Appl Microbiol Biotechnol 98(6):2385–2393

Kang H, Oh YJ, Ahn KS, Eom HJ, Han N, Kim YB, Sohn NW (2009) *Leuconostoc citreum* HJ-P4 (KACC 91035) regulates immunoglobulin E in an ovalbumin-induced allergy model and

induces interleukin-12 through nuclear factor-kappa B and p38/c-Jun N-terminal kinases signaling in macrophages. Microbiol Immunol 53(6):331–339
Kang HK, Yun SI, Lim TY, Xia Y-M, Kim D (2011) Cloning of levansucrase from *Leuconostoc mesenteroides* and its expression in *Pichia pastoris*. Food Sci Biotechnol 20(1):277–281
Keenan T (1968) Production of acetic acid and other volatile compounds by *Leuconostoc citrovorum* and *Leuconostoc dextranicum*. Appl Microbiol 16(12):1881–1885
Kekkonen RA, Kajasto E, Miettinen M, Veckman V, Korpela R, Julkunen I (2008) Probiotic *Leuconostoc mesenteroides* ssp. cremoris and *Streptococcus thermophilus* induce IL-12 and IFN-γ production. World J Gastroenterol 14(8):1192
Kim JE, Eom H-J, Kim Y, Ahn JE, Kim JH, Han NS (2012a) Enhancing acid tolerance of *Leuconostoc mesenteroides* with glutathione. Biotechnol Lett 34:683–687
Kim YJ, Eom H-J, Seo E-Y, Lee DY, Kim JH, Han NS (2012b) Development a chemically defined minimal medium for the exponential growth of *Leuconostoc mesenteroides* ATCC8293. J Microbiol Biotechnol 22(11):1518–1522
Kim JE, Eom H-J, Li L, Yoo K-S, Han NS (2014) Induction of the acid tolerance response in *Leuconostoc mesenteroides* ATCC 8293 by pre-adaptation in acidic condition. Food Sci Biotechnol 23(1):221–224
Klaenhammer TR (1993) Genetics of bacteriocins produced by lactic acid bacteria. FEMS Microbiol Rev 12(1-3):39–85
Kuipers OP, Buist G, Kok J (2000) Current strategies for improving food bacteria. Res Microbiol 151(10):815–822
LeBlanc J, Laiño J, del Valle MJ, Vannini V, Van Sinderen D, Taranto M, de Valdez G, de Giori GS, Sesma F (2011) B-Group vitamin production by lactic acid bacteria–current knowledge and potential applications. J Appl Microbiol 111(6):1297–1309
Leroy F, De Vuyst L (2004) Lactic acid bacteria as functional starter cultures for the food fermentation industry. Trends Food Sci Technol 15(2):67–78
Levata-Jovanovic M, Sandine WE (1997) A method to use *Leuconostoc mesenteroides* ssp. *cremoris* 91404 to improve milk fermentations. J Dairy Sci 80(1):11–18
Li L, Shin S-Y, Lee KW, Han NS (2014) Production of natural antimicrobial compound D-phenyllactic acid using *Leuconostoc mesenteroides* ATCC 8293 whole cells involving highly active D-lactate dehydrogenase. Lett Appl Microbiol 59:404–411
Martinez-Murcia A, Collins M (1991) A phylogenetic analysis of an atypical Leuconostoc: description of *Leuconostoc fallax* sp. nov. FEMS Microbiol Lett 82(1):55–59
Martley F, Crow V (1993) Interactions between non-starter microorganisms during cheese manufacture and repening. Int Dairy J 3(4):461–483
Masuda Y, Ono H, Kitagawa H, Ito H, Mu F, Sawa N, Zendo T, Sonomoto K (2011) Identification and characterization of leucocyclicin Q, a novel cyclic bacteriocin produced by *Leuconostoc mesenteroides* TK41401. Appl Environ Microbiol 77(22):8164–8170
McSweeney PL, Sousa MJ (2000) Biochemical pathways for the production of flavour compounds in cheeses during ripening: A review. Lait 80(3):293–324
Monchois V, Willemot RM, Monsan P (1999) Glucansucrases: mechanism of action and structure–function relationships. FEMS Microbiol Rev 23(2):131–151
Monedero V, Pérez-Martínez G, Yebra MJ (2010) Perspectives of engineering lactic acid bacteria for biotechnological polyol production. Appl Microbiol Biotechnol 86(4):1003–1015
Morishita T, Tamura N, Makino T, Kudo S (1999) Production of menaquinones by lactic acid bacteria. J Dairy Sci 82(9):1897–1903
Nissen-Meyer J, Oppegård C, Rogne P, Haugen HS, Kristiansen PE (2010) Structure and mode-of-action of the two-peptide (class-IIb) bacteriocins. Probiotics Antimicrob Proteins 2(1):52–60
Olson RE (1984) The function and metabolism of vitamin K. Annu Rev Nutr 4(1):281–337
Otgonbayar GE, Eom HJ, Kim BS, Ko JH, Han NS (2011) Mannitol production by *Leuconostoc citreum* KACC 91348P isolated from Kimchi. J Microbiol Biotechnol 21(9):968–971

Papathanasopoulos MA, Krier F, Revol-Junelles A-M, Lefebvre G, Le Caer JP, von Holy A, Hastings JW (1997) Multiple bacteriocin production by *Leuconostoc mesenteroides* TA33a and other *Leuconostoc/Weissella* strains. Curr Microbiol 35(6):331–335

Patra F, Tomar S, Arora S (2009) Technological and functional applications of low-calorie sweeteners from lactic acid bacteria. J Food Sci 74(1):R16–R23

Plengvidhya V, Breidt F, Lu Z, Fleming HP (2007) DNA fingerprinting of lactic acid bacteria in sauerkraut fermentations. Appl Environ Microbiol 73(23):7697–7702

Rhee SK, Song KB, Kim CH, Park BS, Jang EK, Jang KH (2002) Levan. In: Baets S, Vandamme EJ, Steinbuchel A (eds) Biopolymers, polysaccharides I: Polysaccharides from prokaryotes, vol V. Wiley-VCH, Weinheim, pp 351–377

Sanchez JI, Martinez B, Rodriguez A (2005) Rational selection of Leuconostoc strains for mixed starters based on the physiological biodiversity found in raw milk fermentations. Int J Microbiol 105(3):377–387

Sandine W (1996) Commercial production of dairy starter cultures. VCH, New York

Sanz ML, Cote GL, Gibson GR, Rastall RA (2005) Prebiotic properties of alternansucrase maltose-acceptor oligosaccharides. J Agric Food Chem 53(15):5911–5916

Sarbini SR, Kolida S, Naeye T, Einenhand AW, Gibson GR, Rastall RA (2013) The prebiotic effect of α-1,2 branched, low molecular weight dextran in the batch and continuous faecal fermentation system. J Funct Foods 5(4):1938–1946

Schmitt P, Vasseur C, Phalip V, Huang D, Divies C, Prevost H (1997) Diacetyl and acetoin production from the co-metabolism of citrate and xylose by *Leuconostoc mesenteroides* subsp. *mesenteroides*. Appl Microbiol Biotechnol 47(6):715–718

Seo DM, Kim S-Y, Eom H-J, Han NS (2007) Synbiotic synthesis of oligosaccharides during milk fermentation by addition of Leuconostoc starter and sugars. J Microbiol Biotechnol 17(11): 1758–1764

Server-busson C, Foucaud C, Leveay J-Y (1999) Selection of dairy Leuconostoc isolates for important technological properties. J Dairy Res 66(02):245–256

Servin AL, Coconnier M-H (2003) Adhesion of probiotic strains to the intestinal mucosa and interaction with pathogens. Best Pract Res Clin Gastroenterol 17(5):741–754

Sneath PH, Mair N, Sharpe M, Holt J (1986) Bergey's manual of systematic bacteriology. Williams and Wilkins, Baltimore, p 9

Soetaert W (1990) Production of mannitol with *Leuconostoc mesenteroides*. Mededelingen van de Faculteit Landbouwwetenschappen, Rijksuniversiteit Gent 55(4):1549–1552

Speckman R, Collins E (1968) Diacetyl biosynthesis in *Streptococcus diacetilactis* and *Leuconostoc citrovorum*. J Bacteriol 95(1):174–180

Stiles ME, Holzapfel WH (1997) Lactic acid bacteria of foods and their current taxonomy. Int J Food Microbiol 36(1):1–29

Suvarna V, Boby V (2005) Probiotics in human health: a current assessment. Curr Sci 88(11): 1744–1748

Sybesma W, Starrenburg M, Tijsseling L, Hoefnagel MH, Hugenholtz J (2003) Effects of cultivation conditions on folate production by lactic acid bacteria. Appl Environ Microbiol 69(8):4542–4548

Thunell R (1995) Taxonomy of the *Leuconostocs*. J Dairy Sci 78(11):2514–2522

Todorov SD, Dicks LM (2004) Characterization of mesentericin ST99, a bacteriocin produced by *Leuconostoc mesenteroides* subsp. *dextranicum* ST99 isolated from boza. J Ind Microbiol Biotechnol 31(7):323–329

Tolonen M, Taipale M, Viander B, Pihlava JM, Korhonen H, Ryhanen EL (2002) Plant-derived biomolecules in fermented cabbage. J Agric Food Chem 50(23):6798–6803

von Weymarn N, Hujanen M, Leisola M (2002) Production of D-mannitol by heterofermentative lactic acid bacteria. Process Biochem 37(11):1207–1213

Vuyst L, Degeest B (1999) Heteropolysaccharides from lactic acid bacteria. FEMS Microbiol Rev 23(2):153–177

Wiander B, Ryhänen E-L (2005) Laboratory and large-scale fermentation of white cabbage into sauerkraut and sauerkraut juice by using starters in combination with mineral salt with a low NaCl content. Eur Food Res Technol 220(2):191–195

Zhang W, Liu M, Dai X (2013) Biological characteristics and probiotic effect of *Leuconostoc lactis* strain isolated from the intestine of black porgy fish. Braz J Microbiol 44(3):685–691

Fermented Soymilk as a Nutraceutical

Joo-Ann Ewe and Siok-Koon Yeo

Contents

Abstract The increasing health awareness and public interest in nutraceuticals and functional foods are growing tremendously, driven along with the continuous efforts of scientific researches in proving and identifying the properties and potential applications of nutraceutical substances. Nutraceuticals are reported as active natural compounds possessing chemoprotective, antioxidative, anti-inflammatory, and osteogenetic properties, which could be obtained from food or being part of a food. Soy contains phytochemicals such as isoflavones and phytosterols that promote health. Soymilk is considered as an economical substitute for dairy and an ideal nutritional supplement for vegan and lactose-intolerant population. The nutritional content of soymilk supports the growth of beneficial microbes, while the fermentation process enriches the medium with various bioactive components. Fermentation improves the bioavailability of isoflavones, assists in digestion of protein, reduces anti-nutritional factor, enhances calcium solubility and vitamin content, promotes intestinal health, and supports immune system. Fermentation of

J.-A. Ewe (✉) • S.-K. Yeo
School of Biosciences, Taylor's University Lakeside Campus, No. 1 Jalan Taylor's, 47500 Subang Jaya, Selangor Darul Ehsan, Malaysia
e-mail: jooann.ewe@taylors.edu.my

M.-T. Liong (ed.), *Beneficial Microorganisms in Food and Nutraceuticals*,
Microbiology Monographs 27, DOI 10.1007/978-3-319-23177-8_6

soymilk has indeed offered the population a nutraceutical with physiological benefits and/or provides protection against diseases such as cardiovascular disease, bone health, anticancer, obesity treatment, and antidiabetic property. This health potential has granted fermented soymilk to be positioned as a nutraceutical.

1 Introduction

Foods and nutrients play a vital role in normal functioning of the body. They are helpful in maintaining the health of the individual and in reducing the risk of various diseases. Nutraceutical, a contraction of nutrition and pharmaceutical, is created by the US Foundation for Innovation in Medicine as "any substance that may be considered a food or part of a food and have value in health promoting, disease prevention and treatment or semi-medicinal properties" (Rodriguez et al. 2006). The growing awareness of nutraceutical benefits has led a drastic shift of health-care economics towards nonprescription nutraceuticals as self-medication in prevention and disease control. The worldwide nutraceutical market has been projected to reach US$250 billion by 2018 (Global Information Inc. 2012). Annual health-care expenditure of around 20 % has been predicted to be attributable to the widespread consumption of nutraceutical/functional food (Sun-Waterhouse 2011).

Nutraceuticals are reported as active natural compounds possessing chemoprotective, antioxidative, anti-inflammatory, and osteogenetic properties, which could be obtained from food industry, herbal and dietary supplement, pharmaceutical industry and bioengineered microorganisms, agroproducts, or active biomolecules (Sharma 2009). Among others, soy-based products are perceived as healthy food encompassing most of the aforementioned health beneficial effects. Soy (*Glycine max*) is a legume indigenous to East Asia, though the main cultivation areas are now North America (FAO 2007). Soybean is high in protein and oligosaccharides such as raffinose and stachyose, which is known to promote the growth of indigenous microorganisms in the intestinal tract, though they may also cause unpleasant flatulence (Wang et al. 2003). However, the virtues of soybean have been declared, evidenced by the long history of consumption of more than 1000 years. The relatively low cost of soybean protein compared with animal proteins and the simple way it can be fashioned into palatable high protein food such as tofu and soymilk contributes to its role as an alternative protein source (Ang et al. 1985). Soymilk, the aqueous extract of whole soybean, is said to be invented by the legendary An Liu about 2000 years ago in China (He and Chen 2013). It is also low in fat and is an excellent dietary fiber and contains a variety of micronutrients and phytochemicals. In addition, it contains equally abundant amount of protein as dairy milk and is deemed suitable for vegan and lactose-intolerance population.

Fermentation was originally conducted to preserve food. Yet, the fermentation process has indeed caused modification in texture, color, and flavor as well as improved beneficial nutritional content of food. Fermentation has been frequently used as an effective bioprocess to enhance the functional properties of food materials such as the production of bioactive compounds, attributable to the metabolic activity of microorganism involved. Promising probiotics such as *Lactobacillus* and *Bifidobacterium* are often being incorporated in fermented soymilk to improve the quality of the product and health status of consumers. The ability of the microorganisms to utilize the nutrients during growth in the soymilk led to an increase in the concentrations of metabolites. Changes in the concentrations of crude protein, sugars, B vitamins, and organic acids as well as enhancement of isoflavones bioavailability in soymilk can be detected upon fermentation which thus increases the virtue of soymilk as a nutraceutical (Hou et al. 2000). This chapter highlights the research findings on the nutraceutical property of fermented soymilk. At this conjunction, the suitability of soymilk as a vehicle for the delivery of probiotics while improving its nutritional quality before delivery to the host, compositional alterations during microbial fermentation, and the liberation of bioactive compounds in the fermented soymilk to produce a nutritionally enriched nutraceutical are also discussed.

2 Fermentation by Beneficial Microbes to Enhance Nutraceutical Property of Soymilk

Lactose intolerance and cholesterol content are major drawback of dairy beverages which have indeed promoted the development of new probiotics carrier such as vegetable and fruit beverages as well as soy-based drink. Soymilk, the aqueous extract of whole soybean, has been used to emulate the "probiotic delivery role" of dairy beverages and is considered as a suitable economical substitute for dairy milk. This feature is attributable to the high content of indigestible oligosaccharides present naturally in soymilk to serve as carbon source for the fermentation of probiotics. After fermentation, the flatulence-causing oligosaccharides can be reduced to minimum level upon ingestion. Apart from the high α-galactosyl glucosides, the variable amount of soy protein provides pool of peptides and amino acids essential for growth of probiotics (Shihata and Shah 2000).

Probiotics have been reported to produce various glycosyl enzymes simultaneously that enable the microorganisms to hydrolyze various glycosidic bonds (Amaretti et al. 2006). The suitability of soymilk as a probiotic carrier is attributed to the ability of probiotics possessing various glycosyl hydrolases such as α-galactosidase, β-galactosidase, and β-fructofuranosidase that hydrolyze various glycosidic bonds, which enable them to utilize the abundantly found glycosyl sugars such as raffinose and stachyose for growth. The trisaccharide raffinose and tetrasaccharide stachyose in soymilk are hydrolyzed to D-galactose and sucrose by the enzyme α-galactosidase. α-Galactosidase acts upon gal–gal bonds in the tetrasaccharide stachyose, releasing galactose and raffinose, and also acts upon gal–glu bonds with the release of sucrose (Mulimani et al. 1997). Meanwhile, raffinose is hydrolyzed by the α-galactosidase to release galactose and sucrose, and an invertase will then hydrolyze the β-(1,2)-fructofuranoside bond of this disaccharide, releasing fructose and glucose (LeBlanc et al. 2004). The concentration of oligosaccharides such as raffinose and stachyose has been reduced in soymilk upon fermentation by the α-galactosidase activity-possessing probiotics, with changes in the concentration of simpler sugar such as glucose and fructose to support their growth (Ewe et al. 2010; Yeo and Liong 2010). By possessing α-galactosidase activity, probiotics has been found to be able to minimize the content of flatulence-causing oligosaccharides in soymilk, thus improving its acceptability.

Upon hydrolysis by probiotics, the oligosaccharides in soymilk are broken down into simple sugars. The glucose moiety is metabolized via homofermentative or heterofermentative pathway by lactobacilli, while bifidobacteria breakdown glucose via the fructose-6-phosphate shunt. In addition to glucose, lactobacilli, and bifidobacteria are also capable of utilizing fructose as a carbon source (Gomes and Malcata 1999). Caescu et al. (2004) demonstrated that intracellular fructose was metabolized via the fructose-6-phosphate phosphoketolase pathway.

A series of metabolic activities occur during the fermentation of probiotics in the soymilk, such as the production of organic acids and a decrease in pH. Being homofermentative, both *L. acidophilus* and *L. gasseri* ferment glucose predominantly to lactic acid via the Embden–Meyerhof–Parnas pathway, accompanied by the production of minor amounts of acetic acid, while heterofermentative *L. casei* and *L. fermentum* metabolize glucose via pentose phosphate pathway to produce equimolar amount of lactic acid/ethanol and CO_2 as the end products. On the other hand, the metabolism of carbohydrates by *Bifidobacterium* strains follows the fructose-6-phosphate shunt, yielding 2 moles of lactic acid and 3 moles of acetic acid per mole of glucose in pure glucose medium as carbon source. The production of organic acids in soymilk during fermentation is desirable due to their health-promoting potential upon consumption, where study has elucidated that acetate produced by probiotic bifidobacteria can protect the host from enteropathogenic infection (Fukuda et al. 2011).

Probiotics are able to grow in soymilk because they possess a proteolytic system that degrades soy protein, the glycinin and β-conglycinin (Aguirre et al. 2008). The cell-envelope proteins (CEP) of probiotics degrade the soy protein into oligopeptides and are subsequently transported into the cell via specific peptide transporter. The peptides are then further degraded into shorter peptides or amino acids by various intracellular peptidases to support their growth (Kunji et al. 1996; Christensen et al. 1999). The enhanced availability of amino acid in growth medium could increase the viability of probiotics (Kehagias et al. 2008). Apart from degrading the whole protein into amino acid to support growth, the proteinases of various probiotics are capable of releasing angiotensin I-converting enzyme (ACE)-inhibitory peptides, and thus a blood pressure-lowering effect can be derived from the soymilk proteins (Wang and De Mejia 2005). Regardless of the novel beneficial effects exerted by unfermented soymilk, fermentation process of probiotics in soymilk enhances the functional properties of soymilk, covering the advantages granted by probiotics, soy protein, and phytoestrogens and the generated bioactive metabolites, yielding a nutraceutical offering beneficial effects beyond the basic nutritional functions.

In order to improve growth of probiotics in soymilk, soymilk fortified with prebiotics, vitamins, and growth factor-derived components has been studied for their potential growth-stimulating activity. Supplementation of B vitamins (Ewe et al. 2010); prebiotics such as inulin, fructooligosaccharides, mannitol, and maltodextrin (Yeo and Liong 2010); nitrogen sources such as yeast extract, peptone, tryptone, and casitone (Chou and Hou 2000); and carbon sources such as sucrose, fructose, and lactose (Wei et al. 2007) has successfully stimulated the growth of probiotics in soymilk. Therefore, soymilk fermented by probiotics could be a potential nutraceutical with delivery of nutrients and bioactive compounds including vitamins, minerals, antioxidants, as well as probiotics. An intake of the aforementioned fermented soymilk could exert health beneficial effects to host beyond its basic therapeutic benefits.

Mixed culture fermentation has longed been practiced in the production of milk-based products such as yogurt, sourdough, and wine consisting of multiple strains or species (De Vuyst and Neysens 2005; Courtin and Rul 2004; Rodriguez and Manca de Nadra 1995). Heterogeneity of food is one of the major reasons which encourage mixed culture or coculture fermentation of a food product. The physicochemical compositions of food provide diverge niches for specialized strains, leading to distinct flavor and aroma characteristics as well as functional properties. Apart from achieving/enhancing desired organoleptic properties, cocultures also promote productivity from cheaper raw substrates (Lee et al. 2001). The coexisting strains in the growing medium interact physically through physical contact or signaling molecules as well as through nutritional relations, where physiochemical changes of nutritional environment induced by one strain trigger a response in another strain (Sieuwerts et al. 2008). Such interactions could either mutually benefit both party or appear detrimental (Hugenholtz 1986). Sieuwerts et al. (2008) discussed the five main mechanisms of microbial interactions in mixed cultures which include amensalism, competition, commensalism, parasitism, and mutualism. The effects of interactions may either be positive, neutral, or negative, determined by the interactions at the level of substrates, exchange of metabolites, and growth factors or inhibiting compounds.

The organoleptic property of fermented soymilk is one of the main challenges that resisted its wide acceptance despite its emerging role as a new carrier for probiotics. As a result of the lack of economic importance, up to date, only limited study investigated the positive/negative effects of combination strain fermentation on nutraceutical properties as granted upon single culture fermentation, merely on preliminary growth and metabolic activities. Nevertheless, persistent efforts of researchers in search of improving the quality of the fermented product have never ceased. Several coculture fermentations in soymilk have been performed using *Pediococcus acidilactici*, *Lactobacillus acidophilus*, and *Saccharomyces cerevisiae*; *Streptococcus thermophilus*, *Bifidobacterium infantis*, and *B. longum*; and *B. adolescentis* and *Propionibacterium freudenreichii* subsp. *shermanii*. The bacteria selected are of probiotic characteristics as well as dairy sources. Coculture fermentation of the bacteria in soymilk showed higher utilization of carbohydrate (raffinose and stachyose) and amino acids and release of metabolites such as lactic, acetic, and propionic acids than the single culture (Santos et al. 2014; Wu et al. 2012; Wang et al. 2003). The use of yeast cells in co-fermentation gives rise to greater pool of B vitamins and proteins which has enabled a shorter fermentation period while enhancing the growth of probiotics. The strains have been found to possess mutualism characteristics where growth stimulation was observed upon fermentation. Strain interactions involving exchange of metabolites and growth factors could be reasons of the enhanced performance (Sieuwerts et al. 2008).

3 Bioactive Compounds in Fermented Soymilk

Probiotic-fermented food products exert added advantages due to the ability of probiotics to regulate intestinal microflora and impart several bio-therapeutic effects such as antihypertensive, antioxidative, and hypocholesterolemic effects (Jones 2002). It is the enzymic activity of probiotics in enhancing biotransformation of isoflavones and hydrolysis of protein, releasing the bioactive compounds from their native structure that is less bioactive and bioavailable.

3.1 Biotransformation of Isoflavones

Isoflavone phytoestrogens, found abundantly in soymilk, belong to a class of diphenol compounds that have structural and functional similarities to the human estrogen, oestradiol-17β. The parent isoflavones are the aglycone structures of daidzein, glycitein, and genistein, which are conjugated to form malonyl-, acetyl-, and β-glucoside configurations. Gut microflora or probiotics readily hydrolyze the main isoflavonoid glucosides in soymilk, including genistein and daidzein into bioactive aglycones (Cornwell et al. 2004). Probiotics have been reported to possess the ability to biotransform glucosides to aglycones. This is achieved through the enzyme β-glucosidase, enabling probiotics to replace intestinal bacteria in releasing the bioactive aglycone from soymilk. Pyo et al. (2005) reported that an increased concentration of isoflavone aglycones in soymilk was associated with high β-glucosidase activity of lactic acid bacteria. Intestinal β-glucosidases often biotransform conjugated glucosides to bioactive aglycones via hydrolytic cleavage. However, probiotics that are capable of producing β-glucosidase could also liberate such bioactive properties. Consumption of probiotic strains of *Lactobacillus* and

Bifidobacterium was shown to increase β-glucosidase in humans' fecal samples (Marteau et al. 1990).

Past studies have shown that the concentrations of isoflavone aglycones in soymilk are increased upon fermentation by probiotics. The increase in the enzymes from the fermentative probiotics increases the bioavailability of aglycones in fermented soymilk, resulting in increased isoflavone absorption efficiency which may then manifest in greater physiological effects of the aglycone-enriched fermented soymilk compared to glucoside-enriched unfermented soymilk (Kano et al. 2006). In a study performed by Chien et al. (2006) on the transformation of isoflavone phytoestrogens during soymilk fermentation, the concentration of isoflavone aglycones (daidzein, glycitein, and genistein) has increased to 100 %, while a reduction of 50–90 % in the concentration of glucoside counterparts upon fermentation by *S. thermophilus* and *B. longum* was observed. In addition, Pham and Shah (2007) also found that the level of bioactive aglycones increased from 8 % in non-fermented soymilk to approximately 50 % due to fermentation by *Bifidobacterium*, while the concentration of malonyl-, acetyl-, and glucoside isomers decreased by 50 %, 60 %, and 85 %, respectively, in soymilk fermented by *B. animalis*.

Several attempts have also been made to enhance the enzymic transformation of isoflavone glucosides to aglycones of probiotics in soymilk such as establishing favorable growth conditions and application of sublethal physical treatments on probiotic cells. Ewe et al. (2012a, b, 2013) demonstrated that sublethal physical treatments such as ultrasonication, electroporation, and UV radiation have been shown to enhance membrane permeabilization of the probiotic cells. The increased permeability promoted growth of treated cells to beyond 8 $\log_{10}$ CFU/mL after fermentation, associated with enhanced β-glucosidase-specific activities. This led to enhanced bioconversion of isoflavones and higher accumulation of isoflavones aglycones in the fermented soymilk medium.

3.1.1 Bioavailability of Isoflavones

Soymilk contains a reasonable amount of isoflavones which are proclaimed to impart potential therapeutic effects. In general, isoflavones can exist naturally in soymilk as four distinctive chemical forms, namely, aglycones, β-, malonyl-, and acetyl-glucosides. The chemical structures of the isoflavones and its metabolites influence the extent of absorption; aglycone form is more readily absorbed and bioavailable than the highly polar conjugated glucosides (Kano et al. 2006). In general, approximately 80–95 % of isoflavones in unfermented soymilk exist as glucoside conjugates which are less bioactive and non-bioavailable. Therefore, the existing form of isoflavones in unfermented soymilk may seem to be irrelevant in terms of bioactivity and bioavailability.

Among the isoflavones, aglycones are more favorable due to its higher bioactivity and bioavailability. Clinical trials have demonstrated that the aglycone-enriched fermented soymilk is more bioavailable in human than the glucoside-enriched unfermented soymilk (Kano et al. 2006). A total of twelve healthy volunteers (9 men and 3 women) were given three types of soymilk (100 mL), namely, plain soymilk as the control, β-glucosidase-treated soymilk, and probiotic-fermented soymilk to study the absorption and excretion of isoflavones in different soymilk. Consumption of soymilk fermented by *L. mali* YIT 0243 and *B. breve* strain Yakult was found to increase the concentration of total isoflavones in human serum by 118.1 % compared to control soymilk after 4 h of ingestion. The authors suggested that soy isoflavones that have been bio-converted to aglycones by probiotics were absorbed more quickly and efficiently than the glucoside forms in unfermented soymilk. This was mainly due to the lower hydrophilicity and molecular weights of aglycones which allow an efficient intestinal absorption. In addition, probiotics in fermented soymilk might also improve the intestinal metabolism of isoflavones (Kano et al. 2006).

In addition to bioconversion of glucosides to aglycones, probiotics was found to metabolize daidzein into equol when grown in soymilk (Tsangalis et al. 2002). Equol is a secondary isoflavone metabolite that possesses more potent estrogenic metabolite and has a greater affinity for estrogen receptors (ERs). It has been suggested that equol may be the most important feature contributing to the efficiency of soy isoflavones. The chemical structure of equol is strikingly similar to human estradiol where it has been reported to have an activity of approximately 0.2–0.5 % of estradiol (He and Chen 2013). Thus, it is expected that equol could bind to estrogen receptor and regulate various hormone-dependent diseases such as menopausal health and breast and prostate cancers (Setchell et al. 2002). However, equol is not readily available in most processed soy food, and in human, only one third of the population is capable of producing it in sufficient amounts. It is interesting to note that soymilk fermented by probiotics contained an appreciable amount of aglycones and biological potent equol (Tsangalis et al. 2002). This finding suggests that fermentation of soymilk could produce nutraceutical with enriched bioactive aglycones and equol. Figure 1 illustrates the bioconversion of isoflavone and the bioactivity of aglycones and its metabolites.

3.2 Release of Bioactive Peptides During Fermentation

Fermented soy products have been claimed as an excellent source of bioactive peptides (Yang et al. 2004). The enrichment of fermented medium with bioactive peptides is attributed to the proteolytic activity of microorganisms. Probiotics has been reported to exhibit proteolytic activity leading to release amino acids and shorter peptide chains from soy protein to meet their requirement for growth (Liong et al. 2009). Strains of lactobacilli have been found to produce proteinases that could hydrolyze long oligopeptides, liberating angiotensin I-converting enzyme (ACE)-inhibitory peptides (Gobbetti et al. 2000) with antihypertensive properties (Nakamura et al. 1995).

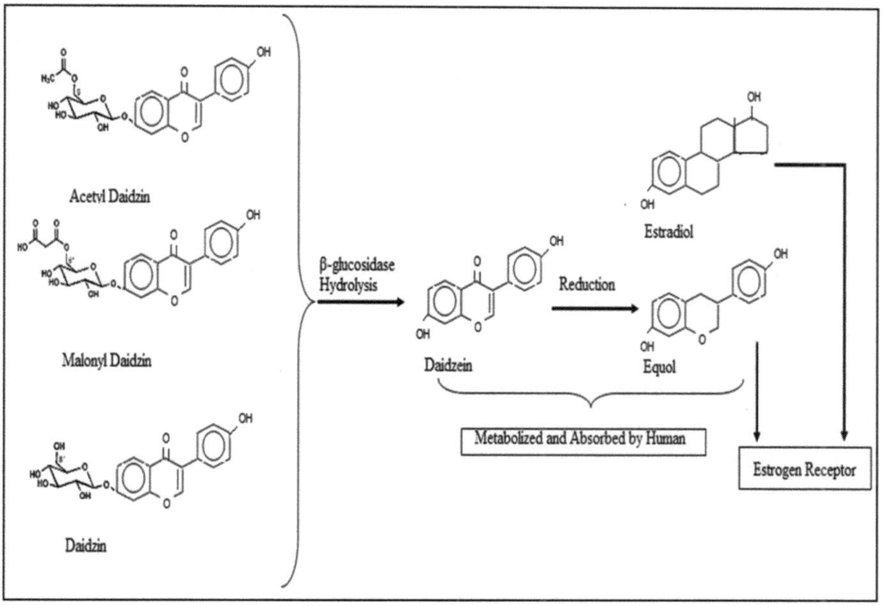

Fig. 1 Bioconversion of conjugated daidzein to equol and structural similarity of equol and estradiol which enables the binding of equol to estrogen receptors

3.2.1 ACE-Inhibitory Peptides

Soy contains a rich source of high-quality protein. Proteolytic activity of probiotics could cleave the soy protein into various amino acid and peptides. Some of the peptides produced are bioactive and could impart antihypertensive effect. These bioactive peptides react by inhibiting the activity of angiotensin I-converting enzyme (ACE), an enzyme that is responsible for the regulation of blood pressure. ACE catalyzes both the production of the vasoconstrictor angiotensin II and the degradation of the vasodilator bradykinin, resulting in overall increase of blood pressure. Therefore, inhibiting the activity of ACE could lead to a reduction of blood pressure.

In a recent study investigating the multifunctional properties of soymilk fermented by *Enterococcus faecium*, the authors found that the fermented soymilk possessed ACE-inhibitory activity and could be a promising strategy in the prevention therapy against cardiovascular disease (Martinez-Villaluenga et al. 2012). Donkor et al. (2007) demonstrated that soymilk fermented by *L. acidophilus* L10, *B. lactis* B94, *L. casei* L26, *B. longum* B1536, and *L. casei* Lc 279 possessed varying degree of proteolytic activity and exhibited ACE-inhibitory activity ranging from 17 % to 43 % inhibition after fermentation at 37 °C for 48 h. Similarly, Ng et al. (2008) demonstrated that *L. bulgaricus* FTCC 0411 and *L. fermentum* FTD 13 could liberate peptides with ACE-inhibitory activity in tofufa. In addition, the

ACE-inhibitory activity was relatively stable and the inhibition remained high (approximately 70–80 % inhibitory) even after 9 days of storage at 4 °C. Probiotic-fermented soy whey (a by-product produced during the manufacturing of tofu) was also found to exert in vitro antihypertensive properties, most probably attributed to the high proteolytic activity observed (Fung et al. 2008). In another study, Donkor et al. (2005) demonstrated that mixed probiotic cultures could produce ACE inhibitors in soy yogurt and the ACE-inhibitory activity increased over the storage periods with an IC_{50} value of 0.28 mg/mL and 69.7 % of ACE inhibition at the end of 28 days of storage at 4 °C. The mixed cultures of probiotic including *L. acidophilus* LAFTI® L10, *B. lactis* LAFTI® B94, and *L. paracasei* LAFTI® L26 substantially increased the ACE-inhibitory activity in soy yogurt as compared to the control yogurt fermented by yogurt culture only. This improvement was partly due to higher proteolytic activity of probiotic cultures. Considering that probiotic-fermented soy products could produce bioactive peptides with ACE-inhibitory activity, it has been strongly suggested that consumption of these functional products could possibly prevent the risk of hypertension in in vivo models. Thus far, only peptide His–His–Leu derived from fermented soy paste was assayed in pure form in SHP, where a decrease of 32 mmHg of SBP was reached at a dose of 100 mg/kg (Shin et al. 2001). Although fermented soymilk has shown promising ACE-inhibitory activity in in vitro studies, the peptides fractions that are responsible for the activity have yet to be characterized from fermented soymilk, and thus further elucidation is needed.

4 Functionality of Bioactive Compounds in Fermented Soymilk

Soybeans and their constituents have been extensively studied for their role in preventing chronic disease, owing to the vast differences in occurrence of metabolic diseases between Chinese and Western women. It has been indicated that dietary factors may account for the different and it should be accredited to soy-containing diet. Consequently, a great amount of scientific reports demonstrating the beneficial effects of isoflavones has been reported in a variety of in vitro, in vivo, and clinical studies. These evidences encompass the function of bioactive compounds of soy towards the reduction of risk factors such as cardiovascular diseases, cancer, hypertension, and osteoporosis. These beneficial effects have been largely ascribed to bioactive peptides and isoflavone aglycones which are released upon fermentation of soy by beneficial microorganisms such as probiotics. The probiotics makeup in the fermented product could always exert their distinctive benefits upon consumption such as modulation of host defense responses and protection against infectious diseases (Saulnier et al. 2009; Mazmanian and Kasper 2006; Sonnenburg et al. 2006). Likewise, the coined health beneficial effects governed by probiotics, fermentation, and soymilk have granted fermented soymilk as a nutraceutical that could be available in the market at a lower budget.

4.1 Cardiovascular Health

High serum cholesterol/triglyceride levels that lead to dyslipidemia, endothelial dysfunction, and hypertension and the inflammatory process have been associated with the induction and perpetuation of cardiovascular diseases (Fraga et al. 2010). The capability of fermented soymilk to confer cardiovascular health effects on the host could be achieved through various mechanisms. For instances, the antihypertensive effects of soy isoflavones have been reported by Nestel et al. (1997), where soy isoflavones could improve systemic arterial compliance. Furthermore, genistein was observed to be able to exert cardioprotective effect, in combination with its ability to lower blood pressure in postmenopausal women (Teede et al. 2001). In a recent study Kim et al. (2014) reported that *Lactobacillus plantarum*-fermented soymilk improved dysregulated lipid metabolism in rats fed with high-cholesterol diet.

4.1.1 Atheroschlerosis

Atherosclerosis, a hallmark of cardiovascular disorders, is always associated with hypercholesterolemia, hypertriacylglyceridemia, and/or low serum HDL-cholesterol (Grundy 1998). Tsai et al. (2009) reported that *Momordica charantia*-supplemented soy skim milk fermented with *L. paracasei* subsp. *paracasei* NTU 101 was found to be effective in preventing and decelerating hyperlipidemia-induced atherosclerosis. Serum cholesterol level and atherosclerotic plaques in the aorta were decreased by 69 % relative to the group of hamsters fed a high-cholesterol diet.

The development of atherogenesis involves the activation of a series of cell signal via stimulation of thromboxane A_2 (TxA_2) receptors. TxA_2 is a pro-atherogenic metabolite of arachidonic acid. Soy isoflavone aglycones such as genistein and equol have been found to be able to act as an antagonist of TxA_2 receptors, which then inhibits TxA_2-mediated platelet responses (Munoz et al. 2009; Nakashima et al. 1991). Furthermore, genistein and daidzein exert protective effect of cardiovascular health through inhibition of platelet adhesion and aggregation as well as the secretory activity of platelets (Munoz et al. 2009; Borgwardt et al. 2008; Guerrero et al. 2005; Gottstein et al. 2003; Sargeant et al. 1993). Thus, with increasing concentration of isoflavone aglycones and equol upon microbial fermentation in soymilk, the product could probably exert a protective effect towards the development of atherosclerosis and thus reduce the risk of cardiovascular-related diseases.

4.1.2 Hypocholesterolemia

Dietary cholesterol, along with fatty acids, has been shown to be a significant contributor to blood cholesterol level (Spady et al. 1993). Studies have suggested there is a negative association between dietary soy compositions with incidence of dysregulated cholesterol metabolism (Taku et al. 2007; Høie et al. 2005; Lin et al. 2004). The regulation of lipid metabolism involves the transcriptional regulation of sterol regulatory element-binding protein (SREBP), as well as transcription factor such as liver X receptor alpha (LXRα) and/or various regulatory steps at the posttranscriptional and/or posttranslational level (Horton et al. 2002; Joseph et al. 2002; Field et al. 2001; Chambers and Ness 1997). Kim et al. (2014) have recently proposed a model about the beneficial effects of fermented soymilk (containing higher aglycones than the unfermented soymilk) on lipid profiles and the related gene expression in the liver and adipose tissues of rats fed on a high-cholesterol diet. Fermented soymilk has been found to downregulate SREBP-1c and SREBP-2 and expression of their target genes, including 3-hydroxy-3-methylglutaryl-CoA reductase (HMGCR), low density lipoprotein receptor (LDLR), fatty acid synthase (FAS), and stearoyl-CoA desaturase (SCD1). Furthermore, fermented soymilk elevated adipose expression levels of genes involved in triglyceride-rich lipoprotein uptake (ApoE, VLDLR, and Lrp1) and fatty acid oxidation (PPARα, CPTIα, LCAD, CYP4A1, UCP2, and UCP3), HDL-biogenesis (ABCA1, ApoA1, and LXRα), and adiponectin signaling (AdipoQ, AdipoR1, and AdipoR2) as well as levels of phosphorylated AMPK and ACC. Taken together, this study provided a better insight of molecular modulation of fermented soymilk with increased levels of isoflavone aglycones in improving lipid metabolism induced by a high-cholesterol diet, via enhancing a reduction of SREBP-dependent cholesterol and triglyceride synthesis in liver, and promoting adiponectin signaling and PPARα-induced expression of genes involved in triglyceride-rich lipoprotein clearance, fatty acid oxidation, and reverse cholesterol transport in adipose tissues.

4.1.3 Antihypertension

One of the most intermediary factors involved in the modulation of hypertension is the action of angiotensin-converting enzyme. Angiotensin I-converting enzyme (ACE) inhibition is a key clinical target for blood pressure control, where ACE inhibitors can lower blood pressure by reducing the production of angiotensin II and subsequently inhibiting the degradation of bradykinin (Saito 2008), a peptide that causes the blood vessels to dilate. The ACE-inhibitory peptides are inactive within the sequence of the parent protein but can be released by microbial activity (Korhonen 2009). Hence, fermentation is considered to be an effective way to produce the bioactive peptides. ACE-inhibitory peptides can be derived from a variety of fermented products including soymilk upon fermentation by various starter microorganisms (FitzGerald and Murray 2006). In addition to yogurt bacteria and cheese starter bacteria, probiotic bacteria have been demonstrated to produce different bioactive peptides in soymilk during fermentation. Probiotics are able to grow in soy products because they possess a proteolytic system that degrades soy proteins, the glycinin and β-conglycinin (Aguirre et al. 2008). Upon fermentation, the proteinases of various probiotics are capable of releasing ACE-inhibitory peptides, and thus a blood pressure-lowering effect can be derived from the soymilk proteins (Wang and De Mejia 2005). Several studies have demonstrated that *Lactobacillus* and *Bifidobacterium* are capable of releasing antihypertensive peptides which are ACE-inhibitory peptides from soy-based protein (Donkor et al. 2005, 2007).

Apart from ACE-inhibitory peptides, the antihypertensive effect of fermented soymilk was also attributable to isoflavone aglycones such as daidzein, genistein, and equol. Despite having weaker binding affinity for estrogen receptors, moderate consumption of fermented soymilk has actually contributed to high concentrations in plasma that sufficiently cause physiological effects (Cornwell et al. 2004). Isoflavone aglycones play a role in inhibition of estrogen sulfotransferase, leading to in vitro increase of circulating estrogens. Furthermore, the isoflavone aglycones such as equol could inhibit the production of contraction factor endothelin-1 via ER-dependent mechanisms, leading to vasodilation and antihypertensive effect.

4.2 Anticancer Effect

Epidemiological studies showed that people with high soy isoflavonoid intake have lower rates of several types of cancers including breast, prostate, and colon cancer (American Institute for Cancer Research 2000). The high intake of isoflavone-rich soy by Asian population is hypothesized to contribute towards their relatively lower incidence of clinical cancers than the Western counterparts (Duncan et al. 1999). The protective effect of soy isoflavones on breast cancer development is suggested to be a hormonal effect involved in lowering circulating levels of unconjugated sex hormones. Isoflavones also could inhibit enzyme activity that would lead to the development of prostate cancer (Cornwell et al. 2004). Guha et al. (2009) revealed in an epidemiological study that increased consumption of daidzein and glycitein could reduce risk of breast cancer compared to no intake of soy isoflavones among postmenopausal women. In fact, soy isoflavones are structurally similar to endogenous estrogens (He and Chen 2013). Soy isoflavones such as daidzein, genistein, and glycitein have been proposed to modulate the occurrence of breast cancer through both hormonally mediated and non-hormonally related mechanism with genistein being the most potent chemoprotective isoflavone.

Genistein has been claimed to act as an agonist of ERβ to compete with endogenous estrogens for binding with ERs in which prolong exposure of cells to endogenous estrogens has been known to increase the risk for the development of breast cancer. Genistein displays weak estrogenic activity mediated by ERs, with a preferential binding to ERβ (Kuiper et al. 1998). It could inhibit estrogen-promoted cell growth without interfere with estrogen-regulated transcription. Thus, by regulating the functions of ERβ, genistein confers protective effects in breast cancer cells (Cappelletti et al. 2006).

Genistein also involves in epigenetic modifications via upregulated expression of tumor suppressors p^{21} and p^{16} whereas downregulated expression of BMI1 and c-MYC complexes in the prevention of cancer (Li et al. 2013b). Furthermore, studies also revealed that genistein promotes reactivation of ERα via remodeling of chromatin structure in ERα promoter. Herewith, genistein can prevent cancer development and reduce the growth of ERα-negative mouth breast tumors, which provide a more effective option in breast cancer therapy (Li et al. 2013a).

Genistein also has an impact on gene expression at a genome-wide scale through regulation of endogenous estrogen-induced genes. Nevertheless, it possesses lower potency resulting in less ERα binding sites and less gene expression changes compared to those induced by endogenous estrogens (Gertz et al. 2012).

Likewise, genistein is capable of binding and activating peroxisome proliferator-activated receptors (PPARs) that play essential roles in the regulation of cellular differentiation and development (Belfiore et al. 2009). It has been known that genistein can stimulate cell growth activity at nanomolar concentration but inhibits cell growth at micromolar concentration. Studies have shown that genistein is capable of inhibiting cell proliferation and triggers apoptosis in human breast cancer cells and triple-negative breast cancer cells (TNBC) through inhibition of NF-κB activity via the MEK5/ERK5 pathway. This indicated that the suppression of cell growth and induction of apoptosis in TNBC can be achieved through inhibition of MEK5/ERK5/NF-κB pathway (Pan et al. 2012). Concomitantly, expression of Bcl-2, a proapoptotic protein, is downregulated, but Bax, an apoptotic protein, is upregulated via downregulation of NF-κB.

Autophagy is a form of programmed cell death which is responsible for the suppression of tumorigenesis (Schwartz et al. 1993). It has been reported that genistein induced both apoptosis and autophagy in several types of cancer cells including ovarian cancer cells, human colon cancer HT-29 cells, and lung cancer cells (Gossner et al. 2007; Singletary and Milner 2008; Nakamura et al. 2009; Pan et al. 2013). In brief, the evidences above demonstrated that consumption of isoflavone aglycone-enriched soy product such as fermented soymilk could exert an anticancer effect to the host.

On top of the physiological functions played by isoflavone aglycones, short-chain fatty acids, such as acetate, lactate, propionate, and butyrate, produced during soymilk fermentation may also have promising uses in the treatment of cancer upon ingestion of fermented soymilk. Short-chain fatty acids are readily absorbed, while butyrate is the major energy source for colonocytes. Among the short-chain fatty acids, butyrate plays a major role in the prevention of cancer of the colon. Butyrate can nourish the colonic mucosa and prevent the development of cancer by promoting cell differentiation, cell-cycle arrest, and apoptosis of transformed colonocytes, as well as inhibiting the enzyme histone deacetylase and decreasing the transformation of primary to secondary bile acids due to colonic acidification (Wong et al. 2006). On the other hand, propionate is largely taken up by the liver, while acetate enters the peripheral circulation to be metabolized by peripheral tissues. Thus, these specific short-chain fatty acids may reduce the risk of developing gastrointestinal disorders and cardiovascular diseases, on top of cancer in various populations (Jones 2002).

4.3 Bone Health

The incidence and risk of osteoporosis is highly associated with postmenopausal women. Estrogen plays an important role in maintaining bone density by regulating the formation and reabsorption of calcium in bone (Nilsson and Gustafsson 2002; Cornwell et al. 2004). Soy isoflavones have been suggested to be effective in maintaining bone mineral density in postmenopausal women (Morabito et al. 2002). Genistein was found to have modest bone-conserving properties when administered in low doses to ovariectomized, lactating rats (Anderson et al. 1995). A double blind, placebo controlled study of postmenopausal women showed significant increase in bone mineral density at the femoral neck after 12 months of daily administration of 54 mg genistein (Morabito et al. 2002). Isoflavones have been found to stimulate osteoblasts and inhibit osteoclast activity in vitro that is essential in reduced bone turnover (Rassi et al. 2002; Chen et al. 2003). Soy foods with isoflavones have been ascribed to be able to prevent lumbar spine bone loss that may otherwise lose 1.5–3 % of bone/year in postmenopausal women (Lydeking-Olsen et al. 2004). This study showed that constant intake of soy food could reduce the risk of osteoporosis and lower fracture rates in the elderly.

Bone remodeling involves the activity of two different cell lines—osteoblasts which are responsible for bone formation and osteoclasts which are the bone-resorbing cells. Estrogen, in particular, is responsible for suppressing osteoclast activity which thus prevents bone resorption in maintaining bone homeostasis. Human osteoblasts contain estrogen receptors ERα and ERβ. The role of genistein as a hormone replacer is significantly important as the expression of ERβ will be increased greatly during bone mineralization, and genistein shows much higher affinity towards ERβ than ERα (Setchell and Lydeking-Olsen 2003). In addition to the coupling of osteoblast and osteoclast activities, the modulation of bone turnover involved a high level of complexity. Genistein and daidzein can suppress osteoclast activity by a number of mechanisms including induction of apoptosis, activation of protein tyrosine phosphatase, inhibition of cytokines, changes in intracellular Ca^{2+}, and membrane depolarization (Okamoto et al. 2001; Gao and Yamaguchi 1999, 2000; Williams et al. 1998; Blair et al. 1996). In a recent study demonstrated by Chiang et al. (2012) to investigate anti-osteoporosis effect of lactobacilli-fermented soy skim milk in 13-month-old female BALB/c aging mice, the trabecular bone volumes in mice fed with fermented soy skim milk increased more than twofold than the control. Furthermore, the network density and thickness of distal metaphyseal trabecular bone in mice fed with fermented soy skim milk also showed higher density than the control. The study suggested that dietary supplement with fermented soy skim milk can attenuate aging-induced bone loss and possibly lower the risk of osteopenia or osteoporosis in aging.

Fermentation of soymilk with probiotics can remove or reduce phytic acid, an anti-nutritional factor that may decrease the absorption of minerals leading to mineral deficiencies. Phytate tends to chelate nutritionally important cations such as Ca^{2+}, Mg^{2+}, Fe^{2+}, and Zn^{2+} which thus decrease the dietary bioavailability of these nutrients (Rekha and Vijayalakshmi 2010). An increase in calcium and magnesium levels accompanied by a decrease in phytic acid levels has been reported in a study involving soymilk fermentation by probiotic and thus reduced the anti-nutritional level of the product.

4.4 Antidiabetic and Obesity

Diabetes and obesity are two of the most challenging public health problems, especially with modernization and industrialization that have led to a reduction in physical activity and increase consumption of convenient foods which are normally high energy diets. With epidemiology studies showing a higher prevalence of diabetes and obesity in the Westerns, diet has been arisen as an important and plausible reason for the lower incidence among Asians (Afridi and Khan 2004). It has been reported that soy-based rich diet played a tight correlation with the prevention and treatment of type 2 diabetes (T2D) and obesity, supported by numerous animal studies and clinical and epidemiological investigations (Kwon et al. 2010). Among others, ingestion of soy protein with isoflavones has been found to improve glucose control and reduce insulin resistance and lipid metabolism. Fermented soymilk that is enriched with bioactive isoflavone aglycones (i.e., genistein) can help prevent and attenuate the progression of T2D and obesity. Particularly, mechanism of action of genistein towards prevention of T2D and obesity includes alleviation of inflammation, oxidative stress, insulin secretion, and β-cell and adipocyte functioning (Behloul and Wu 2013).

Obesity is highly associated with diabetes in which weight gain correlates closely with decreasing insulin sensitivity (Qatanani and Lazar 2007) and eventually develops in resistance. Impaired ability of insulin to inhibit glucose output from the liver and glucose uptake in fat and muscle can be observed in obese individuals who develop resistance to the cellular actions of insulin (Saltiel and Kahn 2001; Hribal et al. 2002). In addition to insulin resistance, the development of T2D is a result of loss of pancreatic islet β-cell mass and function (Gilberta and Liu 2013).

Systemic chronic inflammation has been proposed to have a link to the development of insulin resistance and T2D. Biomarkers of inflammation such as TNFα, IL-6, and C-reactive protein (CRP) are relatively higher in concentrations in insulin resistance and obese individuals, and these biomarkers predict the development of T2D. Genistein has been revealed to ameliorate this inflammatory state by reducing the expression levels of TNFα and proinflammatory cytokines in cerebral endothelial cells (Lu et al. 2009) as well as decreases the plasma levels of TNFα and IL-6 in fructose-fed rats (Palanisamy et al. 2011). Additionally, genistein has also been found effective in dampening diabetes-induced retinopathy by interfering with inflammatory signaling pathways where the release of TNFα was repressed and ERK and P38 MAPK phosphorylation were significantly inhibited in activated microglial cells (Ibrahim et al. 2010).

Besides inflammatory effect, oxidative stress is a causative factor in the development of insulin resistance. Oxidative stress condition has been associated with impaired glucose intolerance, β-cell, and mitochondrial dysfunctions until the development of diabetes (Behloul and Wu 2013). Genistein shows its promising candidacy to combat diabetes and obesity by counteracting the detrimental effects of reactive oxygen species (ROS) that generated during oxidative stress. It has been reported that genistein ameliorates renal inflammation and oxidative stress in streptozotocin-injected mice (Elmarakby et al. 2011). It also induces the expression of antioxidant enzymes manganese superoxide dismutase (MnSOD) and catalase that decreases the reactive oxygen species levels in cancer cells. It counteracts D-galactose-induced oxidative stress by decreasing intracellular ROS and suppressing NF-κB binding activity (Hsieh et al. 2011). Furthermore, genistein has also been found to be able to restore antioxidant enzymes activities and decrease ROS, iNOS, and endothelial NOS (eNOS) contents in sciatic nerve of diabetic mouse (Valsecchi et al. 2011).

Despite offsetting the obesity and diabetes-induced inflammation and oxidative stress, the antiobesity and antidiabetic effect of isoflavone aglycones has extended to be involved in affecting the life stages of adipocyte (Behloul and Wu 2013). Genistein inhibits adipocyte differentiation and lipid accumulation which thus reduce adipose tissue mass while stimulating lipolysis and adipocyte apoptosis. Genistein suppresses differentiation, adipogenesis, and lipid accumulation and increases lipolysis in 3T3-L1 adipocytes (Harmon and Harp 2001). The inhibition of adipogenic differentiation of human adipose tissue-derived mesenchymal stem cell by genistein was reported to be via Wingless and Int/β-catenin (Wnt/β-catenin) signaling pathway, such as extracellular signal-regulated kinase/c-Jun N-terminal kinase (ERK/JNK) signaling (Kim et al. 2010). Genistein inhibit the accumulation of lipid in adipocyte through downregulation of adipocyte-specific transcription factors such as PPARγ, C/EBPα, and glycerol-3-phosphate dehydrogenase (Park et al. 2009). In addition to inhibition of adipocyte differentiation and lipid accumulation, high concentration of genistein was shown to reduce adipocyte viability (Park et al. 2009). Yet, regular genistein intake is much lower than the effective dose which thus a synergistic effect of genistein with other compound such as plant sterol is needed to initiate the apoptotic effect. In a study reported by Rayalam et al. (2007), combination of genistein and resveratrol enhanced their effect on inhibiting adipogenesis, inducing apoptosis, and promoting lipolysis in 3T3-L1 adipocytes via downregulation of adipocyte-specific proteins PPARγ and C/EBPα.

Improving β-cell function and increasing β-cell mass are important for preventing and delaying the progression of type 2 diabetes. β-cells regulate plasma glucose concentration by releasing corresponding amount of insulin. Insulin resistance is a characteristic feature of T2D, and it affects β-cell function where decompensation for insulin resistance is associated with decreased β-cell mass in patients with diabetes (Kulkarni et al. 2004). Thus, the treatment of diabetes involves relieving insulin resistance and potentiating β-cell mass and function. Genistein has been shown to ameliorate glucose and lipid metabolism, via elevation of glucose-stimulated insulin secretion (GSIS) and β-cell regeneration and protection against apoptosis (Jonas et al. 1995; Jones and Persaud 1994; Ohno et al. 1993). Studies reported genistein exerted a dose-dependent response with increase secretion of GSIS was observed from both clonal pancreatic β-cells and cultured islets (Jonas et al. 1995; Sorenson et al. 1994; Ohno et al. 1993). The inducing effect of genistein for β-cell growth is achieved through phosphorylation of ERK1/2 and protein expression of cyclin D1 that is a major cell-cycle regulator (Fu et al. 2010). Additionally, the insulinotropic effect of genistein is found to be as potent as that of incretin hormone glucagon-like peptide-1 (Holst 1994; Thorens 1992), where it activates the cAMP/PKA signaling cascade upon binding to its receptor on the membrane of β-cells. Apart from that, genistein also protects pancreatic β-cells against cytokine-induced toxicity via reduced inducible nitric oxide synthase (iNOS) expression and nitric oxide production. It mitigates cytokine-induced changes in β-cell function by suppressing NFκB, ERK1/2, and JAK/STAT pathways (Kim et al. 2007).

4.5 Gastrointestinal Bacterial Ecosystem

As the organoleptic property of fermented soymilk possesses a great challenge upon consumer acceptance, clinical trial using whole fermented soymilk is scarcely performed by researcher. Nevertheless, effect of fermented soymilk on the intestinal bacterial ecosystem has once been studied by Cheng et al. (2005). The author used a crossover experimental design consisting of 28 healthy adults; each group consumed twice a day of 250 mL of either fermented soymilk or regular soymilk, with 2 weeks crossover for 9 weeks. There were 2 weeks adjustment period and 2 weeks experimental drink consumption followed by 2 weeks washout upon switching of drinks. Fecal samples were collected from subjects every week from week 2, and results showed that there was higher population of *Bifidobacterium* spp. and *Lactobacillus* spp. as well as higher ratio if *Bifidobacterium* spp. and *Lactobacillus* spp. to *Clostridium perfringens* compared to the group ingested normal soymilk. There was a decrease in the population of coliform when the subjects were in the period of fermented soymilk consumption. This study showed that frequent consumption of fermented soymilk could improve the microbial ecosystem in the gastrointestinal tract by increasing the population of probiotics/health beneficial bacteria. Yet, the origin of microorganisms in fecal sample of whether originated from the fermented soymilk or intestinal source was ambiguous as microorganisms used to ferment soymilk were not revealed in the study.

In another in vivo study investigating the effect of ingestion of soy yogurt fermented with *Enterococcus faecium* and *L. helveticus* subsp. *jugurti* on intestinal count of enterococci in 40 SPF Wistar rats (8-week-old male) fed on a beef-based animal diet for 30 days, results of the study showed that the consumption of soy yogurt could alter the composition of *Enterococcus* present in the feces and colon (Bedani et al. 2011). The isolated bacterial population changed from *E. faecalis* that originally dominated the colon to *E. faecium*, the starter that was used for soymilk fermentation. The study suggested that probiotic strains originated from food source could colonize the intestinal tract and thus exert plausible health beneficial effects. Similar outcome was obtained by Cavallini et al. (2011) in a study to investigate the possible correlation between fecal microbiota in rabbits induced hypercholesterolemia that ingested soy extract fermented with *E. faecium* and *L. helveticus* for 60 days. Intake of probiotic soy product was correlated with significant increases on *Lactobacillus*, *Bifidobacterium*, and *Enterococcus* and a decrease in the enterobacteria population. The author suggested that daily ingestion of soy product (supplemented with or not with isoflavones) could contribute to a beneficial balance of fecal microbiota with associated improved cardiovascular disorders. More works need to be done to investigate the interaction of components in fermented soymilk with bacterial population in the gastrointestinal tract.

5 Conclusions

In spite of a myriad of fermented soy products that are currently available in the market; yet, the development of fermented soymilk beverages are still withhold due to the challenges of intolerable organoleptic acceptability. Up to date, there is a lack of clinical study involving the functional attribute of fermented soymilk; yet has only been proven by in vitro and in vivo. Fermented soymilk could confer many nutritional benefits including reducing the risk of cardiovascular diseases, bone loss, cancer, diabetes, and obesity thanks to the soy phytochemicals such as isoflavone aglycones. Addition of suitable beneficial microbes such as probiotics for fermentation has further enriched the nutritional content of soymilk, conjoining the virtue of fermentation, probiotic, and the bioactive metabolites, producing a nutraceutical with health-promoting value.

References

Afridi AK, Khan A (2004) Prevalence and etiology of obesity: an overview. Pak J Nutr 3(1):14–25

Aguirre L, Garro MS, de Giori GS (2008) Enzymatic hydrolysis of soybean protein using lactic acid bacteria. Food Chem 111(4):976–982. doi:10.1016/j.foodchem.2008.05.018

Amaretti A, Tamburini E, Bernardi T, Pompei A, Zanoni S, Vaccari G, Matteuzzi D, Rossi M (2006) Substrate preference of *Bifidobacterium adolescentis* MB 239: compared growth on single and mixed carbohydrates. Appl Microbiol Biotechnol 73(3):645–662. doi:10.1007/s00253-006-0500-9

American Institute for Cancer Research (2000) New American Cancer Society Guidelines on diet, nutrition and cancer prevention. Simple steps to prevent cancer: healthy living and lower cancer risk. American Institute for Cancer Research, Washington, DC

Anderson JW, Johnstone BM, Cook-Newell ME (1995) Meta-analysis of the effects of soy protein intake on serum lipids. N Engl J Med 333:276–282. doi:10.1056/NEJM199508033330502

Ang HG, Kwik WL, Theng CY (1985) Development of soymilk-a review. Food Chem 17(4):235–250. doi:10.1016/0308-8146(85)90033-0

Bedani R, Pauly-Silveira ND, Cano VSP, Valentini SR, de Valdez GF, Rossi EA (2011) Effect of ingestion of soy yogurt on intestinal parameters of rats fed on a beef-based animal diet. Braz J Microbiol 42(3):1238–1247. doi:10.1590/S1517-83822011000300050

Behloul N, Wu G (2013) Genistein: a promising therapeutic agent for obesity and diabetes treatment. Eur J Pharmacol 698(1–3):31–38. doi:10.1016/j.ejphar.2012.11.013

Belfiore A, Genua M, Malaguarnera R (2009) PPAR-γ agonists and their effects on IGF-I receptor signalling: implications for cancer. PPAR Res 2009:830501. doi:10.1155/2009/830501

Blair HC, Jordan SE, Peterson TG, Barnes S (1996) Variable effects of tyrosine kinase inhibitors on avian osteoclastic activity and reduction of bone loss in ovariectomized rats. J Cell Biochem 61(4):629–637. doi:10.1002/(SICI)1097-4644(19960616)61:4<629::AID-JCB17>3.0.CO;2-A

Borgwardt K, Bonifatius S, Gardemann A (2008) Acidic peptides enhanced genistein dependent inhibition of human platelet aggregation: potential protective effect of digestible peptides plus genistein against atherosclerosis. Nutr Res 28(8):523–531. doi:10.1016/j.nutres.2008.03.017

Caescu CI, Vidal O, Krzewinski F, Artenie V, Bouquelet S (2004) *Bifidobacterium longum* requires a fructokinase (Frk, ATP:D-Fructose-6-Phosphotransferase, EC 2.7.1.4) for fructose catabolism. J Bacteriol 186(19):6515–6525. doi:10.1128/JB.186.19.6515-6525.2004

Cappelletti V, Miodini P, Di Fronzo G, Daidone MG (2006) Modulation of estrogen receptor-β isoforms by phytoestrogens in breast cancer cells. Int J Oncol 28(5):1185–1191. doi:10.3892/ijo.28.5.1185

Cavallini DCU, Suzuki JY, Abdalla DSP, Vendramini RC, Pauly-Silveira ND, Roselino MN, Pinto RA, Rossi EA (2011) Influence of a probiotic soy product on fecal microbiota and its association with cardiovascular risk factors in an animal model. Lipids Health Dis 10:126. doi:10.1186/1476-511X-10-126

Chambers CM, Ness GC (1997) Translational regulation of hepatic HMG-CoA reductase by dietary cholesterol. Biochem Biophys Res Commun 232(2):278–281. doi:10.1006/bbrc.1997.6288

Chen X, Garner SC, Quarles LD, Anderson JB (2003) Effects of genistein on expression of bone markers during MC3T3-E1 osteoblastic cell differentiation. J Nutr Biochem 14(6):342–349. doi:10.1016/S0955-2863(03)00056-1

Cheng IC, Shang HF, Lin TF, Wang TH, Lin HS, Lin SH (2005) Effect of fermented soy milk on the intestinal bacterial ecosystem. World J Gastroenterol 11(8):1225–1227. doi:10.3748/wjg.v11.i8.1225

Chiang SS, Lian JW, Pan TM (2012) Effect of bioactive compounds in lactobacilli-fermented soy skim milk on femoral bone microstructure of aging mice. J Sci Food Agric 92(2):328–335. doi:10.1002/jsfa.4579

Chien HL, Huang HY, Chou CC (2006) Transformation of isoflavone phytoestrogens during the fermentations of soymilk with lactic acid bacteria and bifidobacteria. Food Microbiol 23(8):772–778. doi:10.1016/j.fm.2006.01.002

Chou CC, Hou JW (2000) Growth of bifidobacteria in soymilk and their survival in the fermented soymilk drink during storage. Int J Food Microbiol 56(2–3):113–121. doi:10.1016/S0168-1605(99)00201-9

Christensen JE, Dudley EG, Pederson JA, Steele JL (1999) Peptidases and amino acid catabolism in lactic acid bacteria. Antonie Van Leeuwenhoek 76(1–4):217–246. doi:10.1007/978-94-017-2027-4_11

Cornwell T, Cohick W, Raskin I (2004) Dietary phytoestrogens and health. Phytochemistry 65(8): 995–1016. doi:10.1016/j.phytochem.2004.03.005

Courtin P, Rul F (2004) Interactions between microorganisms in a simple ecosystem: yogurt bacteria as a study model. Le Lait 84(1–2):125–134. doi:10.1051/lait:2003031

De Vuyst L, Neysens P (2005) The sourdough microflora: biodiversity and metabolic interactions. Trends Food Sci Technol 16(1–3):43–56. doi:10.1016/j.tifs.2004.02.012

Donkor ON, Henriksson A, Vasiljevic T, Shah NP (2005) Proteolytic strains as starter cultures improve angiotensin-converting enzyme inhibitory activities in soy yogurt. J Food Sci 70(8): M375–M381. doi:10.1111/j.1365-2621.2005.tb11522.x

Donkor ON, Nilmini SLI, Stolic P, Vasiljevic T, Shah NP (2007) Survival and activity of selected probiotic organisms in set-type yogurt during cold storage. Int Dairy J 17(6):657–665. doi:10.1016/j.idairyj.2006.08.006

Duncan AM, Underhill KEW, Xu X, LaValleur J, Phipps WR, Kurzer MS (1999) Modest hormonal effects of soy isoflavones in postmenopausal women. J Clin Endocrinol Metab 84(10):3479–3484. doi:10.1210/jcem.84.10.6067

Elmarakby AA, Ibrahim AS, Faulkner J, Mozaffari MS, Liou GI, Abdelsayed R (2011) Tyrosine kinase inhibitor, genistein, reduces renal inflammation and injury in streptozotocin-induced diabetic mice. Vascul Pharmacol 55(5–6):149–156. doi:10.1016/j.vph.2011.07.007

Ewe JA, Wan Abdullah WN, Liong MT (2010) Viability and growth characteristics of *Lactobacillus* in soymilk supplemented with B-vitamins. Int J Food Sci Nutr 61(1):87–107. doi:10.3109/09637480903334163

Ewe JA, Wan Abdullah WN, Karim AA, Liong MT (2012a) Enhanced growth of lactobacilli and bioconversion of isoflavones in biotin-supplemented soymilk by electroporation. Int J Food Sci Nutr 63(5):580–596. doi:10.3109/09637486.2011.641940

Ewe JA, Wan Abdullah WN, Rajeev B, Karim AA, Liong MT (2012b) Enhanced growth of lactobacilli and bioconversion of isoflavones in biotin-supplemented soymilk upon ultrasound-treatment. Ultrason Sonochem 19(1):160–173. doi:10.1016/j.ultsonch.2011.06.013

Ewe JA, Wan Abdullah WN, Karim AA, Liong MT (2013) Ultraviolet radiation enhanced growth of lactobacilli and their bioconversion of isoflavones in biotin-supplemented soymilk. LWT - Food Sci Technol 50(1):25–31. doi:10.1016/j.lwt.2012.07.042

FAO (2007) Crop prospects and food situations. http://www.fao.org/docrep/010/j9940e/j9940e00.htm. Accessed 18 Dec 2014

Field FJ, Born E, Murthy S, Mathur SN (2001) Regulation of sterol regulatory element-binding proteins by cholesterol flux in CaCo-2 cells. J Lipid Res 42(10):1687–1698

FitzGerald RJ, Murray BA (2006) Bioactive peptides and lactic fermentations. Int J Dairy Technol 59(2):118–125. doi:10.1111/j.1471-0307.2006.00250.x

Fraga CG, Galleano M, Verstraeten SV, Oteiza PI (2010) Basic biochemical mechanisms behind the health benefits of polyphenols. Mol Aspects Med 31(6):435–445. doi:10.1016/j.mam.2010.09.006

Fu Z, Zhang W, Zhen W, Lum H, Nadler J, Bassaganya-Riera J, Jia Z, Wang Y, Misra H, Liu D (2010) Genistein induces pancreatic beta-cell proliferation through activation of multiple signaling pathways and prevents insulin-deficient diabetes in mice. Endocrinology 151(7): 3026–3037. doi:10.1210/en.2009-1294

Fukuda S, Toh H, Hase K, Oshima K, Nakanishi Y, Yoshimura K, Tobe T, Clarke JM, Topping DL, Suzuki T, Taylor TD, Itoh K, Kikuchi J, Morita H, Hattori M, Ohno H (2011) Bifidobacteria can protect from enteropathogenic infection through production of acetate. Nature 469(7331):543–547. doi:10.1038/nature09646

Fung WY, Woo YP, Liong MT (2008) Optimization of growth of *Lactobacillus acidophilus* FTCC 0291 and evaluation of growth characteristics in soy whey medium: a response surface methodology approach. J Agric Food Chem 56(17):7910–7918. doi:10.1021/jf801567j

Gao YH, Yamaguchi M (1999) Suppressive effects of genistein on rat bone osteoclasts: apoptosis is induced through Ca2+ signalling. Biol Pharm Bull 22(8):805–809

Gao YH, Yamaguchi M (2000) Suppressive effect of genistein on rat bone osteoclasts: involvement of protein kinase inhibition and protein tyrosine phosphatase activation. Int J Mol Med 5(3):261–267. doi:10.3892/ijmm.5.3.261

Gertz J, Reddy TE, Varley KE, Garabedian MJ, Myers RM (2012) Genistein and bisphenol A exposure cause estrogen receptor 1 to bind thousands of sites in a cell type-specific manner. Genome Res 22(11):2153–2162. doi:10.1101/gr.135681.111

Gilberta ER, Liu D (2013) Anti-diabetic functions of soy isoflavone genistein: mechanisms underlying its effects on pancreatic β-cell function. Food Funct 4(2):200–212. doi:10.1039/c2fo30199g

Global Information Inc. (2012) Global health movement drives market for nutraceuticals to $250 bn by 2018; Probiotics to touch $39.6 bn and heart health ingredients near $15.2 bn. http://www.ireachcontent.com/news-releases/global-health-movement-drives-market-for-nutraceuticals-to-250-bn-by-2018-probiotics-to-touch-396-bn-and-heart-health-ingredients-near-152-bn-180021771.html. Accessed 31 Dec 2014

Gobbetti M, Ferranti P, Smacchi E, Goffredi F, Addeo F (2000) Production of angiotensin-I-converting-enzyme-inhibitory peptides in fermented milks started by *Lactobacillus delbrueckii* subsp. *bulgaricus* SS1 and *Lactococcus lactis* subsp. *cremoris* FT4. Appl Environ Microbiol 66(9):3898–3904. doi:10.1128/AEM.66.9.3898-3904.2000

Gomes AMP, Malcata FX (1999) *Bifidobacterium* spp. and *Lactobacillus acidophilus*: biological, biochemical, technological and therapeutical properties relevant for use as probiotics. Trends Food Sci Technol 10(4–5):139–157. doi:10.1016/S0924-2244(99)00033-3

Gossner G, Choi M, Tan L, Fogoros S, Griffith KA, Kuenker M, Liu JR (2007) Genistein-induced apoptosis and autophagocytosis in ovarian cancer cells. Gynecol Oncol 105(1):23–30. doi:10.1016/j.ygyno.2006.11.009

Gottstein N, Ewins BA, Eccleston C, Hubbard GP, Kavanagh IC, Minihane AM, Weinberg PD, Rimbach G (2003) Effect of genistein and daidzein on platelet aggregation and monocyte and endothelial function. Br J Nutr 89(5):607–616. doi:10.1079/BJN2003820

Grundy SM (1998) Hypertriglyceridemia, atherogenic dyslipidemia, and the metabolic syndrome. Am J Cardiol 81(4):18B–25B. doi:10.1016/S0002-9149(98)00033-2

Guerrero JA, Lozano ML, Castillo J, Benavente-García O, Vicente V, Rivera J (2005) Flavonoids inhibit platelet function through binding to the thromboxane A2 receptor. J Thromb Haemost 3(2):369–376. doi:10.1111/j.1538-7836.2004.01099.x

Guha N, Kwan M, Quesenberry C, Weltzien E, Castillo A, Caan B (2009) Soy isoflavones and risk of cancer recurrence in a cohort of breast cancer survivors: the Life After Cancer Epidemiology study. Breast Cancer Res Treat 118(2):395–405. doi:10.1007/s10549-009-0321-5

Harmon AW, Harp JB (2001) Differential effects of flavonoids on 3T3-L1 adipogenesis and lipolysis. Am J Physiol Cell Physiol 280(4):C807–813

He FJ, Chen JQ (2013) Consumption of soybean, soy foods, soy isoflavones and breast cancer incidence: differences between Chinese women and women in Western countries and possible mechanisms. Food Sci Hum Well 2(3–4):146–161. doi:10.1016/j.fshw.2013.08.002

Høie LH, Morgenstern EC, Gruenwald J, Graubaum HJ, Busch R, Lüder W, Zunft HJ (2005) A double-blind placebo-controlled clinical trial compares the cholesterol lowering effects of two different soy protein preparations in hypercholesterolemic subjects. Eur J Nutr 44(2):65–71. doi:10.1007/s00394-004-0492-0

Holst JJ (1994) Glucagon-like peptide-1 (GLP-1) - a newly discovered Gl hormone (review). Gastroenterology 107(6):1848–1855

Horton JD, Goldstein JL, Brown MS (2002) SREBPs: activators of the complete program of cholesterol and fatty acid synthesis in the liver. J Clin Invest 109(9):1125–1131. doi:10.1172/JCI15593

Hou JW, Yu RC, Chou CC (2000) Changes in some components of soymilk during fermentation with bifidobacteria. Food Res Int 33(5):393–397. doi:10.1016/S0963-9969(00)00061-2

Hribal ML, Oriente F, Accili D (2002) Mouse models of insulin resistance. Am J Physiol Endocrinol Metab 282(5):E977–81. doi:10.1152/ajpendo.00561.2001

Hsieh HM, Wu WM, Hu ML (2011) Genistein attenuates D-galactose-induced oxidative damage through decreased reactive oxygen species and NF-κB binding activity in neuronal PC12 cells. Life Sci 88(1–2):82–88. doi:10.1016/j.lfs.2010.10.021

Hugenholtz J (1986) Population dynamics of mixed starter cultures. Neth Milk Dairy J 40:129–140

Ibrahim AS, El-Shishtawy MM, Pena A Jr, Liou GI (2010) Genistein attenuates retinal inflammation associated with diabetes by targeting of microglial activation. Mol Vis 16:2033–2042, http://www.molvis.org/molvis/v16/a219

Jonas JC, Plant TD, Gilon P, Detimary P, Nenquin M, Henquin JC (1995) Multiple effects and stimulation of insulin secretion by the tyrosine kinase inhibitor genistein in normal mouse islets. Br J Pharmacol 114(4):872–880

Jones PJ (2002) Clinical nutrition: 7. Functional foods – more than just nutrition. Can Med Assoc J 166(12):1555–1563

Jones PM, Persaud SJ (1994) Tyrosine kinase inhibitors inhibit glucose-stimulated insulin secretion. Biochem Soc Trans 22(2):209S

Joseph SB, Laffitte BA, Patel PH, Watson MA, Matsukuma KE, Walczak R, Colllins JL, Osborne TF, Tontonoz P (2002) Direct and indirect mechanisms for regulation of fatty acid synthase gene expression by liver X receptors. J Biol Chem 277(13):11019–11025. doi:10.1074/jbc.M111041200

Kano M, Takayanagi T, Harada K, Sawada S, Ishikawa F (2006) Bioavailability of isoflavones after ingestion of soy beverages in healthy adults. J Nutr 136(9):2291–2296

Kehagias C, Csapó J, Konteles S, Kolokitha E, Koulouris S, Csapó-Kiss Z (2008) Support of growth and formation of D-amino acids by *Bifidobacterium longum* in cows', ewes', goats' milk and modified whey powder products. Int Dairy J 18(4):396–402. doi:10.1016/j.idairyj.2007.11.014

Kim EK, Kwon KB, Song MY, Seo SW, Park SJ, Ka SO, Na L, Kim KA, Ryu DG, So HS, Park R, Park JW, Park BH (2007) Genistein protects pancreatic beta cells against cytokine-mediated toxicity. Mol Cell Endocrinol 278(1–2):18–28. doi:10.1016/j.mce.2007.08.003

Kim MH, Park JS, Seo MS, Jung JW, Lee YS, Kang KS (2010) Genistein and daidzein repress adipogenic differentiation of human adipose tissue-derived mesenchymal stem cells via Wnt/beta-catenin signalling or lipolysis. Cell Prolif 43(6):594–605. doi:10.1111/j.1365-2184.2010.00709.x

Kim YH, Yoon S, Lee SB, Han HW, Oh H, Lee WJ, Lee SM (2014) Fermentation of soy milk via *Lactobacillus plantarum* improves dysregulated lipid metabolism in rats on a high cholesterol diet. PLoS One 9(2):e88231. doi:10.1371/journal.pone.0088231

Korhonen H (2009) Milk-derived bioactive peptides: from science to applications. J Funct Foods 1 (2):177–187. doi:10.1016/j.jff.2009.01.007

Kuiper GG, Lemmen JG, Carlsson B, Corton JC, Safe SH, van der Saag PT, van der Burg B, Gustafsson JA (1998) Interaction of estrogenic chemicals and phytoestrogens with estrogen receptor beta. Endocrinology 139(10):4252–4263. doi:10.1210/endo.139.10.6216

Kulkarni RN, Jhala US, Winnay JN, Krajewski S, Montminy M, Kahn CR (2004) PDX-1 haploinsufficiency limits the compensatory islet hyperplasia that occurs in response to insulin resistance. J Clin Invest 114(6):828–836. doi:10.1172/JCI200421845

Kunji ER, Mierau I, Hagting A, Poolman B, Konings WN (1996) The proteolytic systems of lactic acid bacteria. Antonie van Leeuwenhoek 70(2–4):187–221

Kwon DY, Daily JW 3rd, Kim HJ, Park S (2010) Antidiabetic effects of fermented soybean products on type 2 diabetes. Nutr Res 30(1):1–13. doi:10.1016/j.nutres.2009.11.004

LeBlanc JG, Garro MS, De Giori GS (2004) Effect of pH on *Lactobacillus fermentum* growth, raffinose removal, α-galactosidase activity and fermentation products. Appl Microbiol Biotechnol 65(1):119–123. doi:10.1007/s00253-003-1532-z

Lee K, Lee J, Kim YH, Moon SH, Park YH (2001) Unique properties of four lactobacilli in amino acid production and symbiotic mixed culture for lactic acid biosynthesis. Curr Microbiol 43(6): 383–390. doi:10.1007/s002840010324

Li Y, Chen H, Hardy TM, Tollefsbol TO (2013a) Epigenetic regulation of multiple tumor-related genes leads to suppression of breast tumorigenesis by dietary genistein. PLoS One 8(1): e54369. doi:10.1371/journal.pone.0054369

Li Y, Zhang Y, Sun S, Zhang A, Liu Y (2013b) Binding investigation on the interaction between methylene blue (MB)/TiO_2 nanocomposites and bovine serum albumin by resonance light-scattering (RLS) technique and fluorescence spectroscopy. J Photochem Photobiol B 128C: 12–19. doi:10.1016/j.jphotobiol.2013.07.027

Lin Y, Meijer GW, Vermeer MA, Trautwein EA (2004) Soy protein enhances the cholesterol-lowering effect of plant sterol esters in cholesterol-fed hamsters. J Nutr 134(1):143–148

Liong MT, Azhar ME, Lim PT, Kang JY (2009) Survival, growth characteristics and bioactive potential of *Lactobacillus acidophilus* in a soy-based cream cheese. J Sci Food Agric 89(8): 1382–1391. doi:10.1002/jsfa.3598

Lu H, Shi JX, Zhang DM, Chen HL, Qi M, Yin HX (2009) Genistein, a soybean isoflavone, reduces the production of pro-inflammatory and adhesion molecules induced by hemolysate in brain microvascular endothelial cells. Acta Neurol Belg 109:32–37

Lydeking-Olsen E, Beck-Jensen JE, Setchell KDR, Holm-Jensen T (2004) Soymilk or progesterone for prevention of bone loss-a 2 year randomized, placebo-controlled trial. Eur J Nutr 43(4):246–257. doi:10.1007/s00394-004-0497-8

Marteau P, Pochart P, Flourié B, Pellier P, Santos L, Desjeux JF, Rambaud JC (1990) Effect of chronic ingestion of a fermented dairy product containing *Lactobacillus acidophilus* and *Bifidobacterium bifidum* on metabolic activities of the colonic flora in humans. Am J Clin Nutr 52(4):685–688

Martinez-Villaluenga C, Torino MI, Martin V, Arroyo R, Garcia-Mora P, Estrella Pedrola I, Vidal-Valverde C, Rodriguez JM, Frias J (2012) Multifunctional properties of soy milk fermented by *Enterococcus faecium* strains isolated from raw soy milk. J Agric Food Chem 60(41): 10235–10244. doi:10.1021/jf302751m

Mazmanian SK, Kasper DL (2006) The love-hate relationship between bacterial polysaccharides and the host immune system. Nat Rev Immunol 6:849–858. doi:10.1038/nri1956

Morabito N, Crisafulli A, Vergara C, Gaudio A, Lasco A, Frisina N, D'Anna R, Corrado F, Pizzoleo MA (2002) Effects of genistein and hormone-replacement therapy on bone loss in early postmenopausal women: a randomized double-blind placebo-controlled study. J Bone Miner Res 17(10):1904–1912

Mulimani VH, Thippeswamy S, Ramalingam S (1997) Enzymatic degradation of oligosaccharides in soybean flours. Food Chem 59(2):279–282. doi:10.1016/S0308-8146(96)00282-8

Munoz Y, Garrido A, Valladares L (2009) Equol is more active than soy isoflavone itself to compete for binding to thromboxane A(2) receptor in human platelets. Thromb Res 123(5): 740–744. doi:10.1016/j.thromres.2008.07.011

Nakamura Y, Yamamoto N, Sakai K, Takano T (1995) Antihypertensive effect of sour milk and peptides isolated from it that are inhibitors to angiotensin-I-converting enzyme. J Dairy Sci 78(6):1253–1257. doi:10.3168/jds.S0022-0302(95)76745-5

Nakamura Y, Yogosawa S, Izutani Y, Watanabe H, Otsuji E, Sakai T (2009) A combination of indole-3-carbinol and genistein synergistically induces apoptosis in human colon cancer HT-29 cells by inhibiting Akt phosphorylation and progression of autophagy. Mol Cancer 8: 100. doi:10.1186/1476-4598-8-100

Nakashima S, Koike T, Nozawa Y (1991) Genistein, a protein tyrosine kinase inhibitor, inhibits thromboxane A2-mediated human platelet responses. Mol Pharmacol 39(4):475–480

Nestel PJ, Yamashita T, Sasahara T, Pomeroy S, Dart A, Komesaroff P, Owen A, Abbey M (1997) Soy isoflavones improve systemic arterial compliance but not plasma lipids in menopausal and perimenopausal women. Arterioscle Thromb Vasc Biol 17(12):3392–3398. doi:10.1161/01.ATV.0000035520.25551.97

Ng KH, Lye HS, Easa AM, Liong MT (2008) Growth characteristics and bioactivity of probiotics in tofu-based medium during storage. Ann Microbiol 58(3):477–487. doi:10.1007/BF03175546

Nilsson S, Gustafsson JA (2002) Estrogen receptor action. Crit Rev Eukaryot Gene Expr 12(4): 237–57. doi:10.1615/CritRevEukaryotGeneExpr.v12.i4.10

Ohno T, Kato N, Ishii C, Shimizu M, Ito Y, Tomono S, Kawazu S (1993) Genistein augments cyclic adenosine 3′5′-monophosphate(cAMP) accumulation and insulin release in MIN6 cells. Endocr Res 19(4):273–285. doi:10.1080/07435809309026682

Okamoto F, Okabe K, Kajiya H (2001) Genistein, a soybean isoflavone, inhibits inward rectifier K + channels in rat osteoclasts. Jpn J Physiol 51(4):501–509. doi:10.2170/jjphysiol.51.501

Palanisamy N, Kannappan S, Anuradha CV (2011) Genistein modulates NF-κB-associated renal inflammation, fibrosis and podocyte abnormalities in fructose-fed rats. Eur J Pharmacol 667(1–3):355–364. doi:10.1016/j.ejphar.2011.06.011

Pan H, Zhou W, He W, Liu X, Ding Q, Ling L, Zha X, Wang S (2012) Genistein inhibits MDA-MB-231 triple-negative breast cancer cell growth by inhibiting NF-κB activity via the Notch-1 pathway. Int J Mol Med 30(2):337–343. doi:10.3892/ijmm.2012.990

Pan X, Zhang X, Sun H, Zhang J, Yan M, Zhang H (2013) Autophagy inhibition promotes 5-fluorouracil-induced apoptosis by stimulating ROS formation in human non-small cell lung cancer A549 cells. PLoS One 8(2):e56679. doi:10.1371/journal.pone.0056679

Park HJ, Della-Fera MA, Hausman DB, Rayalam S, Ambati S, Baile CA (2009) Genistein inhibits differentiation of primary human adipocytes. J Nutr Biochem 20(2):140–148. doi:10.1016/j.jnutbio.2008.01.006

Pham TT, Shah NP (2007) Biotransformation of isoflavone glycosides by *Bifidobacterium animalis* in soymilk supplemented with skim milk powder. J Food Sci 72(8):M316–M324. doi:10.1111/j.1750-3841.2007.00476.x

Pyo YH, Lee TC, Lee YC (2005) Enrichment of bioactive isoflavones in soymilk fermented with β-glucosidase-producing lactic acid bacteria. Food Res Int 38(5):551–559. doi:10.1016/j.foodres.2004.11.008

Qatanani M, Lazar MA (2007) Mechanisms of obesity-associated insulin resistance: many choices on the menu. Genes Dev 21:1443–1455. doi:10.1101/gad.1550907

Rassi CM, Lieberherr M, Chaumaz G, Pointillart A, Cournot G (2002) Downregulation of osteoclast differentiation by daidzein via caspase 3. J Bone Miner Res 17(4):630–638. doi:10.1359/jbmr.2002.17.4.630

Rayalam S, Della-Fera MA, Yang JY, Park HJ, Ambati S, Baile CA (2007) Resveratrol potentiates genistein's antiadipogenic and proapoptotic effects in 3T3-L1 adipocytes. J Nutr 137(12): 2668–2673. http://jn.nutrition.org/content/137/12/2668.full

Rekha CR, Vijayalakshmi G (2010) Bioconversion of isoflavone glycosides to aglycones, mineral bioavailability and vitamin B complex in fermented soymilk by probiotic bacteria and yeast. J Appl Microbiol 109(4):1198–1208. doi:10.1111/j.1365-2672.2010.04745.x

Rodriguez AV, Manca de Nadra CM (1995) Effect of pH and hydrogen peroxide produced by *Lactobacillus hilgardii* on *Pediococcus pentosaceus* growth. FEMS Microbiol Lett 128(1): 59–62. doi:10.1111/j.1574-6968.1995.tb07500.x

Rodriguez EB, Flavier ME, Rodriguez-Amaya DB, Amaya-Farfán J (2006) Phytochemicals and functional foods. Current situation and prospect for developing countries. Segurança Alimentar e Nutricional 13(1):1–22

Saito T (2008) Bioactive components of milk. In: Bösze Z (ed) Advances in experimental medicine and biology. Springer, New York, p 295

Saltiel AR, Kahn CR (2001) Insulin signalling and the regulation of glucose and lipid metabolism. Nature 414(6865):799–806. doi:10.1038/414799a

Santos CCAA, Libeck BS, Schwan RF (2014) Co-culture fermentation of peanut-soy milk for the development of novel functional beverage. Int J Food Microbiol 186C:32–41. doi:10.1016/j.ijfoodmicro.2014.06.011

Sargeant P, Farndale RW, Sage SO (1993) ADP- and thapsigargin-evoked Ca^{2+} entry and protein-tyrosine phosphorylation are inhibited by the tyrosine kinase inhibitors genistein and methyl-2,5-dihydroxycinnamate in fura-2-loaded human platelets. J Biol Chem 268:18151–18156

Saulnier DMA, Spinler JK, Gibson GR, Versalovic J (2009) Mechanisms of probiosis and prebiosis: considerations for enhanced functional foods. Curr Opin Biotechnol 20(2):35–141. doi:10.1016/j.copbio.2009.01.002

Schwartz LM, Smith SW, Jones ME, Osborne BA (1993) Do all programmed cell deaths occur apoptosis? Proc Natl Acad Sci USA 90(3):980–984. doi:10.1073/pnas.90.3.786

Setchell KDR, Lydeking-Olsen E (2003) Dietary phytoestrogens and their impact on bone-evidence from in vitro and in vivo, human observational and dietary intervention studies. Am J Clin Nutr 78(3):593S–609S

Setchell KDR, Brown N, Lydeking-Olsen E (2002) The clinical importance of the metabolite equol-a clue to the effectiveness of soy and its isoflavones. J Nutr 132(12):3577–3584

Sharma R (2009) Nutraceuticals and nutraceutical supplementation criteria in cancer: a literature survey. Open Nutraceuticals J 2(1):92–106. doi:10.2174/1876396000902010092

Shihata A, Shah NP (2000) Proteolytic profiles of yoghurt and probiotic bacteria. Int Dairy J 10(5-6):401–408. doi:10.1016/S0958-6946(00)00072-8

Shin ZI, Yu R, Park SA, Chung DK, Ahn CW, Nam HS, Kim KS, Lee HJ (2001) His-His-Leu, an angiotensin I converting enzyme inhibitory peptide derived from Korean soybean paste, exerts antihypertensive activity in vivo. J Agric Food Chem 49(6):3004–3009. doi:10.1021/jf001135r

Sieuwerts S, de Bok FAM, Hugenholtz J, van Hylckama Vlieg JET (2008) Unraveling microbial interactions in food fermentations: from classical to genomics approaches. Appl Environ Microbiol 74(16):4997–5007. doi:10.1128/AEM.00113-08

Singletary K, Milner J (2008) Diet, autophagy and cancer: a review. Cancer Epidemiol Biomarkers Prev 17:1596–1610. doi:10.1158/1055-9965.EPI-07-2917

Sonnenburg JL, Chen CT, Gordon JI (2006) Genomic and metabolic studies of the impact of probiotics on a model gut symbiont and host. PLoS Biol 4(12), e413. doi:10.1371/journal.pbio.0040413

Sorenson RL, Brelje TC, Roth C (1994) Effect of tyriosine kinase inhibitors on islet of Langerhans: evidence for tyrosine kinases in the regulation of insulin secretation. Endocrinology 134(4):1975–1978. doi:10.1210/endo.134.4.8137766

Spady DK, Woollett LA, Dietschy JM (1993) Regulation of plasma LDL cholesterol levels by dietary cholesterol and fatty acids. Annu Rev Nutr 13:355–381. doi:10.1146/annurev.nu.13.070193.002035

Sun-Waterhouse D (2011) The development of fruit-based functional foods targeting the health and wellness market: a review. Int J Food Sci Technol 46(5):899–920. doi:10.1111/j.1365-2621.2010.02499.x

Taku K, Umegaki K, Sato Y, Taki Y, Endoh K, Watanabe S (2007) Soy isoflavones lower serum total and LDL cholesterol in humans: a meta-analysis of 11 randomized controlled trials. Am J Clin Nutr 85(4):1148–1156

Teede HJ, Dalais FS, Kotsopoulos D, Davis SR, Liang YL, McGrath BP (2001) Soy protein dietary supplementation containing phytoestrogens improves lipid profiles and blood pressure: a double blind, randomised, placebo controlled study in men and postmenopausal women. J Clin Endocrinol Metab 86(7):3053–3060. doi:10.1210/jcem.86.7.7645

Thorens B (1992) Expression cloning of the pancreatic beta cell receptor for the gluco-incretin hormone glucagon-like peptide 1. Proc Natl Acad Sci USA 89(18):8641–8645. doi:10.1073/pnas.89.18.8641

Tsai TY, Chu LH, Lee CL, Pan TM (2009) Atherosclerosis-preventing activity of lactic acid bacteria-fermented milk-soymilk supplemented with *Momordica charantia*. J Agric Food Chem 57(5):2065–2071. doi:10.1021/jf802936c

Tsangalis D, Ashton JF, McGill AEJ, Shah NP (2002) Enzymic transformation of isoflavone phytoestrogens in soymilk by β-glucosidase-producing bifidobacteria. J Food Sci 67(8):3104–3113. doi:10.1111/j.1365-2621.2002.tb08866.x

Valsecchi AE, Franchi S, Panerai AE, Rossi A, Sacerdote P, Colleoni M (2011) The soy isoflavone genistein reverses oxidative and inflammatory state, neuropathic pain, neurotrophic and vasculature deficits in diabetes mouse model. Eur J Pharmacol 650(2–3):694–702. doi:10.1016/j.ejphar.2010.10.060

Wang W, Gonzalez de Mejia E (2005) A new frontier in soy bioactive peptides hat may prevent age-related chronic diseases. Compr Rev Food Sci F 4(4):63–78. doi:10.1111/j.1541-4337.2005.tb00075.x

Wang YC, Yu RC, Yang HY, Chou CC (2003) Sugar and acid contents in soymilk fermented with lactic acid bacteria alone or simultaneously with bifidobacteria. Food Microbiol 20(3):333–338. doi:10.1016/S0740-0020(02)00125-9

Wei QK, Chen TR, Chen JT (2007) Using of *Lactobacillus* and *Bifidobacterium* to product the isoflavone aglycones in fermented soymilk. Int J Food Microbiol 117(1):120–124. doi:10.1016/j.ijfoodmicro.2007.02.024

Williams JP, Jordan SE, Barnes S, Blair HC (1998) Tyrosine kinase inhibitor effects on avian osteoclastic acid transport. Am J Clin Nutr 68(6):1369S–1374S

Wong JM, de Souza R, Kendall CW, Emam A, Jenkins DJ (2006) Colonic health: fermentation and short chain fatty acids. J Clin Gastroenterol 40(3):235–243. doi:10.1097/00004836-200603000-00015

Wu QQ, You HJ, Ahn HJ, Kwon B, Ji GE (2012) Changes in growth and survival of *Bifidobacterium* by coculture with *Propionibacterium* in soy milk, cow's milk, and modified MRS medium. Int J Food Microbiol 157(1):65–72. doi:10.1016/j.ijfoodmicro.2012.04.013

Yang HY, Yang SC, Chen JR, Tzeng YH, Han BC (2004) Soyabean protein hydrolysate prevents the development of hypertension in spontaneously hypertensive rats. Br J Nutr 92(3):507–512. doi:10.1079/BJN20041218

Yeo SK, Liong MT (2010) Effect of prebiotics on viability and growth characteristics of probiotics in soymilk. J Sci Food Agric 90(2):267–275. doi:10.1002/jsfa.3808

Fermented Fish Products in Sudan

Ghada Ahmed El Hag Mohamed

Contents

Abstract Fish is one of the main sources for the provision of animal protein for a growing demand in a world of ever-growing population and increasing consumption. In this respect, Sudan is no exception and the various aquatic resources (marine, freshwater, brackish, groundwater, and others) are tapped in order to fulfill the needs in this direction. To this effect, various methods of fishing are encountered and appropriate gears are employed to match with the requirement of the environs, however, the ultimate goal being the maximization of the products and their utilization for the best and economic satisfaction within the framework of a well-defined exploitation strategy.

G.A.E.H. Mohamed (✉)
Faculty of Sciences, Biology Department, King Abdulaziz University, Jeddah, Saudi Arabia

Faculty of Agricultural Technology and Fisheries Science, Department of Fisheries Science, University of Alneelain, PO Box 12702, Khartoum, Sudan
e-mail: ghahmed@yahoo.com

M.-T. Liong (ed.), *Beneficial Microorganisms in Food and Nutraceuticals*,
Microbiology Monographs 27, DOI 10.1007/978-3-319-23177-8_7

1 Introduction

Fish is one of the main sources for the provision of animal protein for a growing demand in a world of ever-growing population and increasing consumption. In this respect, Sudan is no exception and the various aquatic resources (marine, freshwater, brackish, groundwater, and others) are tapped in order to fulfill the needs in this direction. To this effect, various methods of fishing are encountered and appropriate gears are employed to match with the requirement of the environs, however, the ultimate goal being the maximization of the products and their utilization for the best and economic satisfaction within the framework of a well-defined exploitation strategy.

Fish in Sudan have been a major source of protein and energy for many communities especially among the Niloticus tribes and some of the ethnic group of the far North while some depend on fish in lean months of the year, immediately preceding the rains, when sorghum stores ran low (Jackson 1923). At present, the most important fishing activity of Sudan is that concerned with freshwater sources. Marine fish production (from the red sea) was 1050 metric tons of live fish in 1986, whereas the catch from inland sources was 28.66 metric tons (Yousif 1988). The freshwater fish are distributed in an area of about 100,000 km^2, the primary fishing ground being the group of the reservoirs of the north and the vast swamps area (the Nile Sudd) of the south. The reservoirs occupy about 3000 km^2 and swamps about 17,000 km^2 (Yousif 1988), but these areas are reduced after the separation of the Sudan in 2009.

The current Sudanese fisheries practices are focused on the exploitation of the fishes of the inland waters exemplified by the River Nile and its tributaries, the Red Sea coast, and seasonal water spots. The final product of the inland waters fishes varies from one sector to the other. Equally, handling of the fish and subsequent treatments vary as well. Fish processing in Sudan is dominated by men, but women are also involved. Most of the processed products were salted, fermented, or dried product.

2 Spoilage of Fish

Freshly caught fish have a shining, iridescent surface with bright, characteristic coloration and markings. The surface is covered with a thin, whitish but transparent, smooth, homogeneous layer of mucus or slime. The eyes are clear, bright, and full, with a prominent jet-black pupil and transparent cornea, while the gill is bright pinkish red; skin and flesh are firm, moist, and elastic. Hess (1950) stated that most fish flesh, however, is considered more perishable than meat because of more rapid autolysis by the fish enzymes and because of less acid reaction of fish flesh that favors microbial growth. Fish, in general, usually spoil more rapidly than other muscle foods, particularly when mishandled and such spoilage is primarily bacterial in nature.

A delicate fresh fish odor is typical, although sometimes it is described as a feed odor, once associated with the feed of the fish. In the marketplace, fresh fish odors are masked by the bacterial growth on decomposing refuse and slime. Although the fish begins to deteriorate as soon as it is removed from water, the spoiled condition develops gradually (John 1994).

Fish becomes spoiled within 12 h in tropical regions when a complicated series of chemical and bacterial changes triggered by high temperature take place within the fish (ILO 1989). Rigor mortis is especially important in the preservation of fish, for it retards postmortem autolysis and bacterial decomposition. Therefore, any procedure that lengthens rigor mortis lengthens keeping time. It is longer if the fish have had less muscular activity before death and have not been handled roughly and bruised during catching and later processing, and it is longer in some kinds of fish than in others (Akamatsu 1959). Reducing the holding temperature will lengthen the period. Hess (1950) reported that the bacterial spoilage of fish does not begin until after rigor mortis, and then juices are released from the flesh of fibers.

Fresh whole fish are decomposed by several types of enzymatic and microbial activities. While the fish is still in a state of rigor, autolytic enzymes begin breaking down nitrogenous compounds in areas adjacent to the visceral cavity. Meanwhile, bacterial decomposition of the slime layer, the gills, and the intestinal tract get underway and, after rigor is complete, proceeds rapidly (Evelyn and McDermott 1961). In the stages of spoilage, the highly unsaturated fat which is especially abundant in fatty fish becomes oxidatively rancid.

Spoilage depends on the temperature and water activity of the fish and the presence or absence of the various spoilage microorganisms (FAO 1981). The rate of spoilage is related directly to time and temperature. During postmortem autolytic and microbial activity, numerous chemical changes occur in fish flesh, resulting in liberation of end products of digestion or of metabolism. A number of these end products have been suggested as indicators of spoilage.

Spoilage of fish may take place before, during, or after processing (ILO 1989). There are many ways to minimize the spoilage that occurs during these times such as:

1. Improvement of landing facilities and distribution: these will reduce the period of time that may elapse before the fish can be processed.
2. Maintaining the fish at low temperature will minimize the spoilage when fish is kept in cool condition immediately after catching until processing begins.

   In many areas far from towns especially in tropical area, ice may not be available in sufficient quantities, in order to keep fish at low temperature, fish must be kept in the shade, placing damp sacking over the fish to reduce the evaporates and/or mixing the fish with wet grass or water weeds (FAO 1992), so that the fish could be kept continuously wet.
3. Maintaining a hygienic environment. This is achieved by washing the fish with clean water, keeping it cool, and avoiding breaking the skin by bad handling.
4. Keep all tools, fish boxes, boat holds, and cutting tables clean by washing with clean water.
5. Working area should be cleaned regularly.
6. Use salt during drying to avoid spoilage.
7. After processing, appropriate packaging must be used and cool storage area to avoid dust, insects, etc.

3 Preservation of Fish in Sudan

Artisanal fish processing remains the predominant and most important method of fish preservation in Sudan. The principal methods are salting, fermentation, and drying. These processes may either be used alone or combined in order to achieve the desired product. Preservation by salting, smoking, and drying is called curing of fish. This term is defined as fish preserved without the need for refrigeration or freezing, but excluding sterilized products in airtight containers. Cured fish are consumed mainly in tropical countries (FAO 1981). More than 70 % of the actual fish production is consumed fresh and this fish was basically caught from Gebel Awlia, Sennar, Khashim El-Girba, and Roseiris Reservoir. A total of 25 families and 123 species are recorded for inland water fishes (Ali et al. 1996). In India, for example, only about 5 % of marine fish landings are frozen, but 21 % are preserved by traditional curing (Chakraborty 1978).

Fish is one of the most perishable foods and its preservation is usually accomplished by combination of different techniques. Contamination with spoilage microorganisms is almost unavoidable because fish is a very good culture media. The primary purpose of food preservation methods is the creation of conditions unfavorable to the growth or survival of spoilage organisms. In Sudan, the fish preservation stages involved a combination of processing methods, in order to achieve the desired product and to increase the shelf life so that the product becomes available in shortage season production.

4 The Application of Fermented Fish Products Technology

In Sudan, the fermentation of fish with salting and drying still follows the traditional methods, and remains basically an art or craft. Fisherman can become a food craftsman, and by his experience, he's able to make a good product when his basic materials are normal. But any abnormality in materials or conditions will usually leave him helpless, because he has no knowledge of the fundamentals, which might help him to overcome his difficulties. The evolution of food technology will provide the essential link between fundamental knowledge and empirical practice. One of the criteria that distinguish the technologist from the craftsman is that the former is able to predict the effect of his scientific knowledge, whereas the craftsman can only do this by trial and error.

Work in small-scale (processing of fish) sectors technology should have as its main objective: the development and use of simple, cheap equipment for the processing of food products, to obtain salted-dried fish, salted-fermentative, or any other food commodities on a small scale.

4.1 Salting

An alternative to lowering water activity of fish flesh is by merely extracting water as in simple dehydration, which is to increase the concentration of solutes in the flesh. Common salt is the cheapest solutes used for food preservation (Reid 1976).

Salt is the most important of all curing ingredients. It makes up the bulk of the curing mixture because it is not only a good preservative, but it provides the most desirable flavor and its diffusion in fish meat is by the process of osmosis (Rosenthal 1999). The objective of salting is to ensure that the salt penetration is rapid enough to similarly lower the water activity in the deepest parts of the flesh (Doe et al. 1983). The idea of all traditional processing methods is the water activity, which is a measure of water in the foodstuff available to support the growth of microorganisms, until spoilage processes are prevented or at least greatly slowed. Common salt (sodium chloride) has been used as a preservative for a long time. It acts as a bacteriostatic agent when present in sufficient concentration (Beatty and Fougere 1957). This property of salt has been frequently used in food processing and is the basis for the preservation of salted-dried or salted-fermented fish. Food products containing the required amount of salt can be kept at room temperature for longer time than those without. Salt may also be added to the safety of product. Pearson and Gillett (1996) reported that curing and meat processing can be manipulated by using salts and other additives.

The salt concentrations added to fish product differ according to species of fish. Saisithi (1987) stated that salt concentration differs according to different types and batches of fermentation. It is therefore of interest to identify the optimal salt concentration, which does not inhibit the growth of the fermenting microorganisms and prevent the spoilage products; in addition, it contributes positively to the flavor and texture of the product.

4.1.1 The Salting Process

According to the fish composition and size, salting may be dry, where the fish are stacked in salt and the brine formed is allowed to run away, or wet, where they are immersed in a strong brine, or pickle. The most commonly employed method is a hybrid of the dry and wet methods; the fish are placed in dry salt and eventually become immersed in the liquid pickle formed by solution of the salt in the liquid extracted from the fish. This is sometimes called blood pickle (Doe et al. 1983).

Size of the fish determines whether the fish are salted whole and uneviscerated, eviscerated, and split open, or in smaller pieces ranging from fillets to mince. The barrier presented by fish skins to salt penetration means that only small specimens can be salted whole without gutting. Larger fish salted in this manner would deteriorate at the center before salt could exert its effects there. Sodium chloride diffuses through the fish flesh by a dialysis mechanism and water will diffuse to the outside due to the osmotic pressure between the brine and fish muscle solution.

4.1.2 Salting Methods

Salt is one of the most important food additives in food preservation, as its concentration determines what types of microorganism if any can grow by dehydrating or by lowering the amount of water available for growth. Salting requires a little equipment, but the method used is important. There are four types of salting methods (Kench, Pickling, Brining, and Gaspe):

1. Solid or kench salting: in this method, dry salt is rubbed all over the fish to leave the flesh fairly dry. Fish are split, opened out flat, and placed in layers interspersed with layers of salt, and the liquor which exudates is allowed to drain away, along with the shelf life of the dry product is produced. This method is used for non-fatty fish.
2. Pickling refers to fish that are treated with salt brine (Josephson et al. 1986). This method is similar to kench salting, but here the brine formed is not allowed to drain off. Salt is spread over the surface of fish layer, then another layer of fish is placed, and the processes of salting are repeated. The fish must be kept below the surface of the brine; this is done by a good covering of container used. Pickle curing is recommended in preference to kench salting as it produces a more even salt penetration and provides a better protection of the fish against insects and animals since they are covered with brine. Doe et al. (1983) reported that the immersion in concentrated brines for long periods is generally used for longer-term preservation, mainly of fatty fish. The restriction of oxygen access by immersion retards rancidity reactions, although some rancidity is desirable in the development of characteristic flavor.
3. Brine salting: this is a uniform and controllable salting technique. In this method, salting is done by soaking the fish in light or saturated brine. This method is preferred when a fixed concentration of salt in the final product is needed. Brine salting method is commonly used in developing countries when a smoked product is to be made and the salt concentration required in the final product must be lower than 3 % (ILO 1986). Also, in this method the ultimate concentration of salt in the fish is required to be sufficient only for flavoring purposes that can be affected by other techniques such as smoking applied after salting methods, where fish are treated for several minutes only in less than saturated brine.
4. Gaspe curing: Doe et al. (1977) stated that, instead of allowing the exudates liquor to run away, the dry salting is carried out in tubs, where the brine is formed. Weights are used to keep the fish immersed for 2–3 days, after which they are taken out and dried in the sun.

The use of salt in fish processing may be applied by either dry salting (kenching) or wet salting. In dry salting, the granular salt is applied directly to the fish either in the gills, on the surface, or, in the case of split fish, in the belly. The exudates from the fish may be allowed to drain off or be retained in the latter case; the fish becomes immersed in the exudates and this is often referred to as pickling. In wet salting, the fish immersed in brine for up to 2 days or dipped for few hours. It was observed that some processors who cure fish with brine sometimes reuse the salt solution for a number of times. This may be a potential source of bacterial contamination to fresh fish.

Common salt (sodium chloride) has three major effects on fish and meat products. It enhances taste, solubilizes protein to create desired texture, and controls microbial growth to enhance shelf life and inhibit pathogens (Terrel 1983). Rock-Well and Ebertz (1924) stated that the preserving effect of sodium chloride implies more than its dehydration action on the meat.

4.1.3 Salt Quality

There are four types of salt: solar, brine-evaporated, rock, and manufactured salt may be used in the processing of fish:

- Solar salts are prepared by the evaporation of sea or salt-lake water by sun and wind. Lagoons constructed on the shoreline are flooded, sealed off, and allowed to evaporate to dryness. The salt obtained is very impure due to the presence of salts other than sodium chloride, and when it is dug out, may be further contaminated by sand from the bed of the lagoon.
- Brine-evaporated salts are prepared by the application of heat to evaporate strong brines pumped from deep mines. The purity of such salts depends on the nature of the underground deposit, but they are less likely to be contaminated with sand than solar salts.
- Rock salt is mined from underground deposits of varying purity from 80 % to 90 % sodium chloride.
- Purified manufactured salt may contain 99.9 % sodium chloride and derived from any one of the three types of salt mentioned above, which may contain up to one-fifth of their weight of impurities.

The salt used in the salting and fermentation of fish in the Sudan is solar salt (FAO 1989). It is obtained after the evaporation of seawater. The main constituent of rock salt is sodium chloride, although it may also contain certain amount of calcium, magnesium, and potassium salt, carbonates, sulfates, bromides, and iodides, and even traces of heavy metals and oxides may also be found.

Salt causes temporary fixation of protein in fish, to a certain extent hardening tissues and reducing changing in protein; however, adding too much salt will cause the proteins in fish muscle to swell (Hamm 1994). This leads to a product, which is tough and dry. Too much salt will also mask the taste of the fish and the flesh takes on an unpleasant white color. The occurrence of impurities interferes with salt penetration and acceleration of fat oxidation and consequently leads to low-quality product (FAO 1981).

El-tom (1989) stated that solar salt is used in large scale for production of salting fermenting fish in Sudan. According to Beatty and Fougere (1957), bacteria which contribute to spoilage in fresh fish cannot survive at salt concentration above 12 %

(w/w). Abu Gideiri (2001) found that the number of microorganisms increased rapidly during the first fermentation days and then began to decrease.

The quality and quantity of salt used are important factors that affect the organoleptic qualities of salted fish. Thus, the present random use of salt should be checked in order to determine a satisfactory ratio of salt to fish. Also, the size and shape of the fish should be taken into account.

4.2 Fermentation

Fermentation is derived from the Latin verb "fervere" to boil (Peter and Allan 1984). The boiling appearance is due to the production of carbon dioxide bubbles caused by the anaerobic catabolism of the sugars present in the extract. Fermentation has different meanings to biochemists and to industrial microbiologists. For biochemists, fermentation is related to the generation of energy by the catabolism of organic compounds, whereas it means for industrial microbiologist as any process for the production of product by the mass culture of microorganisms.

Fermentation is considered as an easy and low-energy preservation method for meat that results in distinctive products that have an important part in the diet of people making them (Margy 1992). Such fermented meats contribute both nutritional value and pleasure to meals.

The preservation of fish by fermentation methods differs from place to the other. In Africa especially in Sudan, salting and drying of fish is accompanied with fermentation, but the period is short (a few days) and the products are characterized by strong odor (FAO 1992), but in Asia, the fermented fish products last for several months and the fish flesh may be liquid or turned into a paste. These fermented fish are popular and well liked by the general population and so widely used that the daily diet of the people would not be complete without them. In Philippines, where many fermented food products are known, their popularity is attributed not only to their characteristic flavour but also to the fact that other processing methods are generally expensive (Olympia et al. 1992).

All of fermented fish have a pronounced and distinctive taste and odor and have nutrients important for good health. Watanabe (1982) found that the fermented fish products of Senegal are highly salted and semi-dried fish products with an obnoxious odor and cheesy but rich fish flavor recovery of a sun-dried horse mackerel (*kusaya*) from Japan. Also, Toury et al. (1970) reported that the Guedi is reported as a fermented dried fish product popular in Senegal, where the unsalted fresh fish are piled together for about 24 h in the open air. During this period, the fish undergoes putrefactive fermentation by its own enzymes and endogenous bacteria. Then it is eviscerated; sometimes the big species are filleted to shorten the period of drying and soaked in salty seawater in wooden buckets. The water is changed once a week when it has become too dirty. Eventually, the fishes are spread out on straw mats to dry in the sun for 2–4 days.

In many countries of the world especially in Africa, the fermented foods are popular and well liked by the general populace and so widely used. According to Lopez (1987), fermentation provides an avenue to preserve food products, destroy undesirable factors, make safer product, improve appearance taste of some food, and enhance nutritive value, in order to enable the products to be stored at ambient temperature and used without further preparation.

In developing African countries like Sudan, many fermented food products are known and most of the traditional food fermentation industries are rural, seasonal, and informal with low money invested in these sectors. The processing methods were developed in homes and improvements were based on the observations of the practitioners. Fermentation methods of fish in Sudan are normally transferred from one generation to the others.

In Sudan, the fermented fish products can be divided into two groups.

The first group includes those containing high concentration of salts (20–25 %) in the final products like Fissekh and Muluha. The second group includes fermentation product that used drying methods like Kejeik. Generally, different processing techniques are employed in fish fermentation. In Sudan, the fermentation methods differ from one place to another, and all methods are influenced by factors such as availability of salt and the food habits of the local people. Fermentation is often combined with salting or drying in order to reduce water activity and eliminate the growth of prototypical and putrefying bacteria (FAO 1992). The fermentation methods are divided into:

- Fermentation with salting and drying
- Fermentation and drying without salting
- Fermentation with salting but without drying

As regards the situation of salting and fermentation, attempts have been made by a number of researchers like Mahmoud (1977), Johnson (1994), and Abu Gideiri (2001), who have worked on a number of aspects (post-harvest treatments, meat quality of Nile fish, freezing of fish, salting and smoking of fish, drying, salting, and fermentation of fish) and the main finding is that the proximate composition of whole fresh fish was in the range of 63.3–75.2 %, 15.0–22 %, 0.4–2.5 %, and 0.5–1.94 % for moisture, protein, fat, and ash content, respectively.

4.2.1 Salted Fermented Fish and Fish Type

In Sudan, the salted fermented products without drying are called "fessiekh." The product is also known by the same name in the Middle East (Compbeel-platt 1987) and in Egypt (Hassan et al. 1972; Hamed et al. 1973). Fessiekh is a wet salted product, soft in texture with a strong pungent smell, and a shiny silvery appearance. It can be stored for more than 3 months (Kofi 1992). Fessiekh is a popular fermented product in Sudan when compared with other fish products.

The technique of fessiekh probably entered Sudan during the Turk-Egyptian rule (1821–1885), but its production on a large scale was only well established during the Anglo-Egyptian condominium rule (1898–1956).

The methodology in the preparation of the product was traditionally carried through from one generation to another.

Commonly there are two types of fish used in fessiekh making in Sudan: *Hydrocynus* species (tiger fish) known locally as “Kass” and *Alestes species* (pebbly fish) known as kawara, both belonging to the family Characidae. These two types of fish are considered the first class of fessiekh (Agab and Bashir 1987). The second class consists of other fish.

The totality of fessiekh in the Sudan is made from Nile fish. Only one kind of fish from the Red Sea can be used to make the product. It is the mullet (*Mugil cephalus*) or known as “alarabi.” About 80 % of fessiekh is made from *Alestes baremose* and *A. dentex*, while *Hydrocynus* sp. makes the balance; this is because kawara is always more abundant than Kass in any catch. Fessiekh processing is a seasonal process generally starting in November and ending in June, with a peak in February and sometimes in March. The ratio of salt used depends on the experience of commercial sector and it varies widely. The salted fishes allow fermenting and the fermentation time varies according to the season. Dirar (1993) and EL Hag et al. (2012) mentioned that the first stage of fermentation takes about 3–4 days in the summer and 4–7 days in the winter.

Preparation of Fessiekh

Processing of fesseikh is carried out in temporary sheds for cool environment. During the nineteenth century, the technique of making fessiekh was very simple. After catching, fish was spread on palm-date mats, and then salted them by adding a large quantity of salt. Later after 3–5 days, the product was packaged in an old oil can for around 1 week depending on the temperature, whereas the other method raps up the salted fish in palm-date mats and buries them in deep pits called (Mutmura), for several weeks before packaging in oil cans, and chilies are added in some products. At present, most of the practitioners use plastic containers to prepared salted fermented fish product and about 20–30 % of salt is used (EL Hag et al. 2012). The fish are left to ferment for 3–10 days depending on temperature. Liquid exudates from the fish are allowed to drain off. The product is packaged into polyethylene bags (1–2 kg) for sale in local markets. The Sudanese prepare fessiekh by mincing and boiling with vegetable (tomatoes and spices) to prepare sauce that is eaten with bread (Dirar 1993).

4.2.2 Meluha

Meluha or Terkin is the famous Sudanese liquid or fermented pastry product found. Terkin is defined as a wet pasty mixture of fish muscle and bones which loses moisture during storage and becomes more viscous and dark. The Sudanese salted fermented fish product falls under the category of paste called Terkin, but if the product (Terkin) becomes more liquid, it is called Meluha like nam-pla and bagoong product in south-east Asia (Beddows 1985).

The fermented product uses the same fish which are used in fessiekh (kawara, kass, dibs, and shilbya) and fatty fish were preferred than others. The best product of Terkin is made from small fish with low fat.

After evisceration and removal of scales of fish, salt concentration about 1–4 kg/ body weight is added. Sometimes, salted fish is mixed with pepper and put in suitable covered container (plastic tine) and incubated.

Preparation of Meluha/Terkin

Fermentation occurred at under 37 °C for 10–15 days. These methods will take a longer time in summer than winter season (Kofi 1992; Dirar 1993). Then, the container of the curd fish is placed outdoor under the sun or cooked on an open fire to speed up the fermentation processes. The process continued for about 4–5 days and the cured fish is mixed continuously until the homogenous paste is obtained. The pasty mixture of the muscle and bones produced will have a dark color appearance and a very strong odor.

There are three methods to prepare Meluha/Terkin dish. The first one is known as cooked Meluha, after removal of bones and scales from the product by adding water and straining it; a paste of onion is added to the strained Terkin. Terkin is cooked with tomato sauce until the mixture becomes thicker, then mixed with wheat flour or peanut butter, and eaten with wheat bread. In the second way, chopped onion, cumin, black pepper, and peanut butter, all mixed together to give thick sauce without cooking, are added to the strained Terkin and then eaten with some vegetable and bread. The third method is similar to the second one, where the mixture is cooked for 2–3 min only.

4.3 Drying

Drying preserves food by removing the water that is needed for microbial growth and enzyme activity. According to (Doe et al. 1983) there are three types of processes that can be employed in the drying of fish:

- Air or contact drying, where heat is transferred to the fish from heated air or heated surface, utilizes the air movement above the fish to carry away the moisture.
- Vacuum drying, where advantage is taken of the greater evaporation rate of water from the fish at reduced pressure, utilizes conduction by contact with

heated surfaces or radiation to evaporate the water, which is removed by the vacuum pump.

- Freeze-drying relies upon the attainment of very low pressures by highly efficient vacuum pump in sealed chamber containing the fish.

Fish can be dried with or without salt. The process of drying in tropical countries depends on spreading the fish to sun. Drying allows moisture to evaporate from the flesh surface. In temperate countries, fish can be hung up to dry in the wind. The process highly depends on the wind speed. This process may take longer time than drying under the sun, but the fish take longer time before spoilage. This is due to lower ambient temperature (FAO 1986).

Fish drying rate depends on the speed of water leaving the fish surface and on the rate at which water diffuses from the center of the fish to the surface. The rate of diffusion of small fish is higher compared to large ones. These are due to decreased distance from the center to the surface of the fish (Waterman 1976).

Dried fish is particularly suited for low-income groups who cannot afford expensive fish products. Simple drying of fish is not practical in many of the world's river systems because of the high atmospheric humidity, but in desert or Sahelian Savanna Rivers, the practice is common, especially for smaller species (FAO 1986). In Senegal, basin simple drying is the usual form of treating fish after they have been eviscerated, scaled, and in the case of larger species, cut into strips. In the Niger, only the smaller species such as *Alestes* are sun dried, while in the Chad basin, one of the traditional fish products "salanga" consists of *Alestes dentex* and *Alestes baremose*, which are split open ventrally and laid on mats to dry in the sun. In the Mekong and other Asian rivers, sun drying is also common, although this is sometimes combined with salting. Sun drying is the method used by all fish processors in Sudan. However, it is ineffective as the moisture level of the fish is not reduced to the level that can prevent spoilage of product by microorganisms (Moy 1977). Moreover, the humidity in tropical countries is very high. The average humidity reaches 85–90 % in some countries and thus the salted-dried fish will tend to absorb moisture from the atmosphere when left on the shelf or in storage. This can increase the moisture content of fish and subsequently import a wet appearance to the surface of the fish.

Simple improvement, such as the use of drying racks raised above ground level, can increase drying rates and reduce contamination, thus helping to make products of good quality (FAO 1981). Peter (1997) stated that the dried food can have poorer nutritional and eating quality than the corresponding fresh food, so the correct design and operation of dryers are therefore needed to minimize quality changes to the food.

Tunnel drier have been recommended by FAO for the drying of fish in tropical countries (Waterman 1976). According to this report, tunnel drier has been used in Brazil, Cambodia, and many African countries. The air speed and humidity in the drier can be controlled to suit the drying rates required. The temperature can be adjusted by the use of heaters. In addition to the use of tunnel drier, polythene tent drier can also be used without much capital outlay. They have been used in

Bangladesh, Hawaii, Australia, New Guinea, and some African countries (Doe et al. 1977; Moy 1977). Tent drier can prevent contamination and reduce infestation by flies. The internal temperature is about 45–50 °C. At this temperature, growth of spoilage microorganisms is retarded and the rate of drying is increased. Drying times vary considerably depending on the process and weather, but 3–10 days is generally required to prepare a typical sun dried product (Waterman 1976).

Jan et al. (2001) have found that it is moulds rather than bacteria that cause spoilage during the preparation of dehydrated fish. Open-air drying method is time-consuming. Drying times are considerably different depending on weather condition. Slow drying leads to fish spoilage by microbial contamination.

Attempts to improve open-air drying practices in developing countries have not changed much from traditional open-air drying. Substantial losses can occur in both quality and quantity with subsequent financial losses (FAO 1981). Sun drying as a method of food preservation still has many limitations, even when racks of salt are used (ILO 1986). Long periods of sunshine without rain are required, drying rates are low, products are often of low quality as a result of slow drying, insect damage, contamination of airborne bacteria, dust, etc. It is often too difficult to obtain uniform product.

4.3.1 Kejeik

The Sudanese famous fermented drying fish product is called Kejeik. The most common traditional preservation method used for flesh fish in Sudan is drying. The Niloticus tribes of southern and some part of northern Sudan preserve fish by sun drying of the large fish and the product is called Kejeik.

The product is a hard dried and the color depends on the species of fish used. Black Kejeik is made from Germut (*Clarias anguillaris*, *C. Iazera*), Nauk (*Heterotis niloticus*), Humar el. Hout (*Auchinoglanis biscutatus*, *A. occidentalis*), and Surta (*Heterobranchus bidorsalis*), where all these species of fish are considered as a first class to prepare the best product while the second class product is a white product which is made from Bayad (*Bagrus bayad*), Kharish (*Distichodus niloticus*), Bulti (*Tilapia niloticus*), and Dabs (*Labeo niloticus*) (Omer 1984). The final product can be stored for about 3 months or more.

Preparation of Kejeik

Large fish species are gutted, beheaded, and split longitudinally and washed, whereas small fish are dried as a whole fish after removal of the gut. Then the fish is sprinkled with salt or dipped into salt solution and left to ferment for 3 days before drying. The fermented fish are hung on ropes or tree branches in the open air under the sun for 7 days or more Abu Gideiri (1993).The products are strongly flavored and the odor is different according to product color. Different dishes can be prepared from Kejeik which is commonly consumed as condiment or flavoring agent in certain dishes by adding this product to different cooked that use onion and different vegetables and eaten with bread or cooked sorghum flour.

5 Nutrition and Safety Considerations

The preparation of fermented foods predates the recorded history of Man. Early humans used observation of the apparent effect of microbial alteration of food characteristics to develop processes for food fermentation. Fermented products normally have a different texture and flavor compared to the unfermented starting materials, thus making them more palatable and digestible and prolonging their shelf life (Amano 1962).

Recent advancements in microbiology and biochemistry have improved the safety and efficiency of bio-processing traditional fermented foods.

Microbial fermentation has played an important role in food processing for thousands of years. Fermentation of fish preserves fish products, enhances nutritive value, destroys undesirable factors, makes a safer product, improves the appearance and taste of fish, and reduces the energy required for cooking (Beddows 1985).

During fermentation, the microorganism secretes hydrolytic enzymes into the substrate and assimilates the fatty acids and amino acids while simple sugars are accordingly liberated. These processes are converted into microbial structural components and secondary metabolites. Lactic acid fermentation used for meat is an ancient process whereby a varied group of bacteria ferment carbohydrates producing lactic acid as the major end product (Hall 1989). The organic acid produced during the fermentation of fish in Mali was mainly acetic acid, whereas lactic acid appeared in Asian fermented fish produced (Zakhia and Cua 1991).

Fermentation increases the quantity of soluble protein in foods. It may improve the amino acid profile and reduce the levels of certain antinutritional factors that interfere with digestion. During natural fermentation, food poisoning flora and coliforms may also grow with the lactic acid (Wang and Hesseltine 1981), so the microorganism needs to be eliminated to make fermented foods safe for consumption. There are several factors that contribute to the safety of fermented foods: (1) Soaking and cooking: Soaking and cooking treatment reduce the in situ microbial contamination. (2) Salting: Various fermented foods are made with the addition of salt, which acts as preservative. (3) Acid fermentation: Much fermentation is carried out by acid-producing microorganisms where these organic acids act as preservative or as bacteriostatic agents. (4) Low moisture content: The lower water activity may be important preservative factor.

Safe products are obtained when using hygienic conditions during, handling, and storage (Samson et al. 1987). Kofi (1992) reported that there are various types of salts used for salting and fermentation of fish. Solar salt, which is most widely used in fish preservation, has been found to contain the largest amount of microorganisms. The general bacterial flora of solar salt mostly comprises bacillus type (95 %

with remainder being micrococcus and carcina type). The most important spoilage organism that always presents in solar salt is the red halophilic bacteria. Dry fish can be stored for nearly 6 months, but the soft or semi-dry ones only have a shelf life of up to 3 months (Essuman 1992). One of the main quality parameters that assessed the quality of fermented fish is organoleptic inspection (color and oxidative rancidity).

5.1 Color

The color of fish is an important quality attribute, which affects consumer buying fish. The color of fish flesh is governed by blood and melanin derivatives (produced from skin coloration) or arises directly from its diet. Pink flesh color arises from the carotenoids, which enter food chain originally from photosynthesis (John 1994). The chemical groups involved in the color of fish flesh are heme pigments, carotenoid, and melanin. As in all foods, the specific colors produced are influenced by physical structure. Kropf (1980) reported that the color is the most important factor for consumers when purchasing meat. The color of the product depends on the species of fish used as well as the processing method. For whole products such as fessiekh, a silvery appearance close to the fresh product is considered high quality. Poorly fermented products tend to be grayish or dark, while fermented products which are split and dried are usually light brown in color. Long storage and further drying darkens the product. *Yeet*, for instance, becomes dark brown after weeks of exposure to the sun (Essuman 1992).

The method of catching and handling can be responsible for flesh bruising discoloration. Botta et al. (1987b) found that cod caught by gillnet are significantly more bruised than those caught by trap, hand-line, or long-line. Fish which have struggled extensively before being brought for gutting gives lower color grades (Botta et al. 1987a). On the other hand, blood contained in the vessels of freshly caught fish remains red while holding the fish in ice before filleting results in a darker red or brown color as the hemoglobin oxidizes.

Normally within the fish itself, the flesh is of two types, dark and light on frozen storage; pigments in the darker meat are especially subject to oxidation, becoming deep yellow or brown. The color of pink or red fish is sensitive to frozen storage abuse; the pigments oxidize; and the color can disappear (Hillman 1983). Impurities present in common salt can have an accelerating effect on the oxidative deterioration of frozen fish. Anon (1970) suggested that only high-quality salt should be used for brine dipping of species particularly susceptible to oxidation. Mixtures of tocopherols and ascorbic acid have been used to prevent oxidation of red fish (Wasson et al. 1991).

Salt from the saline deserts to the north contains nitrates and borax as impurities, and these would be converted to the nitric oxide which is necessary for the formation of red complexes heme pigments. Washing minced flesh of fish has a beneficial effect on the color, but significant quantities of water are needed (Martin

1976). The presence of blood, kidney tissue, or the black lining of the belly cavity can also cause darkening of the flesh (King 1973; Dyer 1974).

Oxidation of the blood pigments may be the cause of the yellow and brown color, which develops during storage of fish (Jauegui Carlos and Baker 1980).

Fish oil color varies among species, but the pigments of all fish can become yellow or brown during oxidation. While the whole fish or the fillet is deep frozen, oil can be forced out of the tissue and accumulate on the skin; these fish are rusted (Wasson et al. 1991).

5.2 *Oxidative Rancidity*

The term rancidity is used to describe the task or smell of rank stale fat. Rancidity is associated with a characteristic, unpalatable odor and flavor of the oil. Virtually all fish contain highly unsaturated fatty acids as major components of their lipids. However, the total amount of fat may vary greatly (Lands 1986). Obviously fatty fish such as mackerel and salmon will be more susceptible to appreciable oxidation. Hess (1950) reported that fat and oils of many kinds of fish, especially the fatter ones, such as herring and salmon, are composed of a great extent of unsaturated fatty acids and are subjected to oxidative changes, thus producing oxidative rancidity and sometimes undesirable alteration in color.

Oxidative spoilage of salted fish can occur during processing as well as during storage (Van Arsdel et al. 1973). The rancidity is caused by the changes that occur during the reaction with atmospheric oxygen and is so-called oxidative rancidity. The off flavours are produced by hydrolytic reactions, which are catalyzed by enzymes is the so-called hydrolytic rancidity (Huss and Valimarson 1990). Rancid flavors are chemically very complex, since they are derived from any or all of the unsaturated fatty acids originally present in the oil. Each of these can oxidize through several different mechanisms (Clucas and Ward 1996).

Rancid fat contains a wide variety of chemical substances, whose structures are not all known (Hultin et al. 1982). Human taste buds are highly sensitive to some compounds such as lactones and free fatty acids; therefore, minute amounts of these compounds are sufficient to spoil the taste of food (Harris and Tall 1994). Fresh and saltwater fish contain significant levels of polyunsaturated fatty acids. Consequently, such fish loss their polyunsaturated fatty acids through lipid oxidation during storage (Smith et al. 1990). At ambient temperatures storage, the development of rancidity is not a major problem with fresh fish, because the normal level of microbial flora associated with fresh fish is such that the bacterial spoilage will render it inedible before rancidity has proceeded to any great extent. However, rancidity development is very important in the storage stability of dried or frozen fish.

Fey and Regenstein (1982) stated that the total level of fat in fish varies depending on the species of fish and season. Fish can be divided into low and high fat species. Low fat species usually have less than 4 % total fat and more than

5 % represented high fat species. Any lipid in addition to this level will be made up of triacylglycerides, which are primarily deposited in the liver and under the skin and act as food reserve. Seasonal variation in fat level is due to the availability of food. In addition, environmental conditions, seasonal variations, and sexual dimorphism have also been reported to cause variations in fat deposits of fish. This fat can contain highly unsaturated fatty acids, all of which can make fish very susceptible to rancidity.

Lobuza (1971) found that fish tissue susceptibility to rancidity does not only depend on the amount of lipid present, but also the lipid composition and its location in fish tissue matrix. Different lipid component, such as tocopherol and carotenoid pigments are also associated with autoxidative reaction in fish tissue. These components may be involved in controlling oxidative rancidity, via self-oxidation or co-oxidized with lipid.

Fish lipids are even more susceptible to oxidation and are probably responsible for the rapid spoilage of oily fish such as mackerel and herring. The 20–22 polyunsaturated fatty acids are highly susceptible to peroxidation. The high concentration of 20–22 polyunsaturated fatty acids is found in fish and fish oils; many fish oils contain about 20 % of their total fatty acids as these higher polyunsaturated fatty acids. This explains the susceptibility of fish to oxidative rancidity (FAO 1962). Hence, fish oils appear to be more susceptible to oxidative deterioration than most animal fats (Hess 1950).

5.3 Microbiology of Fermented Fish Products

The most important factors in consumer acceptability are flavor and odor, although the color can be significant. As aseptically a product of fermented fish did not give the typical aroma associated with them, it was anticipated that microorganisms were involved in aroma development (Beddows et al. 1979).

A considerable number of investigators have reported on the microbiology-induced changes in spoiled and fresh fish, but the work on changes in a high salt environment, as found in salt-fermented production, is very limited. Oetterer et al. (2003) stated that the high concentration of salt was used to maintain the product under adequate microbiological control. The contamination comes from the fish, which ordinarily introduces species chiefly consisting of *Pseudomonas*, *Alcligenes*, *Flavobacterium*, and *Corynebacterium* from ice, and from mechanically introduced sources, e.g., dust, which add cocci Eddy (1958).

Boez and Guillerm (1930) isolated an anaerobic spore-bearing bacterium from early stage of fermentation of a fish sauce (nuoc-mam), which was accredited as being a species of *Clostridium*. Aseptic preparation failed to give the typical nuoc-mam aroma, so it was concluded that bacteria must be involved in fermented fish (Beddows 1985).

Krempt (1929) tried to produce a sauce more quickly with a true aroma, on a commercial scale with some degree of success and he concluded that microorganisms were required for aroma production.

Beddows (1985) suggested that the factors (pH of the pickle, temperature, and the salt concentration), which can influence the relative activities of proteolysis enzymes present in the material used, could markedly influence taste of the food.

Saisithi et al. (1966) isolated 10 species of *Bacillus*, 1 *Coryneform*, 2 *Sreptococci* and 1 each of *Micrococcus* and *Staphylococcus* from fish sauce (nam-pla). They stated that these bacteria could have been derived from the solar salt used, since they found that the salt contained an average of 2700 bacteria/g. The total viable count decreased to 2×10^3 cell/cm^3 after 9 months. Crisan and Sands (1975) examined the microflora of four fermented fish. From nam-pla, they isolated *Bacillus cereus* and a strain of *B. licheniformis* after 7 months of fermentation, but at the end of the fermentation period, they isolated another strain of *B. lincheniformis*, *B. megaterium*, and *B. subtilis*. All the isolated species were then grown on a medium containing 10 % salt to which they seem to be halotolerant and not halophilic. No growth was achieved on a medium containing 20 % salt. The primary role of LAB is to ferment the available carbohydrates and thereby cause a decrease in pH. The combination of low pH and organic acids (mainly lactic acid) is the main preservation factor in fermented fish products. Lactic acid bacteria (LAB) are also found as the dominant microorganisms in many fermented fish products (Paludan-Müller et al. 2002).

Crisan and Sands (1975) stated that investigation of patis (fish sauces produced in Philippines) after one month showed no organism growth on 20 % salt, but single strains of *Bacillus pumilus*, *Micrococcus copoyenes*, *M. narians*, and *Candida clausenii* were isolated from 10 % salt medium. Abd-Allah (2011) reported that the microbial load detected in Egyptian fermented salted Mugil cephalus fish (fessiekh) consisted of *Staphylococcus equorum*, *Bacillus subtilis*, *Lactobacillus* sp., *Teratogenococcus halophilus*, *Clostridium bifermentans*, *Clostridium* sp., *Clostridium butyricum*, and *Clostridium cochlearium*. *Staphylococcus* sp. were isolated from thai-fermented fish with salt concentration above 5 % (Tanasupawat et al. 1991, 1992) and from Korean fermented (hydrolyzed) fish with salt concentrations ranging from 8 % to 26 % NaCl. Akanda and Oladosu (1988) stated low incidence of *Staphylococcus aurous* in salted fish at 30 °C storage for 8 weeks. At low water activity in the presence of high salt concentration, microbiological problem should be non-existent (Zapata et al. 1990). Also, Scott (1957) reported that the growth of microorganisms was much dependent on water activity and temperature.

6 Conclusion

Fermented fish product is much appreciated by the local populations as a flavouring agent because of its exceptional flavor and taste. Among the various fermented food processed in Africa, the fermented fish products are one of the oldest and most widespread used as condiments or main sources of protein in Sudan. The fish products were carried out in artisanal way and the processing methods varies from one place to another. Fermentation was also found to be an important method for fish preservation particularly because unpopular species of fish are usually processed in this way and especially used in situations where drying of fish is not possible because the climate is too wet and where cooling and sterilization of the product is too expensive. The preservation methods of fish in Sudan still need efforts and research to be done, for example, to address the magnitude of losses resulting from spoilage that are likely to occur during the post-harvest treatment by provisions that cut them down to the minimum level through the adoption of salting and fermentation methods of preservation, and to obtained progressive promotion of the traditional treatment in the direction of the following scientific approaches based on specific parameters such as the salt quality, salt concentration, fermentation time and impact on the overall nutritive value.

References

Abd-Allah SMS (2011) Bacterial-flora of Egyptian salted Mugil Cephalus Fish (Fessiekh) PCR-identification. Assiut Vet Med J 57(130):116–137

Abu Gideiri OO (2001) Some biochemical and microbiological aspects of fesseikh industry in Sudan. B.Sc. (Hons.) Dissertation, Department of Zoology, Faculty of Science, University of Khartoum, Sudan

Abu Gideiri YB (1993) Fisheries of the Sudan, present and future: food and nutrition in Sudan. I. National food seminar, pp 20–25

Agab MA, Bashir RB (1987) Traditionally salted fermented fish fesseikh. Sudan J Food Sci Technol. Research Centre. 17, Khartoum, Sudan

Ali ME, Babiker SA, Tibin A (1996) Body characteristics, yield indices and proximate chemical composition of commercial fish species of Lake Nubia. In: FAO 6th consultation on fish technology in Africa. Kisumu, Kenya, pp 27–30

Amano K (1962) The influence of fermentation on the nutritive value of fish with special reference to fermented fish products of South-East Asia. In: Fish Nutrition, ed., E. Heen and R. Kreuzer, Fishing News (Books), London. 180. Cited by Mackie, I. M.; Hardy, R. and Hobbs, G. (1971). FAO Fisheries Report No. 100, Fermented Fish Products

Akamatsu M (1959) Bacteriological studies on the spoilage of fish sausage. 1. Number of Bacteria Present in the Meat of Fish Sasuage on the Market. Jpn Soc Sci Fish Bull 25:545–548

Anon (1970) Refrigeration of fish. Part 3. US Department of Interior, Fishery Leaflet No. 429

Beatty SA, Fougere H (1957) The processing of dried-salted fish. Bull. No. 112, Fisheries Research Board of Canada

Beddows CG, Ardeshir AG, Wan JD (1979) Biochemical changes occurring during the manufacture of 'Budu'. J Sci Food Agric 30:1097–1103

Beddows CG (1985) Fermented fish and fish products. In: Wood BJB (ed) Microbiology of fermented food, vol 2. El Sevier Science Publishing, New York, NY, pp 1–40
Boez L, Guillerm J (1930) Le Facteur Microbine Dan La Fabriction De La Saumare Indochinoise ('nuoc-mam'). C R Acad Bulg Sci 190:534–535
Botta JR, Kennedy K, Squires BE (1987a) Effect of methods of catching and time of seasons on composition of Atlantic Cod (*Gadus morhua*). J Food Sci 52:922–927
Botta JR, Bonnell G, Squieres BE (1987b) Effect of methods of catching and time of season on sensory quality of fresh raw Atlantic Cod (*Gadus morhua*). J Food Sci 52:928–931
Chakraborty PK (1978) Technological development of artificial and solar drying of fish in India. Proc IPFC 18(3):322–329
Clucas IJ, Ward AR (1996) Post-harvest fisheries development: a guide to handling, preservation, processing and quality. Natural Resources Institute (NRI), Kent
Compbeel-platt G (1987) Fermented foods of the world: a dictionary and guide. Butter-Worths, London
Crisan EV, Sands A (1975) The microbiology of four fermented fish sauces. Appl Microbiol 29:106
Dirar HA (1993) Indigenous fermented food of the Sudan. Fish product fesseikh. Faculty of Agriculture, University of Khartoum, Khartoum, Sudan, pp 347–366
Doe PE, Ahamed M, Muslemuddin M, Sachithanathan K (1977) Polythene drier for improved sun drying of fish. Food Technol 29(11):437–441
Doe PE, Curran CA, Poulter RJ (1983) Determination of the water activity and shelf-life of dried fish products. J Food Technol 17:202–208
Dyer WJ (1974). Stability of minced fish, an important factor in standard formulation. In: Second technical seminar on mechanical recovery and utilization of fish flesh. National Fisheries Institute, Washington, DC
Eddy BP (1958) Microbiology of fish and meat curing brines. In: 11th International symposium. Food microbiology proceedings, London. AVI Publishing, Westport, CT
EL Hag GA, Abu Gideir YB, Ali ME, AbU Zeid IM (2012) Nutritive value and microflora of salted kawara (Alestes sp.) during storage. Researcher 4(2):69–76
Essuman KM (1992) Fermented fish in Africa: a study on processing, marketing and consumption, FAO Fisheries Technical Paper, 320, 80
Evelyn TPT, Mc Dermott LA (1961) Bacteriological studies of fresh water fish. 1. Isolation of aerobic bacteria from several species of Ontario fish. Can J Microbiol 7:375–382
FAO (1962) Fish in nutrition. In: Heen E, Kreuzer R (eds) Fishing News (Books), London
FAO (1981) The preservation of losses in cured fish. Technical Paper No. 219
FAO (1986) The production of fish meal and oil. Fisheries Technical Paper 142
FAO (1989) Solar drying of fish and paddy. FAO, Environment and Energy Paper, (10)
FAO (1992) Fermented fish in Africa. A study on processing, marketing and composition. FAO Fisheries Technical paper No. 329, 57
Fey MS, Regenstein JM (1982) Extending shelf life of fresh fish wet red hake and Salmon using Co_2–O_2 modified atmosphere and potassium sorbet ice at 0–1 °C. J Food Sci 47:1048–1054
Hall RJ (1989) Application of biotechnology to traditional fermentation. In: Rogers PL, Fleet GH (eds) Biotechnology and the food industry. Gordon and Breach Science Publishers, New York, NY, pp 241–277
Hamed MG, Elias NA, El-Wakeil FA, Foda IO (1973) Studies on the microorganisms in salted mullet fish feseeikh. J Food Sci 1:1–16
Hamm R (1994) Influence of pH on the protein net charge in the myofibrillar system. Reciprocal Meat Conf Proc 47:5–9
Harris P, Tall J (1994) Substrate specificity of mackerel flesh lipoxygenase. In: Allen JC, Hamilton RJ (eds) Rancidity in foods, vol 3. Chapman and Hall, London, pp 257–271
Hassan YM, Abdel-Al A, Awad AT, Awad AA (1972) An Introduction to agricultural research carried out in Egypt (1900–1970). Food and dairy industries. The National Center for Information and Documentation, Egypt

Hess E (1950) Bacterial fish spoilage and its control. Food Technol 4:477–480
Hillman H (1983) Kitchen Science [Boston: Houghton Mifflin] B.H. Himelbloom, J.E. Routledge and S.L. Biede (1983) Colour Changes in blue Carps (*Callinectes sapidus*) During Cooking. J Food Sci 48:652–653
Hultin HO, Mcdonald RE, Kelleher SD (1982) Lipid oxidation in fish muscle microcosms. In: Martine RE, Flick GJ, Hebart CE, Ward DR (eds) Chemistry and biochemistry of marine food products. AVI Publishing, Westport, CT, Chap.1.1
Huss HH, Valimarson G (1990) Microbiology of salted fish. FAO Tech Pap 10(1):190
International Labors Organization (ILO) (1986) Solar drying practical methods of food preservation, Ist edn. Genva, Swiss
ILO (1989) Small-scale processing of fish. Technical Momorandum. Geneva, No. 3:13–41
Jackson HC (1923) The Nure of the Upper Nile Province, Part C. Sudan Notes and Records 6:59189
Jan P, Nedyalka Y, Micheal G (2001) Antioxidants in food. Wood Heat Publishing, England, pp 421–450
Jauegui Carlos A, Baker C (1980) Discolouration problems in mechanically deboned fish. J Food Sci 45:1068–1069
John HB (1994) Food colour and appearance, 1st edn. Blackie Academic and Professional, London, pp 284–318
Johnson WA (1994) Freezing and refrigerated storage in fisheries. FAO, Fisheries Technical Paper No. 340. FSCL. Food Science Laboratory
Josephson DB, Lindsay R, Olafsdotter G (1986) Measurement of volatile aroma constituents as a means of following sensory deterioration of fresh fish and fishery products. In: Kramer DE, Liston J (eds) Seafood quality determination. Elsevier, New York, NY, pp 27–47
King FJ (1973) Improving the supply of minced block for the fish stick trace, a progress report. Mar Fish Rev 38(8):26
Kofi ME (1992) Fermented fish in Africa, FAO Fisheries Technical Paper 329
Krempt A (1929) Traitement Industriel Des Produits De La Peche. Rapport Sr La Fon Ctionnement Du Service Oceonographique Des Peches I'indochine 13:36–43
Kropf DH (1980) Effect of retail display conditions on meat colour. Proceedings of the Annual Reciprocal Meat Conference (RMC) of the American Meat Science Association 33:15–32
Lobuza TP (1971) Kinetics of lipid oxidation in foods. CRC Crit Rev Food Technol 2(3):356
Lopez A (1987) A complete course in canning book: II and III, 12th edn. The Canning Trade, Baltimore, MD
Lands WEM (1986) Fish and human health. Academic, London, pp 1–170
Mahmoud ZN (1977) Studies on meat quality of some common Nile fish. M.Sc. Thesis, Department of Zoology, Faculty of Science, University of Khartoum, Sudan
Margy W (1992) Starter cultures in traditional fermented meats. In: Applications of biotechnology to traditional fermented food, report of an ad hoc panel of the board on science and technology for international development. National Academy Press, Washington, DC, pp 128–159
Martin RE (1976) Mechanically-deboned fish flesh. Food Technol 30(9):64–68
Moy JH (1977) Solar dehydration vs-radiation preservation of foods. In: Proceedings of Conference Food Agriculture, Malaysia, Kuala Lumpur, 2000
Olympia M, Ono H, Shinmyo A, Takano M (1992) Lactic acid bacteria in a fermented fishery product, Burong bangus. J Ferment Bioeng 73:193–197
Oetterer M, Perujo SD, Gallo CR, Ferraz de Arruda L, Borghesi R, Paschoal da Cruz AM (2003) Monitoring the sardine (*Sardinella brasiliensis*) fermentation process to obtain anchovies. Sci Agric 60(3):511–517
Omer AS (1984) Preliminary studies on the chemical composition of the flesh of Hydrocynus Forsklii (Cuvier and Valenciennes, 1986). B.Sc. Honors Dissertation, Department of Zoology, University of Khartoum, Sudan

Paludan-Müller C, Madsen M, Sophanodora P, Gram L, Møller PL (2002) Genotypic and phenotypic characterization of garlic-fermenting lactic acid bacteria isolated from Som-Fak, a Thai low-salt fermented fish product. Int J Food Microbiol 73(1):61–70
Pearson AM, Gillett TA (1996) Processed meat, 3rd edn. Chapman and Hall, New York, NY
Peter F (1997) Traditional food. Intermediated Technology Publication, Southampton Row, London, pp 103–105
Peter SF, Allan W (1984) Principle of fermentation technology. Pergamon Press, Oxford, pp 1–10
Reid DS (1976) Water activity concepts in intermediate moisture foods. In: Davies R, Birch GG, Parker KJ (eds) Intermediate moisture foods. Applied Science, London, pp 54–65
Rock-Well GE, Ebertz EG (1924) How salt preserves. J Infect Dis 35:573
Rosenthal AJ (1999) Flesh foods and their analogues. In: Greaser ML, Pearson AM (eds) Food texture measurement and perception. Aspen Publisher, Gaithersburg, MD, pp 228–233
Saisithi P (1987) Traditional fermented fish products with special reference to Thai products. Asian Food J 3:3–10
Saisithi P, Kasemsarn B, Liston J, Dollar AM (1966) Microbiology and chemistry of fermented fish. J Food Sci 31:105–110
Samson RA, Kooij JA, Boer ED (1987) Microbiological quality of commercial tempe in the Netherlands. J Food Prot 50:92–94
Scott WJ (1957) Water relations of food spoilage microorganisms. Advantage Food Research, Academic Press, New York, NY
Smith G, Hole M, Hanson WH (1990) Assessment of lipid oxidation in Indonesian salted-dried marine catfish (*Arius thalassinus*). J Sci Food Agric 51:193–205
Tanasupawat S, Hashimoto Y, Ezaki T, Kozaki M, Komagata K (1991) Identification of *Staphyococcus carnosus* strains from fermented fish and soy sauce mach. J Genet Appl Microbiol 37:479–494
Tanasupawat S, Hashimoto Y, Ezaki T, Kozki M, Komagata K (1992) *Staphylococcus piscifermentans* sp. Nov., from fermented fish in Thailand. Int J Bacteriol 42:577–581
Terrel RN (1983) The water relation of food borne bacterial pathogens. A review. J Milk Food Technol 36(5):276–288
Toury J, Ware A, Gorgi R, Cros J (1970) Fish in the diet in senegal, quantitative and qualitative aspects, methods of preservation. Food Nutr 8:6–13
Van Arsdel WB, Copley MJ, Morgan AI (1973) Food dehydration, practices and applications. AVI Publishing, New York, NY
Wang HL, Hesseltine CW (1981) Use of microbial culture: legume and cereal products. Food Technol 33(1):79–83
Wasson DH, Reppond KD, Kandianis TM (1991) Antioxidants to preserve rock-fish colour. J Food Sci 56:1564–1566
Watanabe K (1982) Fish handling and processing in tropical Africa. In: Proceedings of the FAO expert consultation on fish technology in Africa, Cosablanca, Morocco, pp 7–11
Waterman JJ (1976) The production of dried fish. FAO Fisheries Technical Paper, No. 160, 52
Yousif OM (1988) Problems and scope of the future development of fisheries in the Sudan, Paper. Presented at the International Symposium on the Development of Animal Resources in the Sudan, University of Khartoum, Faculty of Veterinary Science, Khartoum, Sudan
Zakhia N, Cua JL (1991) A percu surla quality du tilapia seche et commercialise au Mali. FAO Fisheries Report 400:176–181
Zapata JF, Ledward DA, Lawrie RA (1990) Preparation and storage stability of dried salted mutton. Meat Sci 27:109

Consumerism of Probiotics in China

W.L. Hung

Contents

Abstract Probiotics continue to attract considerable interest in both the business and scientific communities. However, the manufacture and applications of probiotics ingredients remain a niche market. Companies with the proprietary probiotic strains, appropriate process technologies, and distribution networks dominate the industry. New information about the genome of existing and emerging probiotics strains provides solid basis for successful commercial applications in a growing range of dietary supplements and food products. Gradually, probiotic-enhanced products are moving further into the mainstream in China. This chapter will be a resource with an interest in the growing popularity of probiotics and presents an informative balance of the scientific and market factors that are critical in the probiotics industry in China.

W.L. Hung (✉)
Department of Biotechnology, Research & Development, Want-Want China Holding Limited, No. 1088, East Hong Son Road, Shanghai 201103, China
e-mail: Hong_WeiLian@want-want.com

M.-T. Liong (ed.), *Beneficial Microorganisms in Food and Nutraceuticals*, Microbiology Monographs 27, DOI 10.1007/978-3-319-23177-8_8

1 Introduction

The United Nation's Food and Agriculture Organization (FAO) defines probiotics as "live microorganisms," which when administered in adequate amounts confer a health benefit on the host. These benefits include the prevention and cure of disorders such as lactose intolerance and inflammatory bowel disease. The major factors driving the growth of probiotics include growing health consciousness and the availability of probiotics in the form of dietary supplements. As shown in Fig. 1, probiotics have an extremely versatile application base. The most important feature of probiotics is their ability to promote the health of the user by strengthening their immune system either directly or indirectly by improving the condition of the gut, digestive process, nutritional value, and others. Using various types of probiotic bacteria, it is also possible to produce tailor-made probiotics according to the applications needed for the end user. Due to such favorable characteristics, probiotics have become part of our daily life and continue to make significant progress around the world in terms of both development and revenues.

Global sales of probiotic ingredients amounted to an estimated US $704 million in 2012. Probiotics of the lactobacillus genus accounted for the largest share, representing an estimated 60.5 % of total sales in 2012. Total global sales of

Fig. 1 Purported health benefits of probiotics and prebiotics: scientific research platforms. *Source:* Euromonitor (2014)

probiotic ingredients are projected to increase at a compound annual growth rate (CAGR) of 6.7 % between 2013 and 2018 (BBC Research 2014).

Probiotics are used in the manufacture of dietary supplements that are sold in the form of capsules, tablets, powders, topical pastes, and gels for human use as well as in animal feed for pets and farm animals. Global sales of probiotic supplements amounted to approximately US $1.0 billion in 2012, with probiotic supplements in capsule form accounting for the largest share of sales (66.3 %). Total global sales of probiotic supplements are projected to reach US $2.1 billion in 2018, representing a CAGR of 11.5 % between 2013 and 2018 (BBC Research 2014).

Food applications for probiotics include dairy-based and nondairy-based products. The main dairy-based categories are yoghurts, cultured drinks, kefir, and cheeses. Other food applications include probiotic-enhanced infant nutrition (formula and cereal), nonalcoholic beverages, breakfast cereal, and snack foods. Global sales of probiotic foods amounted to an estimated US $21.3 billion in 2012. Spoonable yoghurt accounted for the largest share of sales, representing an estimated 34.5 %. Total global sales of probiotic foods are projected to grow at a CAGR of 6 % between 2013 and 2018, to reach $33.5 billion US in 2018 (BBC Research 2014).

Collectively, global sales of probiotic ingredients, supplements, and foods amounted to approximately US $23.1 billion in 2012, with probiotic foods representing an estimated 92.4 % of total sales. Global sales of probiotic ingredients, supplements, and foods are projected to reach US $36.6 billion in 2018, representing a CAGR of 6.2 % between 2013 and 2018 (BBC Research 2014). The following Table 1 and Fig. 2 summarize recent and projected global sales of probiotic ingredients, supplements, and foods through 2018. The market for overall probiotic products was US $26,125.9 million in 2012 and is estimated to grow at a healthy CAGR of 6 % from 2014 to 2019 (Markets and Markets Analysis 2013). One of the key factors contributing to this market growth is the increasing awareness about probiotic products. The Global Probiotics market has also been witnessing the increase in R&D activities. However, government regulation could pose a challenge to the growth of this market.

In terms of regional distribution (Fig. 3), Asia-Pacific accounted for an estimated 38 %, Europe accounted for 32 %, and North America accounted for 15.1 % of global sales of probiotic ingredients, supplements, and foods in 2013. The ROW

Table 1 Global sales of probiotic ingredients, supplements, and foods, through 2018 (US $ Millions)

Probiotics	2012	2013	2018	CAGR% (2013–2018)
Foods	21,313.9	25,090.0	33,505.9	6.0
Supplements	1038.4	1190.1	2051.2	11.5
Ingredients	704.0	798.0	1104.4	6.7
Total	23,056.3	27,078.1	36,661.5	6.2

Source: BBC Research (2014)

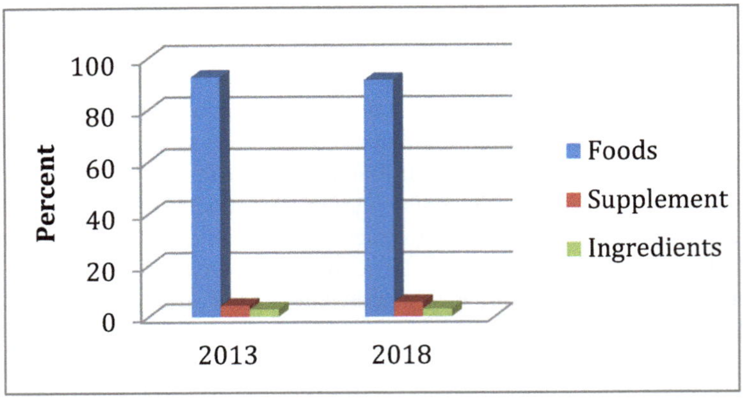

Probiotics	2013	2018
Foods	92.7	91.4
Supplements	4.4	5.6
Ingredients	2.9	3.0
Total	100	100

Fig. 2 Projected shares of global sales of probiotic ingredients, supplements, and foods, 2013 and 2018 (%). *Source:* BBC Research (2014)

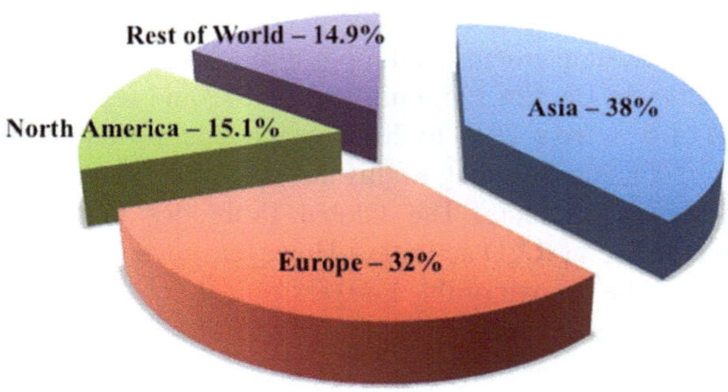

Fig. 3 Global probiotics market value, by Geography, 2013. *Source:* BBC Research (2014), Transparency Market Research (2013)

accounted for the remaining 14.9 % of total global sales in 2013 (Transparency Market Research 2013).

Growing consumption of yoghurt and increasing concerns over digestive and gut health are major driving forces behind probiotic consumption growth. The probiotics market is concentrated in Japan, the USA, and Western Europe, where traditionally probiotic demand has been always strong (BBC Research 2014). In Asia-Pacific, with economic conditions improving at a rapid pace, consumer purchasing power has also gone up multifold (Transparency Market Research 2013).

This has resulted in a sharp rise in revenues for the probiotics industry in the region. Growing demand for probiotics products across the world has translated into increased R&D spending and new product developments over the last 3 years.

The probiotic ingredient market consists of various stakeholders such as suppliers, ingredient processors, research institutes, food and beverage companies, traders, distributors, and consumers. The companies enjoying substantial market share are Yakult Honsha (Japan), Danone (France), Nestle S.A. (Switzerland), etc. At present, Asia-Pacific and Europe dominate the global probiotic consumption. Together, they accounted for over 70 % of the global market in terms of revenue. With over 38 % share in total revenue generated by the global market, Asia-Pacific is the leading market for probiotics (BBC Research 2014).

Asia-Pacific, owing to its high awareness of the benefits of probiotic yoghurts and fermented milk, is the largest market for probiotics, followed by Europe. However, the market is growing rapidly in the USA as well; the general affinity of the US population toward probiotic dietary supplements and the concept of preventive health care (Rephrased from RD) are expected to drive the probiotic ingredient market in future due to which the sales of probiotics yoghurt are already on the rise (Transparency Market Research 2013).

The greatest disadvantage facing the industry is the low level of awareness among consumers about the benefits of probiotics. Moreover, various alternative probiotic strains are available in the market having similar properties of enhancing the immune system and improving gut health, thus creating greater consumer misunderstanding. One of the biggest drawbacks for probiotics is the lack of any kind of insurance cover for probiotic usage.

The probiotics market is currently being driven by the rising popularity of probiotic functional foods and beverages (F&B) among consumers. Age, stress, poor diet, etc., are some of the reasons responsible for digestive ailments, bloating, reduced resistance to infections, etc.; consumption of probiotic-enhanced products helps to alleviate these widespread conditions. These products contain live microorganisms (probiotics) that confer positive health effect on the host (Caselli et al. 2013). Companies such as Yakult Honsha (Japan) and Chr. Hansen (Denmark) have developed patented strains of microorganisms claiming to affect specific health benefits. There has been a proliferation of probiotic ingredient suppliers who develop tailored strains of microorganisms for integrating with a diverse set of products. Awareness, faith on their efficacy, and safety are some of the deciding factors for the success of probiotic products.

Women buyers are the key drivers for the probiotics market. With the probiotic strains becoming a common factor among the manufacturers, taste and convenience continue to be the most important factors to secure the market share. Overall, there is now a flourishing market for functional F&B suppliers and manufacturers, where product innovation will be the key factor to increase the market share.

The issue of counterfeit products making unsubstantiated health claims in the market has diluted the image of the authentic products by making potential consumers wary about making the choice of consuming probiotic products. This is restraining the growth of the probiotic ingredient market. Moreover, the market faces stiff competition from other categories of functional and good-for-health foodstuffs that have significant market share as well as goodwill among consumers, such as products with lesser carbohydrates, or fortified with omega-3, vitamins, etc.

The early movers in the industry are likely to benefit in terms of market share; however, it is important that they focus on innovating probiotic strains that are more efficient in terms of stability and survivability in harsh conditions and are supported by competitively priced production technologies. Extensive research is required to develop cost-efficient manufacturing processes for probiotics. Companies are hence aiming to invest in R&D for the same reasons. In addition, manufacturers have ensured extensive communication with the consumers in terms of legitimate assertions for the health benefits of these products. Such communications, substantiated with scientific publications, are bound to help consumers' faith and in turn gain profits in the market.

2 Macroeconomic Landscape of China

Despite weak and uncertain global conditions, IMF predicated that the Gross Domestic Product (GDP) for China's economic growth (Fig. 4) would slow down to 7.4 % in 2014 and 7.1 % in 2015 (IMF Public Report 2014). And the World Bank trimmed its own 2014 forecast to reflect "the bumpy start to the year," predicting China's GDP to grow 7.6 % this year, with its 2015 figure unchanged at 7.5 % (The World Bank 2014). In China, GDP is divided by three sectors: Primary, Secondary, and Tertiary. The Primary Industry includes farming, forestry, animal husbandry, and fishery and accounts for around 9 % of GDP. The Secondary sector, which includes industry (40 % of GDP) and construction (9 % of GDP), is the most important. The Tertiary sector accounts for the remaining 44 % of total output and consists of wholesale and retail trades; transport, storage, and post; financial

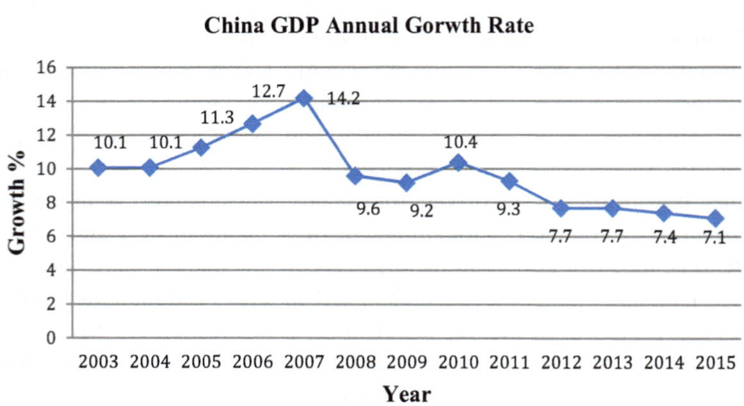

Fig. 4 China GDP annual growth rate data. *Source:* IMF Public Report (2014)

intermediation; real estate; hotel and catering services; and others (Trading Economics 2014).

The GDP in China expanded 7.30 % in the third quarter of 2014 over the same quarter of the previous year. GDP Annual Growth Rate in China averaged 9.10 % from 1989 until 2014, reaching an all-time high of 14.20 % in the fourth quarter of 1992 and a record low of 3.80 % in the fourth quarter of 1990 (The National Bureau of Statistics of China 2014). The National Bureau of Statistics of China (2014) reports GDP Annual Growth Rate of China. In the third quarter of 2014, China's GDP expanded 7.3 % year-on-year, slumping to a 5-year low. The slowdown was driven by lower property investment, dwindling credit growth, and weakening industrial production. But China will launch major investment projects in information networks, environmental protection, and infrastructure and water conservancy. Fiscal and monetary policies would be kept flexible and appropriate targeted adjustments made when needed to support the real economy. In September 2014, China's central bank relaxed lending rules for home buyers and allowed banks to offer a maximum 30 % discount to first-time homebuyers (XinHuaNet 2014). The bank also injected 500 billion CNY (US $81 billion) into five largest banks via a 3-month standing lending facility operation, a move aiming to support credit and growth (Trading Economics 2014).

Figure 5 indicated that consumer confidence index (CPI) in China is kept at 111, and the CPI fell 0.1 substantially by comparing each quarter with the previous quarter, and fast moving consumer goods (FMCG) growth rate% is kept around on average 6.5 % (Neilsen Research 2014). This means that Chinese people on the economic prospect are still full of confidence; therefore, the market also is optimistic about China's employment market, and will be advantageous to help keep the RMB exchange rate. When a country's CPI rose, indicating that the country's inflation rate rose, which means the monetary purchasing power abatement, according to the theory of purchasing power parity, the country's currency should be weakened. On the contrary, when a country's CPI fell, indicating that the country's inflation rate to drop, which means the monetary purchasing power rise,

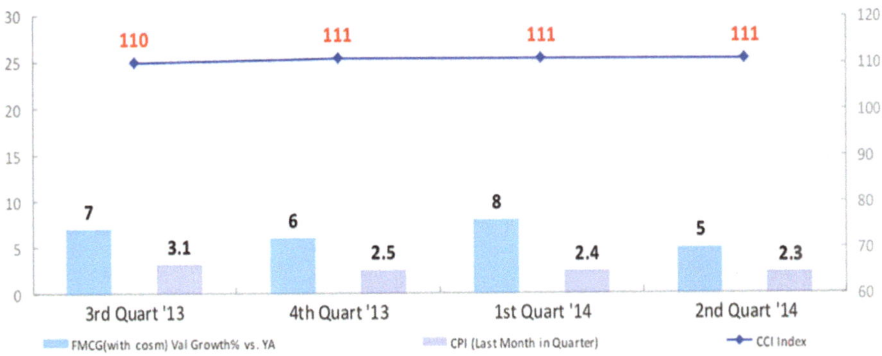

Fig. 5 China consumer confidence index, price index, and fast moving consumer goods growth rate%. *Source:* Neilsen Research (2014)

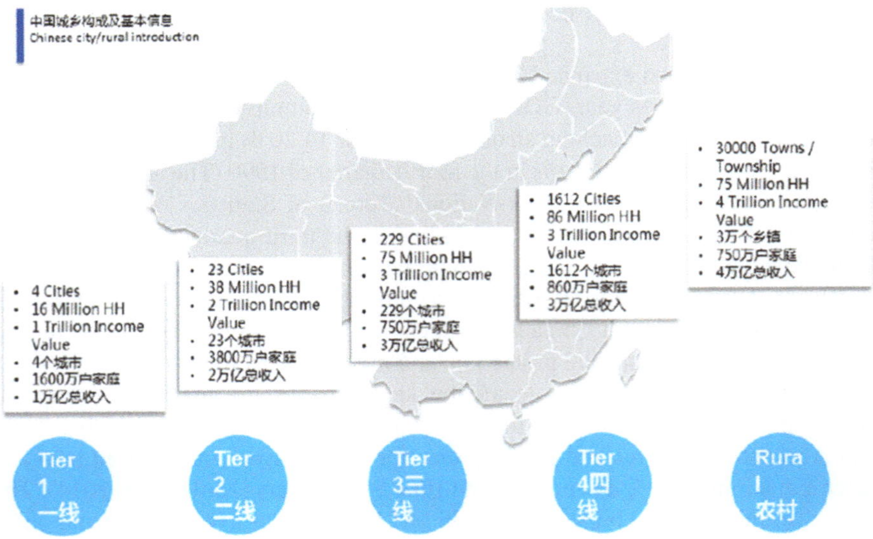

Fig. 6 Chinese city/rural introduction. *Source:* The National Bureau of Statistics of China (2014)

according to the theory of purchasing power parity, the country's currency should be stronger.

Figure 6 shows the Chinese city/rural structure; it is divided into 5 different levels, tier 1, tier 2, tier 3, tier 4, and rural. The new first-tier cities will emerge from the tremendous growth potential of the tier 2 and tier 3 city (Figs. 7 and 8) in the future. The new first-tier cities will bring a new situation in the future consumption mode, because the total population of new first-tier cities will become more than the total population of the original number of first-tier cities, and consumption capacity will also show great strength (The National Bureau of Statistics of China 2014). At the same time, "China Consumer Confidence Survey" study found that (Figs. 9 and 10), the consumers who live in tier 2, 3, and 4 cities are most concerned about the health issue from the amounts factors of health, income, children's education and welfare, health care, job security, parents' welfare and happiness, increasing food prices, personal career, increasing real estate price, environmental protection, and many other issues (China Consumer Confidence Survey 2014). It is said that "health issue" will quickly sweep the Chinese in the future, and related health products will also become the hottest product.

Fig. 7 Potential tier 2 and 3 cities will become the future tier 1 city. *Source:* The National Bureau of Statistics of China (2014)

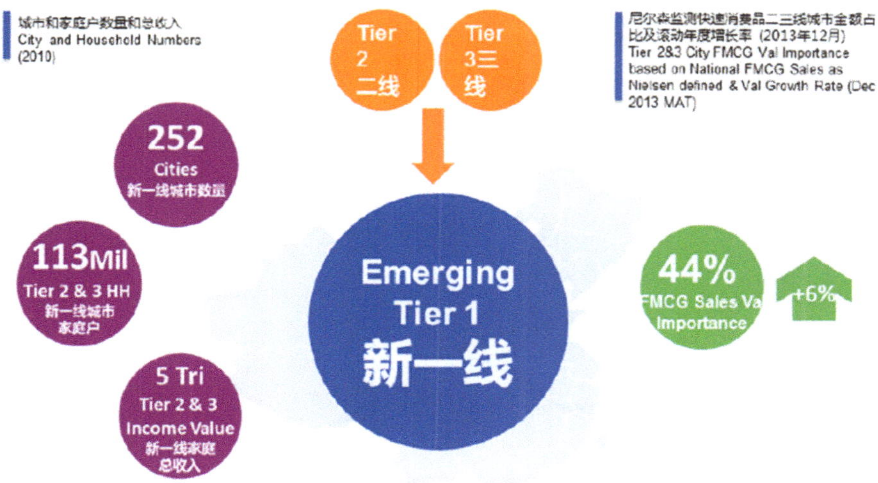

Fig. 8 The future new tier 1 city will be the important new engine in China. *Source:* The National Bureau of Statistics of China (2014)

Fig. 9 Health is the most important issue in China. *Source:* China Consumer Confidence Survey (2014)

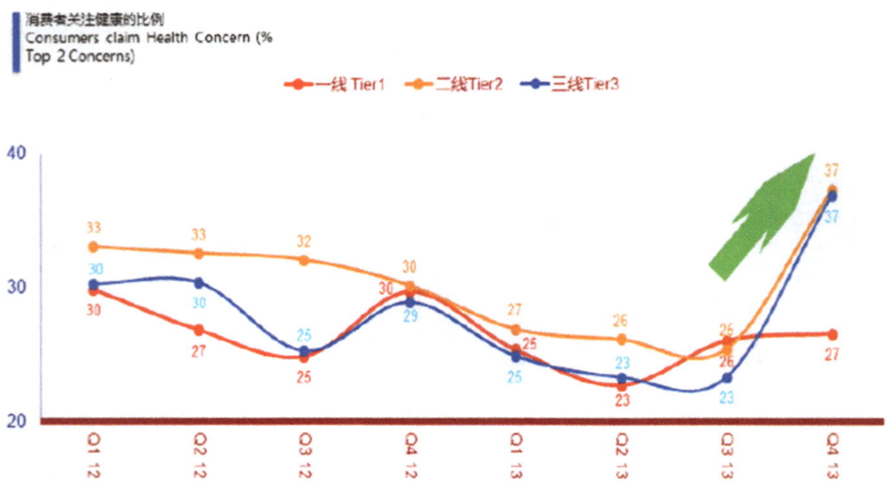

Fig. 10 Consumers claim health concern (%). *Source:* China Consumer Confidence Survey (2014)

3 China Probiotics Market Overview

How to create a successful probiotics market? It is of important experience to learn from the market of European and the USA how they promote the probiotic industry for many years, and the following 5 key points should be considered (Fig. 11): (1) Consumer awareness about the benefits of probiotics. (2) Difference between the buying behavior of consumers in developed and developing nations. (3) Weak immune system of children born from cesarean operations. (4) Women as traditional Food and Beverage buyers. (5) Aging baby boomer population are primary potential consumers. Chinese market will be facing the same situation, and it would be critical to think about these key factors in marketing planning if we were to build a successful probiotic market in China.

Chinese industry information network (2013) pointed out that, along with the deepening research in probiotics, probiotics are increasingly widely applied. In dairy products, it is mainly used in yoghurt, cheese, butter, cheese, and probiotics production. And in the probiotic product market share in China, milk products are 72.4 % and Pharma- & Granule are 10 % (Fig. 12).

The development of traditional probiotics application fields is achieved with innovation. The traditional production of yoghurt fermentation strains by *Lactobacillus bulgaricus* and *Streptococcus thermophilus* was the best combination, but with the development of the manufacturing technology, the "traditional" sense of the yoghurt has a broader connotation, such as dried yoghurt, frozen yoghurt, and yoghurt sterilization. A "new" variety of yoghurt and fermented milk products is constantly emerging (Mugambi et al. 2014). For example, Finland, Norway,

Fig. 11 The keys of booming probiotic market. *Source:* Markets and Markets Analysis (2014)

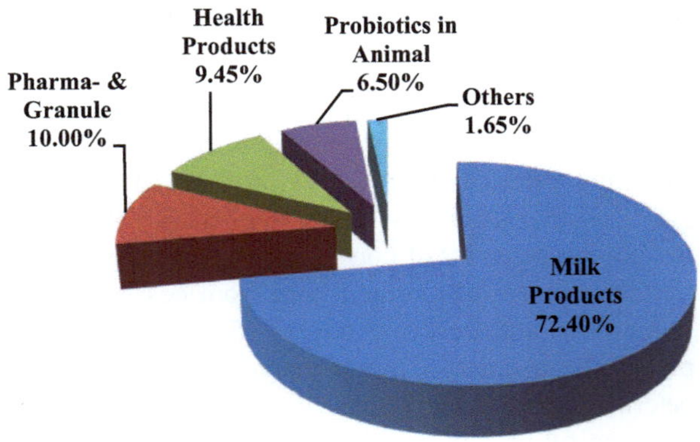

Fig. 12 Probiotics market structure in China. *Source:* Chinese industry information network (2013)

Holland, and other countries have listed new functional yoghurt: *Lactobacillus casei* yoghurt has been listed for sale, Bifidobacterium yoghurt, *Lactobacillus acidophilus* yoghurt has been accepted by consumers. Ymer (Denmark), Nordic ropy milk, and other products; condensed milk (Middle East), Chakka (India) and other types of yoghurt products; Kefir (kephir); Koumiss (cream) yeast fermentation product is becoming a development key point. The annual number of new probiotic dairy products increased 5 times in as short as of 5 years.

At present, the domestic dairy enterprises selected probiotic strains is monopolized by foreign companies; the proportion is about 90 %. Companies such as Danish Hansen Ltd., Danisco, the Holland, and the United States probiotics production enterprises are the main supplier of domestic dairy enterprises of probiotics.

Enterprises have two kinds of choices in their selection of the strain, developing their own or purchasing ready-made bacteria strains; whether it is domestic or foreign isolates strains, they are extracted either from the separation of human body or from nature. If one must say which is "better," from a genetic perspective, it may be only suitable for the population from a group of host's isolated populations of the endemic probiotic species, specific need rigorous determination, and clinical observation. At present, all edible lactic acid bacteria are beneficial to human body, but there must be individual differences in the uptake of probiotics as each intestinal colonies of each person is not identical and individual drinking habits are different (Mugambi et al. 2014).

Chinese government announced the legitimate list of species for infant food (2011) and for normal food usage (2012) (Tables 2 and 3). Only six species that are with substantial clinic research references worldwide could be used for infant food; for normal food, 24 species are deemed to be safe and could be used (Wang and Luo 2011).

According to the report of the Euromonitor (Table 4), pro/prebiotics used in dairy-based yoghurt, amounted from 10,832 million CNY in 2009 to reach 33,011 million CNY in 2014. Currently, pro/prebiotic drinking yoghurt is still the main consumer market, but pro/prebiotic spoonable yoghurt shows rapid growth, from 2079 million CNY in 2009 which rises to 12,579 million CNY in 2014, nearly a

Table 2 The legitimate use of species for infant food in China (2011)

Species name	Strain name
Lactobacillus acidophilus	NCFM
Bifidobacterium animalis	BB-12
Bifidobacterium lactis	HN019
	Bi-07
Lactobacillus rhamnosus	LGG
	HN001

Table 3 The legitimate use of species for food (2012)

Species name	Species name
Bifidobacterium	Lactobacillus
Bifidobacterium adolescentis	*Lactobacillus acidophilus*
Bifidobacterium animalis	*Lactobacillus casei*
Bifidobacterium lactis	*Lactobacillus crispatus*
Bifidobacterium bifidum	*Lactobacillus delbrueckii* subsp. *Bulgaricus*
Bifidobacterium breve	*Lactobacillus delbrueckii* subsp. *Lactis*
Bifidobacterium infantis	*Lactobacillus fermentum*
Bifidobacterium longum	*Lactobacillus gasseri*
	Lactobacillus helveticus
Streptococcus	*Lactobacillus johnsonii*
Streptococcus thermophilus	*Lactobacillus paracasei*
Leuconostoc	*Lactobacillus plantarum*
Leuconostoc mesenteroides subsp. *mesenteroides*	*Lactobacillus reuteri*
Propionibacterium	*Lactobacillus rhamnosus*
Propionibacterium freudenreichii subsp. *Shermanii*	*Lactobacillus salivarius*

Table 4 The market of pro/pre biotic yoghurt between 2009 and 2014

	2009 ▼	2010 ▼	2011 ▼	2012 ▼	2013 ▼	2014 ▼
Pro/Pre Biotic Drinking Yoghurt—CNY mn	8753.4	10,302.2	12,648.7	14,544.9	17,063.5	20,432.3
Pro/Pre Biotic Spoonable Yoghurt—CNY mn	2079.3	2888.1	4227.8	5856.9	8775.1	12,579.2
Pro/Pre Biotic Flavored Spoonable Yoghurt—CNY mn	236.0	340.3	465.3	597.8	769.8	971.8
Pro/Pre Biotic Fruited Spoonable Yoghurt—CNY mn	422.1	594.2	929.8	1236.6	1619.9	2080.7
Pro/Pre Biotic Plain Spoonable Yoghurt—CNY mn	1421.2	1953.6	2832.8	4022.5	6385.4	9526.6

Source: Euromonitor (2014)

6 times growth. In terms of CAGR% between 2013 and 2014 of pro/prebiotic drinking yoghurt, the sales volume is increasing by 12 % and the sales value is increasing by 19.7 % (Fig. 13). However, the retail value RSP (retail sales price) of probiotic supplements is just starting to growth, from 196 million CNY in 2009 to 358 million CNY in 2014, and is estimated to reach 553 million CNY in 2018 (Fig. 14).

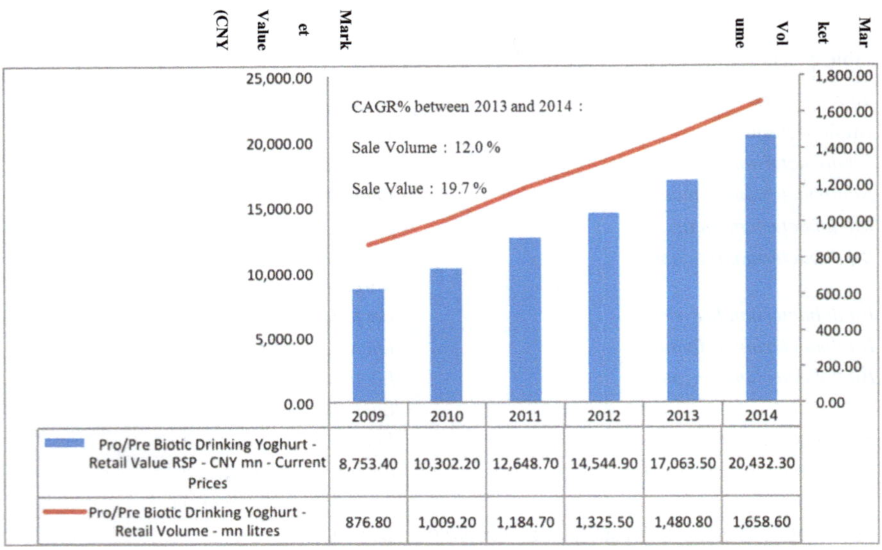

	2009	2010	2011	2012	2013	2014
Pro/Pre Biotic Drinking Yoghurt - Retail Value RSP - CNY mn - Current Prices	8,753.40	10,302.20	12,648.70	14,544.90	17,063.50	20,432.30
Pro/Pre Biotic Drinking Yoghurt - Retail Volume - mn litres	876.80	1,009.20	1,184.70	1,325.50	1,480.80	1,658.60

Fig. 13 Pro/pre biotic drinking yoghurt retail value and volume between 2009 and 2014 in China. *Source:* Euromonitor (2014)

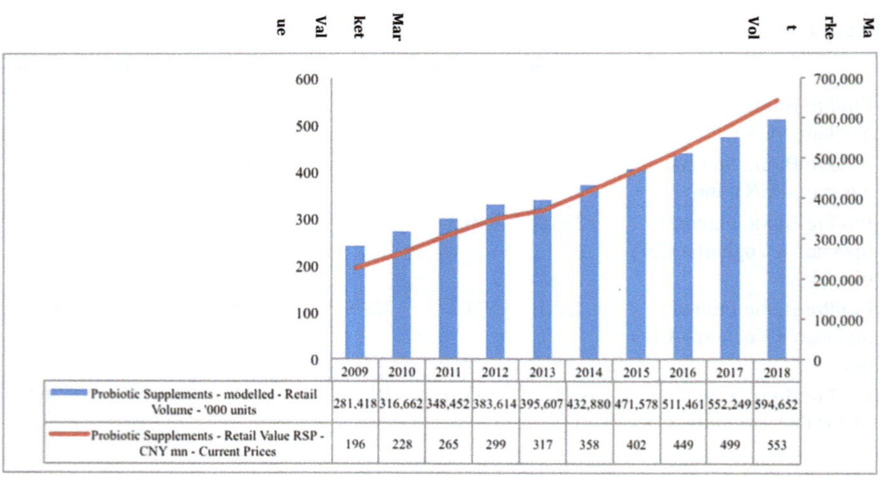

	2009	2010	2011	2012	2013	2014	2015	2016	2017	2018
Probiotic Supplements - modelled - Retail Volume - '000 units	281,418	316,662	348,452	383,614	395,607	432,880	471,578	511,461	552,249	594,652
Probiotic Supplements - Retail Value RSP - CNY mn - Current Prices	196	228	265	299	317	358	402	449	499	553

Fig. 14 Probiotic supplements market value and volume in China. *Source:* Euromonitor (2014)

Table 5 Probiotic yoghurt product companies markets share in China

Companies	2009	2010	2011	2012	2013
Hangzhou Wahaha Group Co. Ltd.	45.90	46.40	43.60	38.90	33.70
China Mengniu Dairy Co. Ltd.	5.50	12.50	14.70	16.70	17.90
Yakult Honsha Co. Ltd.	10.00	11.90	13.10	14.90	17.20
Inner Mongolia Yili Industrial Group Co. Ltd.	8.20	9.50	10.90	11.70	12.70
Bright Food (Group) Co. Ltd.	8.40	9.00	8.30	8.90	10.30
Wei Chuan (BVI)	1.70	1.70	1.60	1.60	1.60
Guangzhou Zhujiang Meileduo Beverage (HK) Co. Ltd.	2.10	2.10	1.90	1.70	1.50
Wonder Sun Dairy Co. Ltd.	1.20	1.20	1.10	1.10	1.00
Sichuan New Hope Agribusiness Co. Ltd.	0.90	1.00	0.90	1.00	0.90
Guangzhou Yantang Dairy Co. Ltd.	0.70	0.70	0.80	0.80	0.80
Green's Bioengineering (Shenzhen) Co. Ltd.	0.60	0.60	0.50	0.50	0.50
Others	8.30	3.40	2.70	2.30	2.00
Total	100.00	100.00	100.00	100.00	100.00

Source: Euromonitor (2014)

Table 6 Sales comparison of Yoghurt and Yoghurt drink in China

China market	Annual sales	2010 and 2006 sales comparison	2013 and 2006 sales comparison
Yoghurt	+19.1 %	+2.2 times	+3.5 times
Yoghurt drink	+37.1 %	+2.5 times	+9 times

Source: Neilsen Research (2014)

Table 5 shows that market sharing of Probiotic yoghurt product companies in China from 2009 to 2013, the Hangzhou Wahaha group Co. Ltd. stays ahead for many years, China Mengniu Dairy Co. Ltd. is now No. 2, and Yakult Honsha Co. Ltd. No. is No. 3. China's yoghurt market is constantly changing and growing. According to the Neilsen Research Report (2014), the annual sales of Yoghurt in China increased 19.1 %, and Yoghurt drink increased 37.1 % (Table 6). If 2010 and 2006 sales are compared, yoghurt part increased 2.2 times and yoghurt drinks increased 2.5 times, but if 2013 and 2006 sales are compared, yoghurt part increased 3.5 times and yoghurt drinks increased 9 times.

4 Application and Prospect of Probiotics in China

4.1 *The General Situation of the Development of Probiotic Products*

Probiotics market of China started late, along with China's consumer eating habits changing to finer health-oriented foods! Probiotic products' market is currently mainly in the food and dietary supplements on foreign markets. A healthy probiotic with continuous development of new technology and new products to meet consumer tastes and ideas and to combine with dietary fiber, oligosaccharide dairy products, and healthcare products, etc., will further expand market share.

The main product types are liquid milk and yoghurt with probiotic. Normally, dietary supplement products contain probiotics in the form of a capsule, powder, oral liquid, or tablet added with other ingredients such as milk, non-dairy, sheep milk powder, cranberry extract carrier, fructo-oligosaccharides (FOS), immunoglobulin, fermentation by-products, and other biologically active substances (Mugambi et al. 2014; Wang et al. 2013; Fen et al. 2010). Probiotic product is either a single strain or combined multiple strains, including lactobacillus, Bifidobacterium, and less use of Enterococcus, bacillus, *Escherichia coli*, and yeast. Dietary supplements are primarily a health food or natural salt crystal. In the past few years, the probiotics market has grown steadily with the growth of overall natural products' market. Study on the physiological effects of daily intake of 1×10^9 CFU~1×10^{10} CFU of probiotics (such as the treatment of diarrhea, lactose intolerance, increased fecal enzyme activity test research) proved effective. Now the dietary supplement with active probiotics is easy to reach this effective level (Wang and Luo 2011; Yuan et al. 2010).

4.2 *Health Function and Application of Probiotics Products*

Probiotics, the new healthcare doctor, has clinical value and wide application prospect. With in-depth research, one starts to understand the relationship between probiotics and intestinal microflora. It promotes health by improving immunity and balancing host and intestinal microbial. Enterprises endeavors to increase the rate of utilization and expand the space for probiotics development. Probiotics could adjust the imbalance of intestinal flora and improve micro-ecological environment, thus providing the effective results of prevention and treatment on the various causes of acute/chronic diarrhea, constipation, and other digestive diseases (Wang et al. 2013). Probiotics also have the antitumor effect which is mainly reflected in the optimal combination of metabolism products of intestinal flora to enhance the immune function of organism (Caselli et al. 2013; Yuan et al. 2010). At the same time, anticancer probiotics could eliminate the degradation of carcinogenic nitrosamines. In addition, probiotics could promote intestinal peristalsis and help to flush the harmful bacteria out of the body.

Probiotics is mainly used in the food industry in China, and 90 % of them are used in dairy products that are mostly in yoghurt, milk drinks, yoghurt drinks, infant milk powder, milk, and the activity of neutral flavored milk products. Especially in fermented milk, the development and utilization effect of probiotics is more significant, and has become a hot selling product. The global annual production of fresh yoghurt reached 1600 million tons. The yoghurt production of China is nearly 310 thousand tons in 2001, which increased to 1620 thousand tons in 2005. It has increased more than 4 times during 5 years. Yoghurt drinks beverage is more than 140 thousand tons in 2001 and increased to nearly 1080 thousand tons in 2005; this has increased more than 6 times. Fermented dairy products have become a new growth point in the dairy industry (Yun 2013; Yuan et al. 2010).

Recently, functional food of probiotics has become a hot research and development topic, including functional yoghurt in improving lactose intolerance, lowering cholesterol, preventing cancer, reducing diarrhea, and playing a very important role in improving the immunity. Common method for production of functional yoghurt is to add probiotics after yoghurt fermentation is completed, to complete the subsequent fermentation. Chang et al. (2006) have reported the beneficial effect on the basis of this, according to their experimental results; xylitol could completely replace the sucrose and the used low fat skim milk. So functional yoghurt would be produced and could help diabetic patients and older consumers' consumption.

At present, Probiotics is not only used in dairy products, such as yoghurt, yoghurt drink, and infant milk powder, etc., but also gradually being used for the development of various functional foods, such as infant food, beverage, candy, baked goods, and snack foods (Yun 2013). With the development of science and technology: micro-capsule technology in recent years, which greatly improves the resistance to storage and the processing resistance to a large extent, the technical advance has been applied in the processing of probiotics for probiotics products and developing enhanced function of probiotics (Wang et al. 2013).

4.3 *The Problem of Probiotics in Food Application*

At present, every country is vigorously developing probiotic foods to expand the scope of application of probiotics. But many probiotic food problems still need to be addressed urgently in China. First, the application of probiotics in dairy products is very common, but compared with foreign countries, China's consumption of dairy products is still relatively low. Second, application of probiotics in food is not suitable for Chinese taste, so we should work hard on strain breeding as soon as possible, through screening and cultivating the excellent probiotics to develop suitable products for Chinese consumers' tastes. Third, application of probiotics should use multiple methods to add into food; probiotics also should be diversified to be suitable for all kinds of people. Therefore, scientific researchers need to try high-tech biotechnology to develop the application of probiotics in food (Wang et al. 2013; Yun 2013).

When probiotics are widely used in the food industry and medicine fields in China, only few studies have been conducted to evaluate the actual microbial amounts and species in probiotic products, which may conflict with the labels and mislead consumers to choose inappropriate foods or medicines. The combination of culture-dependent and culture-independent methods was proven to quickly and conveniently detect the microbial diversity in probiotic products, and more effort is required to regulate the probiotic market in China (Chen et al. 2014).

In order to promote the health of the human body and the improvement of living standard, we should increase the investment on the existing basis, expand the range of application of probiotics in food and other industries, to develop new products, and give full focus to the function of health care of probiotics.

5 Conclusion

With the continuous improvement of people's living standards and growing health awareness, people pay more and more attention to probiotic products containing probiotic foods. Nowadays, healthy and beneficial microorganisms were found in many different types of food like traditional foods, folk foods, and strong regional features foods, and probiotic products of industrial scales have been formed in China where the market changes rapidly. The natural probiotics existence and current application within our familiar foods signify the safety of the probiotics, which fit perfectly for our daily consumption to keep the body health. Deep understanding of the value of probiotics and impelling the probiotics foods consumption during our daily life is our main target.

References

BBC Research (2014) Global sales of probiotic ingredients, supplements and foods between 2013 and 2018. http://www.bbcresearch.com/

Caselli M, Cassol F, Galo G, Holton J, Zuliani G, Gasbarrini A (2013) Actual concept of "probiotics": is it more functional to science or business? (Review). World J Gastroenterol 19:1527–2540

Chang Y, Sun Y, Wang C (2006) The research and application of functional yoghurt. Dairy Guide 2:25–27

Chen T, Wu Q, Li S, Xiong S, Jiang S, Tan Q, Zhang Z, Zhu D, Wei H (2014) Microbiological quality and characteristics of probiotic products in China. J Sci Food Agric 94:131–138

Chinese industry information network (2013) 2013–2017 year Chinese probiotics market supply and demand forecast and the prospect of the Industry Research Report. www.chyxx.com

China Consumer Confidence Survey (2014) Chinese consumer satisfaction index hits record high. http://finance.ce.cn/rolling/201402/20/t20140220_2342595.shtml

Euromonitor (2014) Analysis of the current situation of Chinese probiotic industry demand in 2014. www.euromonitor.com
Fen Y, Qu D, Wang D, Shen W (2010) The application and towards of probiotics. Food Industry 3: 86–88.
IMF Public Report (2014) People's Republic China and the IMF—China: fastest growing consumer market in the world. www.imf.org
Markets and Markets Analysis (2013) Probiotics Market by Products (Functional Foods, Dietary Supplements, Specialty Nutrients, Animal Feed), Applications (Regular, Therapeutic, Preven tive Health Care) & Ingredients (Lactobacilli, Bifidobacteria, Yeast)—Global Trends & Fore casts to 2019. www.marketsandmarkets.com/
Mugambi M, Young T, Blaauw R (2014) Application of evidence on probiotics, prebiotics and synbiotics by food industry: a descriptive study. BMC Res Notes 7:754–762
Neilsen Research (2014) China consumer confidence index, price index, and fast moving consumer goods growth rate% and Sales comparison of Yoghurt and Yoghurt drink in China. www.nielsen.com/cn/zh.html
The National Bureau of Statistics of China (2014) Annual GDP Accounting Instructions of China. http://www.stats.gov.cn/english/
The World Bank (2014) Global Economic Prospects—Forecasts. http://www.worldbank.org/
Trading Economics (2014) China GDP Annual Growth Rate & China Consumer Confidence 1991–2014. http://www.tradingeconomics.com/
Transparency Market Research (2013) Probiotics Market (Dietary Supplements, Animal Feed, Foods & Beverages)—Global Industry Size, Share, Trends, Analysis, Growth And Forecast 2012–2018. www.transparencymarketresearch.com/
Wang J, Luo X (2011) Key points of management on foods added with probiotics. Chin J Food Hygiene 23:160–162
Wang R, Zheng L, Tong Y (2013) Application and prospect of probiotic products. Food Res Dev 34:128–130
XinHuaNet (2014) China Focus: China's eased mortgage policies spur sales, sector pressure remains. http://news.xinhuanet.com/english/china/2014-10/31/c_133757213.htm
Yuan X, LV L, Han D, Xiao J (2010) Research progress of probiotic products. China Dairy Ind 38: $32#46–49
Yun L (2013) The latest progress of research and application on probiotics and its products. J Gansu lianhe Univ (Nat Sci) 27:114–116

Probiotics in Dairy Products

Sejong Oh

Contents

Abstract At the beginning of the twentieth century, Ellie Metchnikoff (1845–1916), Ernest Moro (1874–1951), and Leo Rettger (1874–1954) made their first scientific contributions to the research on probiotics. In humans, lactic acid bacteria (LAB) have a strong influence on the host's health because LAB are an important biodefense factor for preventing colonization by and subsequent proliferation of pathogenic bacteria in the intestine. Probiotics have been defined as beneficial microorganisms by Lilly and Stillwell since 1965. Some species of probiotic LAB have been called probiotics, such as *Lactobacillus acidophilus*, *L. casei*, *L. rhamnosus*, *L. plantarum*, and *L. reuteri*. There have been hundreds of publications describing the use of *Lactobacillus* strains to prevent and treat a variety of gastrointestinal disorders. However, only a few have contributed convincingly to our knowledge of the health effects of *Lactobacillus* species as starter cultures. *Lactobacillus* strains in dairy products represent an exciting prophylactic and therapeutic advance, although further investigations must be undertaken before their role in intestinal health can be delineated clearly.

S. Oh (✉)
Division of Animal Science, Chonnam National University, 77 Yongbong-ro, Buk-gu, Gwangju 500-757, South Korea
e-mail: soh@jnu.ac.kr

M.-T. Liong (ed.), *Beneficial Microorganisms in Food and Nutraceuticals*, Microbiology Monographs 27, DOI 10.1007/978-3-319-23177-8_9

1 Introduction

Lactobacillus species have been claimed as probiotics and closely associated with the human's health because these are an important biodefense factors. *Lactobacillus* is the largest genus within the group of lactic acid bacteria (LAB). It contains a very large number of species, isolated mainly from humans, animals, plants, and foods, and the genus shows large phenotypic, biochemical, and physiological variability (Laura et al. 1999). Several health benefits are also reported for *Lactobacillus* species that colonize the gastrointestinal tracts. These include stimulation of immunoglobulin production, induction of interferon expression in macrophages, acidification of the local environment, hypocholesteraemic effects, binding of mutagenic compounds, production of bacteriocins, and prevention of the adhesion of pathogenic bacteria to epithelial cells (Fuller and Gibson 1997; Tannock 2003; Oh et al. 2012).

Lactobacillus species are used as a starter in the fermentation of foods. Therefore, the use of *Lactobacillus* to enhance intestinal health in dairy products has been proposed for many decades. There have been hundreds of publications describing the use of *Lactobacillus* species to prevent and treat a variety of gastrointestinal disorders (Holzapfel and Schillinger 2002). However, the scientific basis of *Lactobacillus* in dairy products has been firmly established only recently, and the clinical studies have begun to be published. Currently, the best-studied *Lactobacillus* species are *Lactobacillus acidophilus*, *L. casei*, *L. rhamnosus*, and *L. plantarum*.

To understand the general mechanism of probiotics completely, this chapter explores the early studies relating to probiotics and the intestinal microbiota. Furthermore, we briefly describe the commercial probiotics used in dairy products at present.

2 History of LAB and Probiotics

Studies on lactic acid bacteria (LAB) as probiotics started with the research on the fermentation process. These studies made a substantial contribution to the fledgling field of microbiology. Although LAB are found in dairy products, humans, animals, and insects, it was Pasteur (Louis Pasteur, 1822–1895) who first discovered LAB; he found them to be an influencing factor in the fermentation of wine. Escherich (Theodor Escherich, 1857–1911) was the first to conduct systematic studies on the intestinal microbiota. He was able to identify a gram-positive bacillus that was dominant in a healthy baby's stool. Unfortunately, with the knowledge and the state of technology back then, he failed to isolate the microorganisms. In 1892, Döderlien (Albert Döderlien, 1860–1941) found an acid bacillus within a healthy woman's vaginal secretion. In 1895, a microorganism, isolated by Boas and Oppler, had characteristics very similar to those of LAB isolated from a patient with stomach cancer. In the nineteenth century, most research was focused on the discovery of LAB species; in the twentieth century, the research on LAB was focused more on the health and the concept of probiotics. LAB, as a subject of research on probiotics, take their origin from Tissier's research into *Bacillus bifidus* (currently, *Bifidobacterium* sp.), Moro's studies on *Bacillus acidophilus* (currently, *Lactobacillus acidophilus*), and Grigoroff's work on *Bacillus bulgaricus* (currently, *Lactobacillus bulgaricus*).

In the thesis that Tissier (Henry Tissier, 1866–1926) published in 1900, he mentioned *B. bifidus* for the first time; it was isolated from a healthy infant's feces (Tissier 1900). This microorganism appeared to belong to a small group of Y-shaped bacteria, and Tissier named it "bifurcated forms." In the subsequent research, Tissier found these intestinal bacteria in a 3-day-old infant's feces who was fed only breast milk. Tissier argued that, because these bacteria were common in healthy infants, it was possible to transplant *B. bifidus* and thereby generate healthy intestinal microbiota. Diarrhea was treated with healthy intestinal microorganisms; this concept of probiotics is used in present-day studies just as in 1900.

In the same year, Moro (Ernst Moro, 1874–1951) found a microorganism in the stool of a baby who was breast-fed and named it *B. acidophilus* (*L. acidophilus*). Dr. Moro believed that this microorganism originated from the mother's milk. Furthermore, he stated that this bacterium was easily detectable in the baby's mouth, stomach, and intestines and that it was a gram-positive rod 1.502 μm long with a pointy tip; this bacterium could survive under highly acidic conditions (Moro 1900). In 1901, Cahn reported that *B. acidophilus* was found in bottle-fed infants and in breast-fed infants (Cahn 1901). Weiss (1904) reported that when a person drinks a lot of milk, the number of *B. acidophilus* cells increases (Weiss 1904). *Bacillus acidophilus* was renamed *Lactobacillus acidophilus* in 1920.

Fermented dairy products are traditionally known to alleviate constipation and diarrhea; because of these characteristics, various countries in Europe undertook numerous studies on fermented milks. In 1905, Grigoroff isolated the bacteria from the fermented milk that was used in Bulgaria and named it as *B. bulgaricus*. In 1907, a Ukrainian scientist, Ellie Metchnikoff (1845–1916), argued that lactic acid could suppress the intestinal putrefaction and the formation of poisonous substances that occur in the intestines. Due to the suppression of the decaying process within the gut, he believed that lactic acid produced by LAB could prolong human life (Metchnikoff 1908). His theory was accepted around the world because of the connections among lactic acid, LAB, health, and prolongation of life (Oh 2009). It is an undeniable fact that Metchnikoff's research was revolutionary and accelerated the studies of microorganisms in general and LAB in particular. Many of the journals cited above frequently cite Metchnikoff's work.

Nonetheless, Metchnikoff was not the first to conduct studies on intestinal decomposition and lactic acid. In 1887, Poehl reported that when a person ingests fermented milk, it inhibits decomposition of refuse in the intestines. Furthermore, Rovighi, Embden, and Brundinki confirmed this finding in 1892, 1894, and 1900, respectively. Tissier and Martelly (1902) reported that lactic acid could inhibit the decomposition of refuse in the intestines (Rettger and Cheplin 1921). Orla-Jensen (1919) was the first to notice the industrial value of LAB then; he collected hundreds of LAB isolates around Europe. On the basis of this collection, Orla-Jensen started to classify LAB, and his dedication was the starting point for the research on fermented dairy products such as cheese.

Studies on health-promoting effects of LAB were not conducted in Europe alone. As an up-and-coming nation, the United States of America saw numerous studies of LAB by its scientists. In the Rockefeller Institute, Kendall was using *B. bifidus* that was isolated by Metchnikoff. Kendall injected *B. bifidus* into a monkey's small intestine, and the decomposition of intestinal contents was noticeably inhibited. He reported that the selection criteria for LAB should be as follows: microorganisms that produce lactic acid should be selected by (1) their ability to settle down in the intestines, (2) their harmlessness to the host (the bacteria do not produce any harmful substances), and (3) their ability to produce sufficient quantities of lactic acid. Such selection criteria for probiotics are still used today.

In 1921, Professor Rettger of Yale University published a book, showing that *L. acidophilus* has characteristics that are completely different from those of *L. bulgaricus* (Rettger and Cheplin 1921). Rettger argued that *L. bulgaricus* was never found in human feces, and that this organism does not take up residence in human (or animal) intestines. He also proved that when *L. acidophilus* is injected into rat intestines, it dominates the population of microorganisms there (Rettger and Cheplin 1922) (Fig. 1).

3 Probiotics and Dairy Products

In 1965, the word "probiotics" first appeared in a scientific article, "*Probiotics: growth promoting factors produced by microorganisms*" by Lily and Stillwell. In this article, probiotics were defined as the opposite of antibiotics: as microorganisms that stimulate other microorganism's growth. Although another article, "*Anti- und Probiotika,*" was already published by Vergin in 1954. Lily and Stillwell were the first authors to define probiotics as bacteria that facilitate another microorganism's growth. In 1974, Parker defined probiotics as "a substance and/or a microorganism that balances intestinal microbiota in livestock." In 1989, Fuller redefined probiotics as "a live microbial supplement that enhances beneficial effects of intestinal micro biota in host." Fuller's definition is mentioned in many research articles because he redefined the meaning of probiotics by implying administration of live microorganisms.

Since then, many scientists have changed the definition of probiotics. Using Fuller's concept, in 2002, the Food and Agriculture Organization of the United Nations (FAO) and WHO defined a probiotic as "live microorganisms that when administered in an adequate amount confer a health benefit on the host." In this definition, the FAO and WHO emphasized the amount that should be administered (or injected) to promote enhancement in health. The focus in this definition was on strains available as probiotics in food, which are mainly members of the genera *Lactobacillus* and *Bifidobacterium*. The definition excluded the reference, but is not

Fig. 1 The pioneers of research on probiotics. Ellie Metchnikoff (*top*; Metchnikoff 1921), Ernst Moro (*bottom left*; Weirich and Hoffmann 2005), and Leo Rettger (*bottom right; http://www.asmbranches.org/brCtValley/HISTORY.htm*). The photographs were reproduced with permission

limited, to the term biotherapeutic agents and beneficial microorganisms not used in food (Pineiro and Stanton 2007). Currently, the FAO/WHO's definition of probiotics is widely accepted.

As shown in Table 1, the basic concepts in probiotics and antibiotics are subtle yet distinct. Both have similar roots to each other by using microorganisms. The first recorded history of our ancestral use of probiotics date back to 4500 BC. Probiotics is largely categorized as food and the first commercial production with the use of it was in 1935 by a Japanese company named Yakult. However, the first recorded history of the use of antibiotics goes back to 2500 BC. Unlike probiotics, antibiotic's mode of action is direct. Antibiotics are categorized as drugs and it's first public appearance as a commercial product was in the United States in 1942 by the Merck Company.

Probiotics are used most frequently as live microorganisms in food, drugs, and animal feed. Commercial products that are based on probiotics can be classified into

Table 1 Comparison between probiotics and antibiotics

	Probiotic	Antibiotic
Term's meaning	For life	Against life
Denominator	Lilly and Stillwell (1965)	Waksman and Woodruff (1942)
First record in history	4500 BC	2500 BC
Mode of action	Indirect	Direct
Product category	Food	Drug
First commercial production	Japan (1935), Yakult Co.	USA (1942), Merck Co.

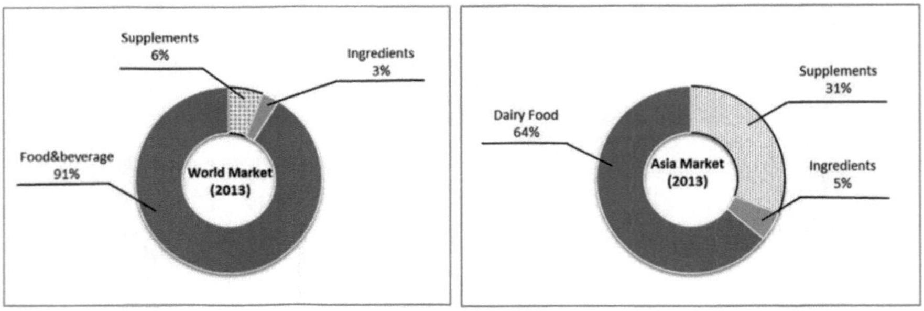

Fig. 2 The global and Asian markets of probiotics (Nutrition Business Journal 2014)

four categories: probiotic agents, animal feed supplements, health foods, and yogurt. In some countries, probiotic agents should be sold only at pharmacies. Although many other microorganisms are used in dairy foods, LAB plays a major role in this market sector.

As shown in Fig. 2, most people in Asia consume >64 % of probiotics in the types of yogurt. Thus, fermented milk is an important probiotic product in the Asian market. Various types of fermented milk including yogurt have become well-established food products worldwide (Table 2).

Yogurt, the most widely used fermented dairy product containing probiotics, should contain more than 6–8 $\log_{10}$ colony-forming units (CFU) of LAB per milliliter (Table 3). LAB that are used to make yogurt are known as "yogurt starter cultures." These cultures ferment lactose and produce lactic acid, which acts on milk proteins and gives yogurt its texture and gustatory characteristics. No one knows where or how yogurt originated, but approximately 10,000 BC, the human way of life changed from food gathering to food production, and this change included domestication of cows, sheep, goats, buffaloes, camels, yaks, horses, and even reindeer. Although the primary purpose of fermentation is to extend the shelf life of milk products, this process also improves the taste of milk and enhances the digestibility of this food product.

Yogurt products available around the world can be classified into five common categories. Set-type yogurt is incubated and cooled in the final packaging and is characterized by firm jellylike texture. Stirred-type yogurt is incubated in a tank and the final coagulum is broken up by stirring prior to cooling and packaging. The

Table 2 Microorganisms in fermented dairy products

Name	Origin of milk	Microorganisms
Yogurt	Buffalo, cow, goat, sheep	*Lactobacillus bulgaricus* *Streptococcus thermophilus*
Cultured buttermilk	Buffalo, cow, goat, sheep	*Streptococcus lactis* *Streptococcus cremoris*
Acidophilus milk	Cow, goat	*Lactobacillus acidophilus*
Lassi	Buffalo, cow	*Lactobacillus bulgaricus*
Kefir	Cow, goat, sheep	*Lactobacillus kefir* *Streptococcus lactis* *Leuconostoc* spp. (yeast)
Koumiss	Mare	*Lactobacillus acidophilus* *Lactobacillus bulgaricus* (yeast)
Leben	Goat, sheep	*Streptococcus lactis* *Streptococcus thermophilus* *Lactobacillus bulgaricus*

Table 3 Composition of fermented milk in accordance with Codex requirements (Codex Standard 243-2003)

	Fermented milk	Yogurt, alternate culture yogurt, and acidophilus milk	Kefir	Koumiss
Milk protein[a] (%)	min. 2.7 %	min. 2.7 %	min. 2.7 %	–
Milk fat (%)	<10 %	<15 %	<10 %	<10 %
Titratable acidity, expressed as % lactic acid	min. 0.3 %	min. 0.6 %	min. 0.6 %	min. 0.7 %
Ethanol (% vol/w)	–	–	–	min. 0.5 %
All microorganisms constituting starter culture (CFU/g)	min. 10^7	min. 10^7	min. 10^7	min. 10^7
Microorganisms[b] on the label (CFU/g)	min. 10^6	min. 10^6	–	–
Yeast (CFU/g)	–	–	min. 10^4	min. 10^4

Source: FAO, with permission (www.codexalimentarius.org/input/download/standards/400/CXS_243e.pdf. Accessed 9 January 2013)
Abbreviations: CFU colony-forming units, *vol/w* volume/weight
[a]Protein content is 6.38 multiplied by the total Kjeldahl nitrogen
[b]Applies where a content claim is made in the labeling (mentions the presence of a specific microorganism)

texture of stirred yogurt is less firm than that of set yogurt: the former resembles very thick cream. Drinking-type yogurt is similar to stirred yogurt: the coagulum is broken up prior to cooling. In drinking yogurt, to break up the coagulum, they use potent agitation. Little if any re-formation of the coagulum will occur after packaging. Frozen-type yogurt is inoculated and incubated in the same manner as stirred

Table 4 Regulatory requirements on viable-cell counts of lactic acid bacteria (LAB) in yogurt

Country	Viable cells of LAB (CFU/g)
USA[a]	$\geq 10^7$
Australia[b]	$\geq 10^7$
Europe[c]	$\geq 10^7$
China[d]	$\geq 10^6$
Japan[e]	$\geq 10^8$
Korea[f]	$\geq 10^8$

[a]US FDA Code of Federal Regulations, parts 131–203
[b]Australia & New Zealand Food Standard 2.5.3
[c]European Union Council Regulation No. 178/2002
[d]China National Dairy Standard GB2746-1999
[e]Ministry of Health and Welfare Ordinance No. 52, Japan
[f]Livestock Product Processing Control Act 2.1.4, Korea

yogurt is. Nonetheless, cooling is achieved by pumping through a whipper, chiller, and freezer in a fashion similar to the production of ice cream. Concentrated-type yogurt is inoculated and fermented in the same way as stirred yogurt is. After resuspension of the coagulum, the yogurt is concentrated by boiling off some of the water; this process is often performed under vacuum to reduce the temperature required. Each country has its own standards with respect to the regulations governing the production of fermented dairy products. Some of these standards are similar, but others differ: e.g., milk protein content and milk fat level, the content of solids-not-fat, etc. The regulations and standards pertaining to fermented milk in various countries are listed in Tables 3 and 4.

4 Major Lactobacillus Species in Dairy Products

As shown in Table 5, some species of LAB are called probiotics around the world. *Lactobacillus* species are beneficial microorganisms that have several probiotic effects on humans and animals, such as alleviation of acute diarrhea and allergy (Szajewska et al. 2001; Ouwehand 2007), of inflammatory bowel disease (Ewaschuk and Dieleman 2006; Limdi et al. 2006), and of antibiotic-associated gastrointestinal symptoms (Lenoir-Wijnkoop et al. 2007; Maslowski et al. 2009); a reduction in the number of potentially pathogenic bacteria (Savard et al. 2011); and immunomodulatory effects (Bahrami et al. 2011).

In particular, *L. acidophilus*, *L. casei*, and *L. rhamnosus* GG are most popular probiotics used in dairy products such as yogurts, lactic drinks, and probiotic agents because these bacteria are a resident species of the human digestive tract.

L. acidophilus is a widely used probiotic microorganism because it is known to have beneficial effects on the host. The name "*Lactobacillus acidophilus*" first appeared in the 1920s because of an international agreement on the nomenclature of microorganisms (the original name was "*Bacillus acidophilus*"). *L. acidophilus* is one of 64 species of Lactobacillus with the characteristic inability to ferment

Table 5 Commercial probiotics in dairy products

Strain	Supplier
Lactobacillus acidophilus DDS-1	Nebraska Culture
Lactobacillus acidophilus LA-5	Christian Hansen
Lactobacillus acidophilus NCDO 1748	Arla Foods
Lactobacillus acidophilus NCFM	Danisco
Lactobacillus casei DN 114 001	Danone
Lactobacillus casei Shirota	Yakult
Lactobacillus gasseri OLL2716	Meiji Dairies
Lactobacillus paracasei 431	Christian Hansen
Lactobacillus paracasei F-19	Arla Foods
Lactobacillus plantarum 299v	Pro Viva
Lactobacillus reuteri	Bio Gaia
Lactobacillus rhamnosus GG	Valio

5-carbon sugars. *L. acidophilus* is a genetically heterogeneous species, and the classification of the strains has been a difficult task. With the technological advances in molecular biology, in 1992, *L. acidophilus* was split into six major species—*L. crispatus*, *L. amylovorus*, *L. gallinarum*, *L. gasseri*, *L. johnsonii*, and *L. acidophilus*—which have clearly observable differences but constitute the same *L. acidophilus* group (Hammes and Vogel 1995).

In 1921, Rettger and Cheplin hypothesized that if a person consumed fermented milk made with *L. bulgaricus*, gram-positive bacteria counts would not increase within the fecal matter. Nonetheless, when fermented milk is produced by *L. acidophilus*, the numbers of gram-positive bacteria increase significantly (Fig. 3). Thus, the conclusion was drawn that *L. bulgaricus* cannot survive in the large intestine but *L. acidophilus* could.

In 1928, Dr. Shahani of Nebraska State University was the first to study the production of fermented milk by means of *L. acidophilus*. Dr. Shahani then established Nebraska Culture Inc. to mass-produce the *L. acidophilus* DDS-1 strain for preparation of various kinds of probiotics. Although *L. acidophilus* was discovered in Europe, scientists in the USA had already been analyzing numerous kinds of LAB and went even further commercializing them (Oh 2009).

The strain NCFM of *L. acidophilus* was then commercialized by North Carolina State University. This strain was first isolated from human feces; this microbe can produce bacteriocin and reduces the serum level of cholesterol. In 2005, the genome of this strain was first sequenced. Currently, Danisco has a patent on the use of this strain for production of yogurt. Although the strains DDS-1 and NCFM were both described in the USA, La5 was first isolated and studied in Europe. A Danish bacterial-culture company named Christian Hansen first commercialized a starter culture containing the La5 strain. La5 can prevent diarrhea and significantly improve the intestinal health of humans. Nowadays, this strain is used in yogurt and food supplements worldwide.

L. acidophilus was the first group of bacteria reported to have a cholesterol-lowering effect in humans and animals. Several studies involving rats (Akalin

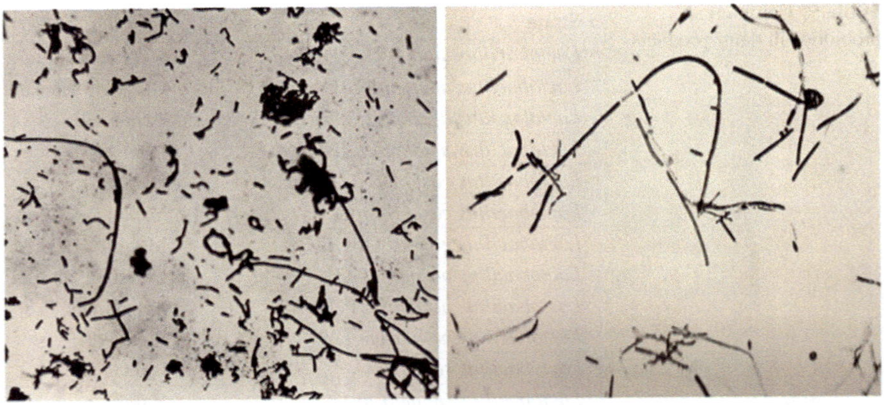

Fig. 3 Gram stained *Lactobacillus acidophilus* (*left*) and *Lactobacillus bulgaricus* (*right*). The photographs were reproduced from Rettger and Cheplin (1921)

et al. 1997), pigs (De Rodas et al. 1996), and humans (Schaafsma et al. 1996) who were fed *L. acidophilus*-fermented products (or a diet containing *L. acidophilus*) demonstrated a significant reduction in blood cholesterol levels. Bile is produced in the liver from various substances, including cholesterol. The liver turns cholesterol into cholic and deoxycholic acids that are combined with glycine and taurine (Brandt and Bernstein 1976). Gilliland et al. (1985) reported that when *L. acidophilus* is grown in the presence of cholesterol, some of the cholesterol is incorporated into the *L. acidophilus* cells while they are growing.

There are recent reports showing that *L. acidophilus* growth may be regulated by the changes in the AI-2 (autoinducer 2) signaling system (Kim et al. 2008; Yun et al. 2014). Quorum sensing is a bacterial cell-to-cell communication mechanism that involves production, detection, and a response to extracellular signaling molecules called autoinducers. Autoinducers accumulate in the environment as the bacterial-population density increases, and bacteria detect a threshold concentration of these autoinducers; this event leads to changes in gene expression. Because of these signal response systems, bacteria can act as a collective unit, that is, a multicellular entity, as opposed to individual cells performing individual functions. Quorum sensing is involved in bioluminescence, sporulation, competence, antibiotic production, biofilm formation, and secretion of virulence factors (Rutherford and Bassler 2012). Many pathogens synthesize an extracellular signal called LuxS or AI-2, for example, *Escherichia coli* O157:H7, *Clostridium perfringens*, and *C. difficile*. Thus, probiotics such as *L. acidophilus* have been implicated in the control of virulence of pathogenic bacteria via downregulation of AI-2 (Kim et al. 2008; Yun et al. 2014) (Fig. 4).

In 1930, Dr. Minoru Shirota succeeded in isolating and culturing a strain of *Lactobacillus* that can survive gastric juice and bile salts. This strain is now known as *L. casei* strain Shirota. Dr. Shirota then developed a fermented-milk beverage, called Yakult in 1935. *L. rhamnosus* GG (ATCC 53103) was isolated from the stool

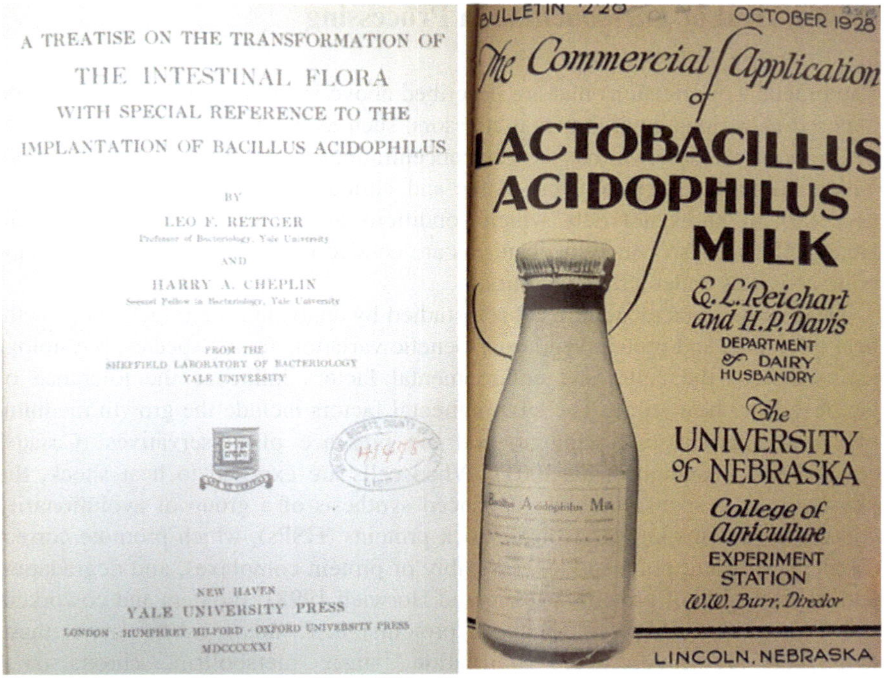

Fig. 4 The cover page of the book written by Rettger and Cheplin (1921; *left*) and by Reichart and Davis (1928; *right*)

of a healthy person in 1985 (Gorbach 2000), and subsequent studies showed beneficial effects on patients with colitis (Gorbach et al. 1987). This strain has been shown to adhere to the colonic mucosa in humans (Alander et al. 1999) and can be isolated from the colonic mucosa and feces. It survives for 1–3 days in most people and for up to 7 days in 30 % of people. *L. rhamnosus* GG has most of the characteristics that are generally desirable in a good probiotic strain, including the ability to survive passage through the human gastrointestinal tract after ingestion and the ability to transiently colonize the ileum and colon (Doron et al. 2005).

L. casei Shirota and *L. rhamnosus* GG stimulate mucosal IgA responses and enhance antigen uptake by Peyer's patches (Gorbach 2000). Consumption of *L. casei* Shirota and *L. rhamnosus* GG has been shown to be beneficial for human health. Accordingly, both microbes are used worldwide as probiotics in various milk products.

5 Survival of *Lactobacillus* in Processing

The practical applications that are described above imply that lactobacilli may be exposed to various environmental stressors, such as the extremes of temperature, pH, osmotic pressure, various oxygen concentrations, high pressure, and starvation, which may affect physiological status and various properties of the cells. It is necessary to know not only which conditions are favorable or detrimental to lactobacilli but also which mechanisms are conducive to their survival and metabolic activities under stress conditions.

Heat stress in lactobacilli has been studied by analyzing the effects on growth, heat tolerance, and protein synthesis. Genetic variation among species, physiological status of the cells, and environmental factors influence the tolerance of lactobacilli to heat stress. The environmental factors include the growth medium, pH, water activity, salt content, and the presence of preservatives (Casadei et al. 2001; Desmond et al. 2001). When cells are exposed to heat shock, the physiological response involves enhanced synthesis of a group of evolutionarily conserved proteins known as heat shock proteins (HSPs), which promote correct folding of nascent polypeptides, assembly of protein complexes, and degradation and translocation of proteins (Bukau and Horwich 1998). Wouters and coworkers (2000) identified CIPs (cold-induced proteins) in *L. lactis* subsp. *lactis*; these molecules are involved in translation, sugar metabolism, chromosomal restructuring, and signal transduction. Low-molecular-mass (approximately 7 kDa) CIPs differ from other inducible proteins because CIPs putatively belong to the cold shock protein (CSP) family. These CSPs are thought to help the cell to survive at temperatures lower than the optimal growth temperature, in contrast to heat shock proteins, which help the cell to survive at temperatures higher than the optimal level, possibly because of condensation of chromosomes and formation of the prokaryotic nucleoid (Obokata et al. 1991).

Acidity is a major environmental stressor that affects LAB during fermentation of foods and beverages. Probiotic lactobacillus strains are exposed to extreme acid stress when they reach the stomach, where hydrochloric acid is abundant in gastric juice. A number of mechanisms behind the regulation of intracellular pH homeostasis have been identified in gram-positive bacteria, for example, the mechanisms involving F_1F_0-ATPase proton pumps, amino acid decarboxylation, general stress proteins, and molecular chaperones (which repair and degrade damaged DNA and proteins).

Lactobacillus species are also exposed to osmotic stress. A sudden increase in the osmolarity of the environment causes water to move from the cell to the outside; this process causes a detrimental loss of cell turgor and changes the intracellular solute concentrations and the cell volume. During hyperosmotic stress (18 % NaCl for 2 h), the survival of mid-exponential-phase cells of *L. acidophilus* decreased by 46 % (Kim et al. 2001). Exposure of the cells for 1 h to a sublethal concentration of NaCl (2 %) slightly increases the rate of survival, and the cells adapted to this level of NaCl salinity show increased tolerance of bile salt. The detrimental concentration of NaCl for *L. alimentarius* is 8 % (Lemay et al. 2000). Hyperosmotic conditions induce transcription of the *dnaK* and *htrA* operons in *L. sakei* LTH681 (Schmidt et al. 1999). In general, hyperosmotic conditions that are based on sugar stress are much less detrimental and have only transient effects because lactobacilli can equilibrate the extracellular and intracellular concentrations of lactose and sucrose. As shown for *L. plantarum*, the uptake of these sugars most likely proceeds via facilitated diffusion with the help of a transport system with a low affinity for the substrates; this notion is consistent with the inability of the sugars to serve as compatible solutes (Glaasker et al. 1996).

Oxidative stress means conditions where the production of reactive oxygen species (ROS) results in adverse effects on cellular functions. The 4-electron reduction of O_2 to two molecules of H_2O gives rise to ROS: superoxide radical anion (O_2^-), hydrogen peroxide (H_2O_2), and hydroxyl radical ($HO^\cdot$). O_2^- and H_2O_2 may contribute to the formation of the highly reactive oxidant $HO^\cdot$ via the Fenton and Haber–Weiss reactions. O_2^- can diffuse across a considerable distance to reach a potential target and is more reactive with intracellular proteins than H_2O_2 or $HO^\cdot$ is. Cellular components such as hemoproteins, and especially lipids and DNA, are $HO^\cdot$ targets. Lactobacilli are fermentative aerotolerant microorganisms that do not use a proton-translocating electron transport chain but possess oxidases that use O_2 to oxidize substrates such as pyruvate or NADH. The NADH oxidase/NADH peroxidase system is an alternative way to regenerate NAD^+ along with the conversion of pyruvate to lactate or ethanol. Pyruvate is then available for conversion to acetate, yielding an extra mole of ATP. The activity of these oxidases can produce ROS that cause oxidative stress in the cell. It is therefore expected that the presence of O_2 will induce a specific cellular response. Bacteria may use enzymes (catalase, NADH oxidase, NADH peroxidase, or superoxide dismutase) or nonenzymatic compounds (Mn^{2+}, ascorbate, tocopherols, or glutathione) to reduce oxygen radicals (De Angelis and Gobbetti 2004).

Another stressor, high pressure, causes denaturation of proteins and damage to membranes with detrimental effects on microbial functions and viability. Microbial growth is generally inhibited at a pressure of 20–130 MPa, and cell death occurs in the range 130–800 MPa. Tolerance to high pressure varies depending on the species, strain, and the culture medium used. Milk suspensions of three *L. lactis* subsp. *lactis* strains that were subjected to 400 MPa for 20 min showed a decrease in cell viability of log 1.87, 2.98, or 3.18 CFU/mL. Similar differences were found among the strains of *L. casei* subsp. *casei*. The stress induced by high pressure negatively affects the survival of bacteria during food storage under adverse conditions. Hydrostatic pressure influences pH_i by enhancing dissociation of weak organic acids and by increasing permeability of the cytoplasmic membrane; these changes limit the efficiency of pH homeostasis (De Angelis and Gobbetti 2004).

In a recent report (Bang et al. 2014), investigators measured viability of commercial probiotics during long-term storage. Figure 5 shows the viability of starter cultures from three commercially available types of yogurt during a 12-month storage period. The counts of viable cells were maintained at 10^7–10^8 CFU/mL

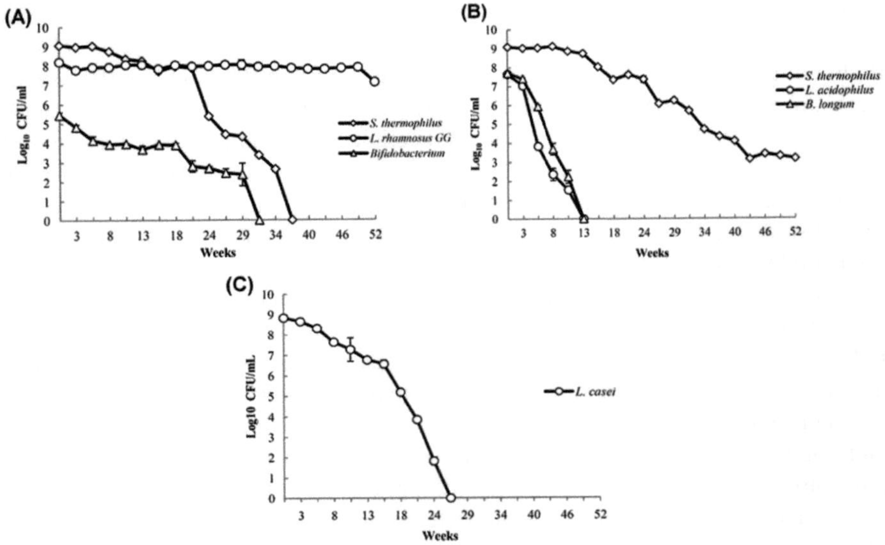

Fig. 5 Survival of lactic acid bacteria (LAB) in yogurt A (*S. thermophilus*, *L. bulgaricus*, *L. rhamnosus* GG, and *B. animalis* BB12), yogurt B (*S. thermophilus*, *L. acidophilus* La5, and *B. animalis* BB12), and yogurt C (*L. casei* Shirota) during storage for up to 52 weeks at 5 °C. The counts of viable cells of LAB were measured at 15-day intervals. The figures were reproduced from Bang et al. (2014) with permission

for *Lactobacillus* sp. and 10^9 CFU/mL for *S. thermophilus* for up to 5 weeks. Similarly, the initial counts of *Bifidobacterium animalis* BB12 were ca. 10^5–10^7 CFU/mL in both yogurt A and yogurt B. The *L. rhamnosus* strain GG exhibits the highest viability in samples “A” during the storage period, with counts of 10^7 CFU/mL after 52 weeks of storage. These results indicate that commercial probiotics such as *L. rhamnosus* GG are tolerant of the acidic conditions for 1 year at 5 °C.

6 Future Perspectives

One hundred fifteen years have passed since Ernest Moro made his first scientific contribution to *Lactobacillus* study. Many researches have been conducted that *Lactobacillus* species in dairy products enhance the intestinal health including immune-stimulations and pathogen-inhibitions. And also, there are many proposed mechanisms by which *Lactobacillus* may protect the host from gastrointestinal diseases. Therefore, more clinical trials must be conducted to define clearly the mechanisms of specific *Lactobacillus* strains. Much work remains to classify the mechanisms of action of particular *Lactobacillus* strains for health benefits in dairy products.

References

Akalin AS, Gonc S, Duzel S (1997) Influence of yogurt and acidophilus yogurt on serum cholesterol levels in mice. J Dairy Sci 80(11):2721–2725. doi:10.3168/jds.S0022-0302(97)76233-7

Alander M, Satokari R, Korpela R, Saxelin M, Vilpponen-Salmela T, Mattila-Sandholm T, Von Wright A (1999) Persistence of colonization of human colonic mucosa by a probiotic strain, *Lactobacillus rhamnosus* GG, after oral consumption. Appl Environ Microbiol 65(1):351–354

Bahrami B, Macfarlane S, Macfarlane GT (2011) Induction of cytokine formation by human intestinal bacteria in gut epithelial cell lines. J Appl Microbiol 110(1):353–363. doi:10.1111/j.1365-2672.2010.04889.x

Bang M, Oh S, Lim KS, Kim Y, Oh S (2014) The involvement of ATPase activity in the acid tolerance of *Lactobacillus rhamnosus* strain GG. Int J Dairy Technol 67(2):229–236. doi:10.1111/1471-0307.12123

Brandt LJ, Bernstein LH (1976) Bile salts: their role in cholesterol synthesis, secretion and lithogenesis. Am J Gastroenterol 65(1):17–30

Bukau B, Horwich AL (1998) The Hsp70 and Hsp60 chaperone machines. Cell 92(3):351–366. doi:10.1016/S0092-8674(00)80928-9

Cahn A (1901) Ueber die nach Gram färbbaren Bacillen des Säuglingsstuhles. Centralbl F Bakt Orig 30:721–726

Casadei MA, Ingram R, Hitchings E, Archer J, Gaze JE (2001) Heat resistance of *Bacillus cereus*, *Salmonella typhimurium* and *Lactobacillus delbrueckii* in relation to pH and ethanol. Int J Food Microbiol 63(1–2):125–134. doi:10.1016/S0168-1605(00)00465-7

De Angelis M, Gobbetti M (2004) Environmental stress responses in *Lactobacillus*: a review. Proteomics 4(1):106–122. doi:10.1002/pmic.200300497

De Rodas BZ, Gilliland SE, Maxwell CV (1996) Hypocholesterolemic action of *Lactobacillus acidophilus* ATCC 43121 and calcium in swine with hypercholesterolemia induced by diet. J Dairy Sci 79(12):2121–2128. doi:10.3168/jds.S0022-0302(96)76586-4

Desmond C, Stanton C, Fitzgerald GF, Collins K, Ross RP (2001) Environmental adaptation of probiotic lactobacilli towards improvement of performance during spray drying. Int Dairy J 11(10):801–808. doi:10.1016/S0958-6946(01)00121-2

Doron S, Snydman DR, Gorbach SL (2005) *Lactobacillus* GG: bacteriology and clinical applications. Gastroenterol Clin North Am 34:483–498

Ewaschuk JB, Dieleman LA (2006) Probiotics and prebiotics in chronic inflammatory bowel diseases. World J Gastroenterol 12(37):5941–5950. doi:10.3748/wjg.v12.i37.5941

FAO/WHO (2002) Guidelines for the evaluation of probiotics in food. Report of a joint FAO/WHO working group on drafting guidelines for the evaluation of probiotics in food. London, Ont., Canada

Fuller R (1989) Probiotics in man and animals. J Appl Bacteriol 66(5):365–378. doi:10.1111/j.1365-2672.1989.tb05105.x

Fuller R, Gibson GR (1997) Modification of the intestinal microflora using probiotics and prebiotics. Scand J Gastroenterol Suppl 222:28–31

Gilliland SE, Nelson CR, Maxwell C (1985) Assimilation of cholesterol by *Lactobacillus acidophilus*. Appl Environ Microbiol 49(2):377–381

Glaasker E, Konings WN, Poolman B (1996) Osmotic regulation of intracellular solute pools in *Lactobacillus plantarum*. J Bacteriol 178(3):575–582

Gorbach SL (2000) Osmotic regulation of intracellular solute pools in *Lactobacillus plantarum*. Am J Gastroenterol 95:S2–S4

Gorbach S, Chang TW, Goldin B (1987) Successful treatment of relapsing *Clostridium difficile* colitis with *Lactobacillus* GG. Lancet 330(8574):1519

Grigoroff S (1905) Étude sur un lait fermente comestible. Le "Kisselomleko" de Bulgarie. Rev Med De la Suisse Romande 25:714–720

Hammes WP, Vogel RF (1995) The genus *Lactobacillus*. In: Wood BJB, Holzafel WH (eds) The lactic acid bacteria, vol 2, The genera of lactic acid bacteria. Blackie Academic and Professional, Glasgow, pp 19–54

Holzapfel WH, Schillinger U (2002) Introduction to pre-and probiotics. Food Res Int 35 (2):109–116

Kim WS, Perl L, Park JH, Tandianus JE, Dunn NW (2001) Assessment of stress response of the probiotic *Lactobacillus acidophilus*. Curr Microbiol 43(5):346–350

Kim Y, Oh S, Park S, Seo JB, Kim SH (2008) *Lactobacillus acidophilus* reduces expression of enterohemorrhagic *Escherichia coli* O157: H7 virulence factors by inhibiting autoinducer-2-like activity. Food Control 19(11):1042–1050

Laura JF, Fuller R, Gibson GR (1999) Prebiotics, probiotics and human gut microbiology. Int Dairy J 9:53–61

Lemay MJ, Rodrigue N, Garièpy C, Saucier L (2000) Adaptation of *Lactobacillus alimentarius* to environmental stresses. Int J Food Microbiol 55(1-3):249–253

Lenoir-Wijnkoop I, Sanders ME, Cabana MD, Caglar E, Corthier G, Rayes N, Sherman PM, Timmerman HM, Vaneechoutte M, Loo JV, Wolvers DAW (2007) Probiotic and prebiotic influence beyond the intestinal tract. Nutr Rev 65:469–489

Lilly DM, Stillwell RH (1965) Probiotics: growth-promoting factors produced by microorganisms. Science 147(3659):747–748

Limdi JK, O'Neill C, McLaughlin J (2006) Do probiotics have a therapeutic role in gastroenterology? World J Gastroenterol 12(34):5447–5457

Maslowski KM, Vieira AT, Ng A, Kranich J, Sierro F, Di Y, Schilter HC, Rolph MS, Mackay F, Artis D, Xavier RJ, Teixeira MM, Mackay CR (2009) Regulation of inflammatory responses by gut microbiota and chemoattractant receptor GPR43. Nature 461:1282–1286

Metchnikoff E (1908) The prolongation of life. Putnam, Mahopac, NY

Metchnikoff O (1921) Life of Elie Metchnikoff, 1845–1916. Houghton Mifflin

Moro E (1900) Ueber den Bacillus acidophilus. Jahrb F Kinderh 52:38–55

Nutrition Business Journal. Nuance in the probiotics market (2014) http://newhope360.com/node/924421

Obokata J, Ohme M, Hayashida N (1991) Nucleotide sequence of a cDNA clone encoding a putative glycine-rich protein of 19.7 kDa in Nicotiana sylvestris. Plant Mol Biol 17(4):953–955

Oh S (2009) Probiotics and prolongation of life. Korean J Dairy Sci Technol 26:31–37

Oh S, Chai CH, Kim SH, Kim YJ, Kim HS, Worobo RW (2012) Comparison of acid and bile tolerances, cholesterol assimilation, and CLA production in probiotic *Lactobacillus acidophilus* strains. Korean J Food Sci Anim Resour 32:409–413

Orla-Jensen S (1919) The Lactic acid bacteria. Anhr Fred Høst and Søn Copenhagen Andr Fred Host & Son Copenhagen

Ouwehand AC (2007) Antiallergic effects of probiotics. J Nutr 137(3):794S–797S

Parker RB (1974) Probiotics. The other half of the antibiotics story. Anim Nutr Health 29:4–8

Pineiro M, Stanton C (2007) Probiotic bacteria: legislative framework-requirements to evidence basis. J Nutr 137:850S–853S

Reichart EL, Davis HP (1928) The commercial application of *Lactobacillus acidophilus* milk. The University of Nebraska, College of Agriculture Experimental Station, Lincoln, Nebraska. Bulletin 228:1–19

Rettger LF, Cheplin HA (1921) A treatise on the transformation of the intestinal flora with special reference to the implantation of *Bacillus acidophilus*. Yale University Press, New Haven, CT

Rettger LF, Cheplin HA (1922) *Bacillus acidophilus* and its therapeutic application. Arch Intern Med 29:357–367

Rutherford ST, Bassler BL (2012) Bacterial quorum sensing: its role in virulence and possibilities for its control. Cold Spring Harb Perspect Med 2:a012427

Savard P, Lamarche B, Paradis ME, Thiboutot H, Laurin É, Roy D (2011) Impact of *Bifidobacterium animalis* subsp. *lactis* BB-12, and Lactobacillus acidophilus LA-5-containing yoghurt, on fecal bacterial counts of healthy adults. Int J Food Microbiol 149(1):50–57

Schaafsma G, Meuling WJA, van Dokkum W (1996) Nieuw zuivelprodukt gefermentteerd met *Lactobacillus acidophilus*. Zuivel kan serumcholesterol verlagen Voeding 57:12–14

Schmidt G, Hertel C, Hammes WP (1999) Molecular characterisation of the dnaK operon of *Lactobacillus sakei* LTH681. Syst Appl Microbiol 22:321–328

Szajewska H, Kotowska M, Mrukowicz JZ, Armánska M, Mikolajczyk W (2001) Efficacy of *Lactobacillus* GG in prevention of nosocomial diarrhea in infants. J Pediatr 138(3):361–365

Tannock GW (2003) Probiotics: time for a dose of realism. Curr Issues Intest Microbiol 4 (2):33–42

Tissier H (1900) Rechèrches sur la flore du intestinale normale et pathologique du nourrison. Thése Paris 85–96

Tissier H, Martelly A (1902) Recherches sur la putréfaction de la viande de boucherie. Ann Inst Pasteur 16:865–903

Vergin F (1954) Anti- und Probiotika (Anti- and probiotics). Hippokrates 25:116–119

Waksman SA, Woodruff HB (1942) Streptothricin, a new selective bacteriostatic and bactericidal agent, particularly active against Gram-negative bacteria. Exp Biol Med 49(2):207–210

Weirich A, Hoffmann GF (2005) Ernst Moro (1874–1951)—A great pediatric career started at the rise of university-based pediatric research but was curtailed in the shadows of Nazi laws. Eur J Pediatr 164(10):599–606

Weiss H (1904) Zur Kentniss der Darmflora. Centralbl F Bakt Abt 1 Orig 36: 13–28

Wouters JA, Kamphuis HH, Hugenholtz J, Kuipers OP, De Vos WM, Abee T (2000) Changes in glycolytic activity of *Lactococcus lactis* induced by low temperature. Appl Environ Microbiol 66(9):3686–3691

Yun B, Oh S, Griffiths MW (2014) *Lactobacillus acidophilus* modulates the virulence of *Clostridium difficile*. J Dairy Sci 97(8):4745–4758

Current Trends and Future Perspectives on Functional Foods and Nutraceuticals

Eric Banan-Mwine Daliri and Byong H. Lee

Contents

Abstract The rising awareness of consumers toward the health benefits of foods and their nutritional benefits for potential disease prevention and health enhancement is the driving force of the global functional food and nutraceutical market. Functional foods are the medicinal foods that provide health benefits beyond energy and essential nutrients. Many studies, including several European Commission (EC) funded projects, have led to the understanding of the potential mechanisms of biologically active components in food, which could improve health and

E.B.-M. Daliri
School of Biotechnology, Jiangnan University, 1800 Lihu Avenue, Wuxi 214122, Jiangsu, China

B.H. Lee (✉)
School of Biotechnology, Jiangnan University, 1800 Lihu Avenue, Wuxi 214122, Jiangsu, China

Department of Food Science and Agricultural Chemistry, McGill University, Ste-Anne-de-Bellevue, QC, Canada, H9X 3V9
e-mail: byong.lee@mail.mcgill.ca

M.-T. Liong (ed.), *Beneficial Microorganisms in Food and Nutraceuticals*, Microbiology Monographs 27, DOI 10.1007/978-3-319-23177-8_10

probably reduce the risk of disease while enhancing our overall well-being. Functional foods and nutraceutical products are helping to improve health, reducing the healthcare costs, and supporting the economic development in rural areas. Growing demand for functional foods is also helping the producers to diversify their agriculture and marine-based crops and promoting research and innovation. There is growing demand for functional foods, especially in developed economies due to increasing awareness toward health benefits of functional foods and an increase in disposable income. The USA is the world's largest market for functional foods and is expected to witness 21 % growth in the coming years owing to the growing demand for functional foods and expected to reach the value of USD 8.62 billion by 2015. This growth is mainly driven by the continuously growing demand for energy drinks and fortified dairy products. Meanwhile, country-specific regulations and health claim substantiation are some of the challenges the functional food and nutraceutical market continues to face. This chapter gives an overview of the functional food and nutraceutical market, motivations, challenges, and future perspectives of the market.

1 Introduction

The statement "Let food be thy medicine, thy medicine shall be thy food," made by Hippocrates, the father of modern medicine, is receiving a lot of interest today as food scientists and consumers realize the health benefits of certain foods (El Sohaimy 2012). The concept of functional foods was invented in Japan. In the 1980s, health authorities in Japan recognized the need for an improved quality of life as well as increasing life expectancy for the expanding number of elderly people in the population in order to control healthcare costs. The concept of foods that were developed specifically to promote health or reduce the risk of disease was introduced and was called Foods for Specified Health Use (FOSHU). According to the European Food Information council, functional foods must contain biologically active components that have the potential to optimize physical and mental well-being and which may also reduce the risk of disease (http://www.eufic.org/article/en/nutrition/functional-foods/expid/basics-functional-foods/). Such foods include fortified food, enhanced food, enriched food, dietary supplements, and health food. The term "nutraceutical" was coined from the words, "nutrition" and "pharmaceutical" (Fig. 1) in 1989 by Stephen DeFelice, founder and chairman of the Foundation for Innovation in Medicine (FIM), Cranford. He defined nutraceuticals as *"a food (or part of a food)* that provides medical or health benefits, including the prevention and/or treatment of a disease" (El Sohaimy 2012). They are usually in the form of pills, tinctures, and capsules. Nutraceuticals cover a broad range of products including beverages, dietary supplements, isolated nutrients, genetically engineered designer foods, herbal products, and processed foods.

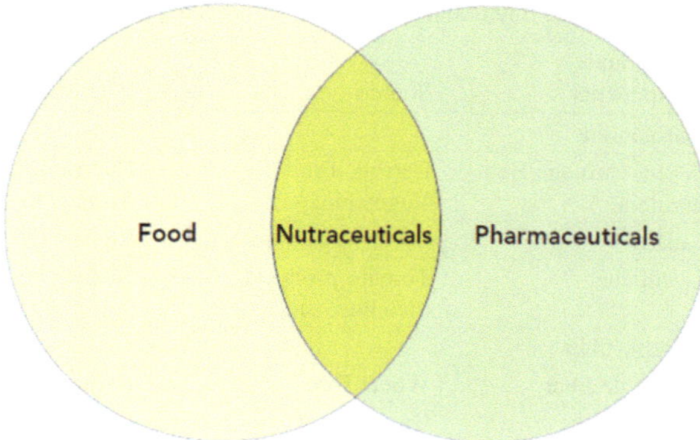

Fig. 1 Nutraceuticals: a blend of pharmaceuticals and food

The basic categories of nutraceuticals are (1) dietary supplements, (2) functional foods and beverages, and (3) nutraceutical ingredients (raw minerals or oils). The classical definition of nutraceutical and functional food states that both functional foods and nutraceuticals are derived from food or part of a food (Shahidi 2009). However, it seems inapplicable to the modern nutraceuticals since a number of nonfood-derived ingredients are now recognized as safe and are used in nutraceutical formulations to prevent cancer, neurodegenerative diseases, and other chronic diseases (Dewapriya and Kim 2014). Various kinds of natural plants (Harbourne et al. 2013; Johanningsmeier and Harris 2011), animals (Galanakis 2013), and marine sources (Freitas et al. 2012) possess biologically active metabolites that can be used as novel sources for developing modern nutraceuticals and functional foods. Several foods have been shown to help fight against or reduce the risk of cardiovascular diseases, obesity, diabetes, hypertension dyslipidemia, and other chronic ailments (Table 1), and an extensive review on such foods has been published (Mohamed 2013).

2 Some Bioactive Components in Food and Health Benefits

It is generally known that foods do not only provide nutrients for nourishment but may also confer additional health benefits for the prevention and treatment of various types of diseases. Recently, there is a pool of evidence substantiating the health claims of certain foods due to the bioactive compounds they contain (Palozza et al. 2010). These health-promoting foods or compounds are either classified as functional foods or nutraceuticals.

Table 1 Some functional foods, their active ingredients, and potential benefits

Functional components	Source	Potential benefits
Carotenoids		
Alpha-carotene/Beta-carotene	Carrots, Fruits, Vegetables	Neutralize free radicals, which may cause damage to cells
Lutein	Green vegetables	Reduce the risk of macular degeneration
Lycopene	Tomato products (ketchup, sauces)	Reduce the risk of prostate cancer
Dietary fiber		
Insoluble fiber	Wheat Bran	Reduce risk of breast or colon cancer
Beta-glucan	Oats, barley	Reduce risk of cardiovascular disease. Protect against heart disease and some cancers; lower LDL and total cholesterol
Soluble fiber	Psyllium	Reduce risk of cardiovascular disease. Protect against heart disease and some cancers; lower LDL and total cholesterol
Fatty acids		
Long-chain omega-3 fatty acids—DHA/EPA	Salmon and other fish oils	Reduce risk of cardiovascular disease. Improve mental, visual functions
Conjugated linoleic acid (CLA)	Cheese, meat products	Improve body composition. Decrease risk of certain cancers
Phenolics		
Anthocyanidins	Fruits	Neutralize free radicals; reduce risk of cancer
Catechins	Tea	Neutralize free radicals; reduce risk of cancer
Flavonones	Citrus	Neutralize free radicals; reduce risk of cancer
Flavones	Fruits/vegetables	Neutralize free radicals; reduce risk of cancer
Lignans	Flax, rye, vegetables	Prevention of cancer, renal failure
Tannins (proanthocyanidins)	Cranberries, cranberry products, cocoa, chocolate	Improve urinary tract health. Reduce risk of cardiovascular disease
Plant sterols		
Stanol ester	Corn, soy, wheat, wood oils	Lower blood cholesterol levels by inhibiting cholesterol absorption
Prebiotics/probiotics		
Fructo-oligosaccharides (FOS)	Jerusalem artichokes, shallots, onion powder	Improve quality of intestinal microflora; gastrointestinal health
Lactobacillus	Yogurt, Other dairy	Improve quality of intestinal microflora; gastrointestinal health
Soy phytoestrogens		
Isoflavones: DaidzeinGenistein	Soybeans and soy-based foods	Reduce menopause symptoms, such as hot flashes. Protect against heart disease and some cancers; lower LDL and total cholesterol

Source: International Food Information council (IFIC, www.ific/org, 2009)

2.1 Probiotics and Prebiotics

The human body is colonized by over a trillion microbes (Martín et al. 2013), some of which are beneficial and others harmful. It is believed that an imbalance between the beneficial and harmful bacteria leads to diseases such as obesity and urogenital infections (Vujic et al. 2013). Therefore, probiotic and prebiotic supplements may be effective in preventing as well as combating such conditions. The World Health Organization (WHO) defines probiotics as "Live microorganisms which, when administered in adequate amounts, confer a health benefit on the host." Prebiotics are nondigestible food ingredients that stimulate the growth and/or activity of bacteria in the digestive system in ways claimed to be beneficial to health (FAO/WHO 2002). Many foods including chicory roots, banana, tomato, and alliums are rich in fructo-oligosaccharides, while beans and peas contain raffinose and stachyose (Das et al. 2012). For a microbe to be classified as probiotic, it must be (1) from human origin, (2) nonpathogenic (3), resistant to technological processes, (4) stable in acid and bile, (5) adhere to target epithelial tissue, (6) able to persist within the GI tract, (7) produce antimicrobial substances, (8) able to modulate the immune system, and (9) able to influence metabolic activities (Baek and Lee 2009). By far, the most commonly selected probiotic strains are from the *Lactobacillus* and *Bifidobacterium* genera. Probiotics and prebiotics are found in various foods such as yogurt and sauerkraut and they have been reported to have many health effects such as weight reduction (Dewulf et al. 2013), cholesterol reduction (van Baarlen et al. 2010), cardiovascular health (Ramchandran and Shah 2011; Kerry et al. 2011), and hypertension (Lin et al. 2013). Prebiotics are fermented by probiotics to release short-chain fatty acids such as acetate, butyrate, and propionate, which activate peroxisome proliferator-activated receptors (PPARγ) to regulate fatty acid oxidation in muscle and adipocytes (Sheril et al. 2013; Korecka et al. 2013). Though *Lactobacillus reuteri cardioviva* NCIMB 30242 has recently been reported to significantly lower cholesterol levels in humans and commercialized on the market (Jones et al. 2012, 2013), more research is still needed to validate the safety on secondary bile acids and other adverse effects of long-term consumption (Ishimwe et al. 2015). Well-designed human trials are needed to substantiate the health claims of prebiotics and probiotics as well as their mechanisms of action. Also, information on the optimum dose, age group, the specific effect of each probiotic when put in food matrix, and adverse effects will aid food researchers to design specific functional foods to meet specific needs (Delphine et al. 2009).

2.2 Proteins and Peptides

Proteins are long-chain polymers of amino acids, while peptides are short amino acid chains. Proteins in foods may be digestible or indigestible. Indigestible proteins trap and expel toxins and bile through feces as well as aid in cholesterol reduction (El Sohaimy 2012) in the large intestine. Wheat and soya beans are known to contain good amounts of indigestible proteins and eating them is claimed to enhance a healthy gut. Digestible proteins are however broken down into peptides during digestion and absorbed into the blood circulatory system. The peptides from the digestible proteins of soya beans have been reported to be effective in reducing blood cholesterol levels (Beecher 1999). In fact, bioactive peptides have been isolated from various sources such as marine organisms (Ngo et al. 2012), milk proteins (Hafeez et al. 2014), meat (Ryan et al. 2011), and other sources for application in nutraceuticals. The activities of bioactive peptides are based on their amino acid composition and sequence. These short chains of amino acids are inactive within the sequence of the parent protein, but can be released during gastrointestinal digestion, food processing, or fermentation (Ngo et al. 2012). Bioactive peptides have been reported to have many health benefits such as antioxidant activity (Bordbar et al. 2013), antihypertensive activity (Ryan et al. 2011), anti-human immunodeficiency virus activity (Chen et al. 2012), anti-cancer (Hsu et al. 2011), and many others (Table 2). Many angiotensin-I-converting enzyme (ACE) inhibitor peptides have been isolated from milk proteins (Hayes et al. 2006), beef (Jang et al. 2008), chicken (Saiga et al. 2006), and microalgae (Byun et al. 2009). The beneficial effects of food-derived antioxidants in health promotion and disease prevention are attracting attention. Recently, there has been a particular focus on milk-derived peptides. As a source of antioxidants, these

Table 2 Some bioactive peptides, sources, and activities

Source of peptide	Sequence/Name	Activity	Reference
Microalga	VECYGPNRPQF	Antioxidant	Byun et al. (2009)
Rice endosperm	FRDEHKK	Antioxidant	Zhang et al. (2010)
Rotifer	LLGPGLTNHA	Antioxidant	Byun et al. (2009)
Scorpion venom	Kn2–7	Anti HIV	Chen et al. (2012)
Sponge	Theopapuamide B	Anti HIV	Senevirathne and Kim (2012)
Tuna muscle	LPHVLTPEAGAT	Antiproliferative	Hsu et al. (2011)
Microalga	VECYGPNF	ACE inhibition	Sheih et al. (2009)
Styelaplicata	MLLCS	ACE inhibition	Ko et al. (2011)
Blacktip shark	Gelatin hydrolysate	Antihypertensive, cholesterol reduction	Kittiphattanabawon et al. (2013)
Mactraveneriformis	NGAVMILTH	Calcium binding	Wang et al. (2012)
Fish backbone	NKDRG	Antiproliferative	Naqash and Nazeer (2012)
Tilapia	TNTLHYT	Ca-binding	Charoenphun et al. (2013)
Beef	VLAQYK	ACE inhibition	Jang et al. (2008)
Chicken	FKGRYYP	ACE inhibition	Saiga et al. (2006)
Milk	GLSDGEWQ and GFHI	Antimicrobial	Hayes et al. (2006)

peptides remain inactive within the sequence of the parent protein but can be released during enzyme hydrolysis. Once released, the peptides show radical scavenging, metal ion chelation properties, and the ability to inhibit lipid peroxidation (Power et al. 2012). Lactoferrin in milk has been reported to inhibit intestinal tumors. This peptide acts by inducing apoptosis, inhibiting angiogenesis and modulating carcinogen-metabolizing enzymes (El Sohaimy 2012). Parodi (2007) reported that the addition of selenium to cow milk increased selenoprotein levels and these proteins were found to inhibit colon tumorigenesis in rats. Though much is known about bioactive peptides, greater understanding of the biological mechanisms surrounding how these peptides control the human body is necessary in order to successfully design and produce new functional foods and nutraceutical agents.

2.3 Lipids and Fatty Acids

Essential polyunsaturated fatty acids (PUFA) such as omega-3 (n-3) and omega-6 (n-6) are essential in regulating metabolic processes. Docosahexaenoic acid (DHA; 22:6) and eicosapentaenoic acid (EPA; 20:5) which are N-3 PUFA have attracted a lot of attention due to their potential health benefits (Balogun et al. 2013). N-3 PUFA prevent atherosclerosis (Wassall and Stillwell 2009), regulate nuclear transcription factors involved in the gene expression of inflammatory markers, stimulate cognitive development (Cottin et al. 2011), and reduce high triglyceride (TG) levels, endothelial dysfunction, inflammation, and cardiac arrhythmia which are markers of cardiovascular disease (Guttler et al. 2012). DHA has been shown to be a structural component of the brain which contributes to memory functions. This has therefore caused an increased incorporation of DHA into margarines and baby foods to enhance brain memory development and reduce the severity of Alzheimer's disease (Cole and Frautschy 2010). Other foods such as fish oil, vegetable oils, and nuts such as peanuts and almonds are rich in linoleic and linolenic acids and therefore provide cardiovascular benefits as well as antioxidant activities. Conjugated linoleic acid (CLA) is another important lipid found mostly in dairy products or meat products derived from ruminants. Foods based on milk are naturally enriched with *cis*-9, *trans*-11-CLA (c9, t11-CLA) and are reported to prevent cancer, heart health, obesity, diabetes, and bone health in animal models (Koba and Yanagita 2014). Also *trans*-10, *cis*-12 CLA have a predominant effect on weight management and prevention of athrogenesis (Gaullier et al. 2004) and thus anti-obesity CLA pills or syrups are on the market. Recently, genetic engineering for the production of trasn-10, cis-12 CLA in a yeast, *Yarrowia lipolytica*, was successfully overproduced (Zhang et al. 2013). Although many reports have been made on the potential health effects of CLA, further substantiation of the scientific evidence relating to CLA and human health benefits is still required before health claims can be confirmed. Soybean oil has high tocopherol content and therefore has a good antioxidant capacity. Also wheat germ oil contains high levels of Vitamin E, an antioxidant, and therefore used in lotions and creams to heal

Table 3 Fatty acid composition of food oils

Food	Stats (%)	Omega-9 MUFA (oleate)	Omega-6 linoleate	Omega-3 a-linolenate
Avocado oil	13.4	72	9	0.5
Olive oil	17.3	66.8	11.8	0.6
Peanut butter	21.6	51.6	26.2	–
Maize oil (corn oil)	17.2	28.7	47.8	1.5
Canola oil	7.8	58.2	20.8	10.1
Wheat germ oil	20.1	15.4	53.6	10.4
Almonds	8.3	70.9	19.1	0.5

Adapted from El Sohaimy (2012)

and rejuvenate the skin. Table 3 shows some food oils and their fatty acid composition.

2.4 *Catechins (Proanthocyanides)*

The recent interest of polyphenols in green and black tea has increased due to their antioxidant activities and their possible roles in the prevention of cancer, cardiovascular disease, renal disease, and intestinal bacterial flora (Fukuzawa et al. 2014). A majority of phenolic compounds (catechins) belong to flavonoids. There are four main types of catechins: epicatechin (EC), epigallocatechin (EGC), epicatechingallate (ECG), and epigallocatechingallate (EGCG) (Fig. 2). Zhang et al. (2012) reported the effects of catechin-enriched green tea beverage on visceral fat loss in adults with a high proportion of visceral fat, and Fukuzawa et al. (2014) reported the ability of green tea catechins to treat nonalcoholic steatohepatitis (NASH) in human subjects. Also, DeRossi et al. (2014) observed a restoration of salivary functions in xerostomia human subjects. Cranberry juice is reported to contain high levels of epicatech in polymers which prevent adhesion of viruses and bacteria to the urinary tract (Hisano et al. 2012). Regular consumption of cranberry juice or cranberry concentrate tablets has been shown to reduce antibiotic requirements in women experiencing urinary tract infections (El Sohaimy 2012). These health claims have increased the application of catechins in the nutraceutical industry. Using the tea creaming effect and by enhancing it as a phase separation via precipitation makes it possible to recover a large amount of polyphenols from the cream phase without using toxic solvents. The addition of aluminum chloride (Wu and Bird 2010) to green tea aqueous solutions and the addition of a precipitation agent such as polyvinylpyrrolidone (PVP) are other methods used for recovering catechins from tea leaves (Dong et al. 2011). Water can also be used to extract high amounts of polyphenols and proanthocyanidins from pomegranate peels at 50 °C for 20 min in a buffer at pH 3.5 (Wissam et al. 2012). Since water is nontoxic

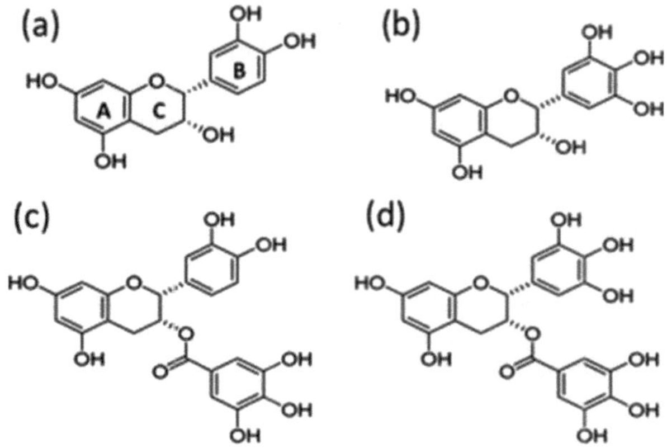

Fig. 2 Catechins chemical structure; (**a**) epicatechin (EC), (**b**) epigallocatechin (EGC), (**c**) epicatechingallate (ECG), and (**d**) epigallocatechingallate (EGCG). From Monsanto et al. (2014)

and gives a good recovery, this method could represent an environmentally friendly method for producing antioxidants from pomegranate peels.

Catechins are not very stable in products due to preparation, food processing conditions, brewing conditions, as well as storage conditions; hence, nutraceuticals containing them must be analyzed to ensure their bioavailability to consumers (Abd-ElSalam et al. 2014).

2.5 *Lycopene*

Lycopene is the red pigment of tomato and other red fruits and vegetables such as papaya, watermelon, carrot, pink guava, and pink grapefruit. It is a carotenoid with high antioxidant potential. Though the biological function of lycopene has not been fully known, its positive impact in modulating the gene expressions of prostate cancer is well known (Lee and Foo 2014). By a nutrigenomic approach, Lee and Foo found lycopene to be a unique candidate in the regulation of DNA repair system, testosterone signaling, inflammatory cytokine secretion, intercellular gap junction interaction, PPARγ–LXRα–ABCA1 pathway, cell cycle and integrin, IGF-IR-activated endogenous antioxidant enzyme, and the migration and invasion of tumor signaling in prostate cancer. Lycopenes are also reported to prevent cardiovascular diseases and oxidative stress in the liver, kidney, and testes (Boeiraa et al. 2014) and also have an antiproliferative activity. Sonmez et al. (2011) suggested lycopene as an alternative treatment for sperm toxicity after chemotherapy.

Lycopene extraction methods include solvent extraction, hydrostatic pressure processing, enzymatic treatment, supercritical fluid extraction (SCFE) with CO_2, ultrasonic extraction, and soxhlet extraction. Conventional spectrophotometric and HPLC assays for quantifying lycopene from plant tissue use organic solvents to extract the compound. However, if the solvents used for solvent extraction are toxic, the product becomes unwholesome for human consumption. The selective inclusion complex method has also been reported as promising for lycopene extraction and purification (Mahmoud et al. 2013). Neagu et al. (2014) have recently reported a possibility of enhancing the yield of carotenoids and phenolics' extraction from tomato tissue by using crude cell wall degrading enzymes, such as pectinases and cellulases. This method prevents the use of high quantities of organic solvents and also gives a low production cost, thereby making this enzyme-assisted hydrolysis a good biotechnological process.

The nutraceutical effects of lycopenes have been extensively reviewed by Waliszewski and Blasco (2010); however, knowledge about the mechanism by which lycopenes exhibit health effects in humans is needed to promote the design of specific nutraceuticals.

2.6 Herbs and Spices

Spices are esoteric food adjuncts that are used throughout history to enhance the sensory quality of foods. They impart characteristic flavor, aroma, and color to foods, stimulate appetite, as well as modify food texture. Dietary spices and herbs, even in minute quantities, have immense influence on the human health by their antioxidative, chemopreventive, antimutagenic, anti-inflammatory, and immune modulatory effects on cells and a wide range of beneficial effects on human health (Zhang et al. 2015). Most of the spice components are terpenes and other constituents of essential oils (Das et al. 2012). Curcumin has been shown to protect neuronal cells against toxicity, reduce apoptosis (Jayaraj et al. 2013), protect against multiple sclerosis (Lian et al. 2013), as well as antiarthritic properties (Chandran and Goel 2012). 6-Shogaol and 6-gingerol derived from ginger showed significant inhibition against downregulation of adiponectin expression mediated by TNFα in 3T3-L1 adipocytes (Isa et al. 2008), while the addition of 1.5 % w/w of lemon balm and marjoram herbs was found to increase antioxidant capacity of a portion of salad by 150 % and 200 %, respectively (Ninfali et al. 2007) (Table 4).

Veda and Srinivasan (2009) have reported that species such as black pepper, capsaicin, red pepper, and ginger enhance the uptake of β-carotene in the intestines. Piperine and ginger increased the uptake of β-carotene by 147 % and 98 %, respectively. While an increase in absorption was 59 % and 27 % in black pepper and red pepper fed animals, respectively, dietary capsaicin also increased the uptake by 50 %. This could therefore be a promising way of improving β-carotene bioavailability in the body, thereby reducing vitamin A deficiency. As the potential of spice-derived nutraceuticals against various neurodegenerative diseases becomes more evident, some of these nutraceuticals may be developed as new nutraceuticals against brain tumors, Parkinson's disease, and other neurodegenerative diseases.

Table 4 Experimentally documented potential health benefits of some spices

Potential health benefits	Spices observed to exert
Lowering of blood cholesterol	Garlic, Onion, Fenugreek, Turmeric/Curcumin, Red pepper/ Capsaicin
Prevention and dissolution of cholesterol gallstones	Curcumin, Capsaicin
Protection of erythrocyte integrity in hypercholesterol-emic condition	Curcumin, Capsaicin, Garlic
Hypoglycemic potential	Fenugreek, Garlic, Onion, Turmeric, Cumin
Amelioration of diabetic nephropathy	Curcumin, Onion
Antioxidant effect	Turmeric/Curcumin, Capsaicin, Eugenol
Anti-inflammatory and antiarthritic effect	Turmeric/Curcumin, Capsaicin, Eugenol
Antimutagenic effect/Cancer preventing	Turmeric/Curcumin, Garlic, Ginger/Gingerol, Mustard
Digestive stimulant action	Curcumin, Capsaicin, Piperine, Ginger, Cumin, Ajowan, Fennel, Coriander, Onion, Mint
Antimicrobial	Turmeric/Curcumin, Garlic, Asafoetida
Antidepressant	Black pepper, Cloves, ginger, All-spice, Cloves
Schizophrenia	Onion
Meningitis	Garlic
Spongiform encephalopathy	Turmeric

Adapted from Das et al. (2012) and Kannappan et al. (2011)

3 Nutraceutical Market Overview

The global population of individuals over 60 years of age is expected to reach 1 billion by 2020, 70 % of which will be living in developed nations driving anti-aging products to higher product visibility (Global Industry Analysts 2012). The global functional food and nutraceutical market revenue for the year 2013 was approximately $175 billion (www.researchandmarkets.com/research/m9qvsw/global_functional). The market is expected to grow from $221.58 billion in 2014 (www.reportlinker.com/p01990122/Functional-Foods-Nutraceuticals-Market-2014-) to $424 billion by 2017 (Ken Research 2014) owing to the aging population and robust over-the-counter market in different countries. The nutraceutical beverages' market alone is expected to experience the highest growth, at a compound annual growth rate (CAGR) of 8.8 %. The sector is expected to be worth nearly $87 billion in 2016. Nutraceutical food market is the second largest market, expected to reach $67 billion in 2016, for a CAGR of 6.4 % (Fig. 3).

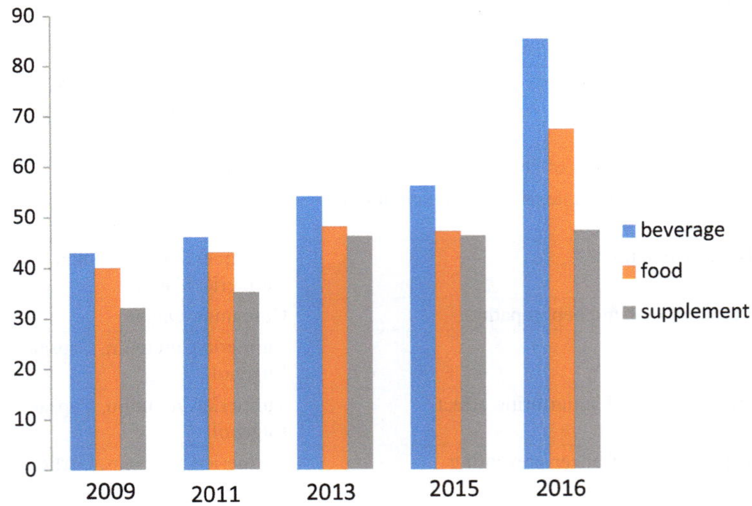

Fig. 3 Global nutraceutical market—foods, beverages, and supplements, 2009–2016 ($ MILLIONS). *Source:* Euromonitor, March 2012 and BCC Research (2011)

The Asia Pacific Nutraceutical product market is an emerging market in dietary supplements and functional food segment. Japan is the largest consumer of nutraceuticals in Asia, while China is the second largest consumer of nutraceutical products since they have the largest population in the world and the people are more mindful of their food habits. The Japanese nutraceutical industry features a wide assortment of products in various categories including functional foods and beverages and nutritional and dietary supplements. The rapidly aging demography along with healthy dietary habits among the population has been the driving force of the about three-decade market (Shimizu 2014). According to a report by Ken Research (2014), the highest percentage of sales has been recorded by the fortified and sports products. The dietary supplement market has also shown a healthy growth over the years. Energy-boosting foods and beverages have the highest demand with vitamins leading the dietary supplements. Major players such as Yakult, Pocari Sweat, and Aquarius are the giants in the Japanese functional food and beverage market, while Otsuka pharmaceuticals, Miki, Amway, and Taisho pharmaceuticals are the giants in the dietary supplement market. Due to the highly developed nature of the Japanese nutraceutical market, the nutraceutical industries are expected to produce specific products that will meet various age requirements in the country. The revenue from the nutraceutical market is expected to grow at a CAGR of 1.5 % by 2017.

The functional food and beverage market in India is expected to have 70.74 % growth compared to the dietary supplement by 2017 (Keservani et al. 2014). Over the last 7 years, the nutraceuticals market in India has been growing rapidly. An increase in health consciousness, increasing awareness about the various types of nutraceuticals available in the market, and willingness of people to spend on health-fortifying food and additives have been reported to be the factors propelling the market (Keservani et al. 2014). In India, dietary supplements hold the largest share, while the functional foods and beverages segment is relatively smaller (Rajagopal 2014). However, the industry has witnessed a rapid growth in the dietary supplements market, due to increasing awareness among people to protect themselves against chronic diseases. According to a report by Rajagopal (Frost and Sullivan) in 2014, the sales of vitamins have dominated the dietary supplements market of the Indian nutraceutical market over the years, with mineral supplements closely following behind. There is an increasing consumer interest in energy and sports drinks in recent years, since there is a growing demand for improved physical endurance, mental alertness, and enhanced activity among the young and affluent professionals. The Indian nutraceutical market has in the past been viewed as an export-focused industry, but with the changing market trends, most local companies have started to launch products in India and expand their product line according to Indian consumer needs. The nutraceuticals market of India is promising, with a major presence in urban parts of the country. The vast population base of India presents wide prospects for the nutraceuticals market to flourish in the near future. Busier lifestyles of the urban middle class population have encouraged increased consciousness about health and fitness, which are likely to be the major driving force behind the growth of the nutraceuticals industry in the future. According to Rajagopal (2014), Indian nutraceutical market is projected to grow to US $4 billion by 2018 at a CAGR of 17 %. The fortified foods market is as well projected to grow at a rapid pace with a CAGR of 21.7 % by 2018. Vitamins and minerals occupy 36 % of the total Indian nutraceutical market, followed by probiotic with a 9 % share, while omega-3 fatty acids occupy 5 %. DSM Nutritional Products dominates in the vitamins, minerals, and omega-3 fatty acids markets, while BASF and Merck are also key participants in the vitamins and minerals market. Danisco, Chr. Hansen, and Yakult, on the other hand, dominate the probiotics segment.

In 2013, the European nutraceutical market was valued at $6.4 billion and is estimated to grow at an annual rate of 7.2 % between 2013 and 2018, to reach a projected $9.0 billion by 2018. In Eastern Europe, however, the nutraceutical market is growing on account of high growth in functional food and dietary supplement segments. Russia is recognized as the largest consumer of nutraceuticals in the region and this is due to its increasing middle age population. The Russia and Hungary nutraceutical markets are expected to have a market share of 24.4 % and 20.2 %, respectively, in 2017. The increase in disposable income of Eastern European consumers over the evaluation period (2012–2017) would also help in the faster adoption of nutraceuticals in the region. The market for Europe nutraceuticals can be segmented on the basis of applications, regions, and types (Fig. 4).

According to a report by Research and Markets, functional foods remained the fastest-growing segment of North America nutraceutical market with a 6.5 % CAGR during 2007–2011. Omega fatty acid fortified food segment of functional food market is expected to have a moderate growth rate. Protein and peptide supplements will support the growth of the global nutraceutical market, while the

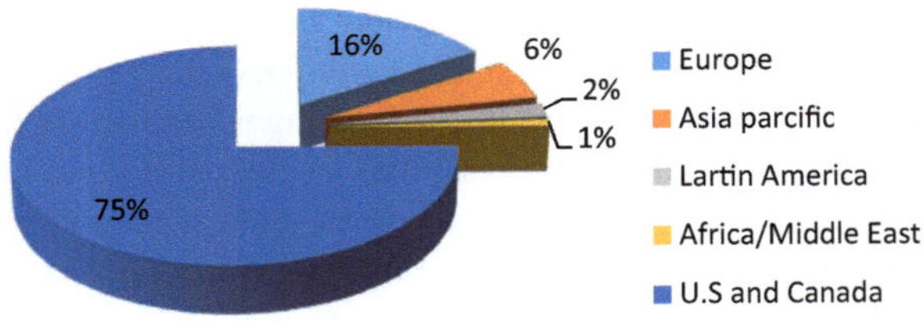

Fig. 4 Geographic distribution of mergers and acquisitions (M&A) deals by region. *Source:* Bourne Partners Proprietary Research, Capital IQ (January 2013)

nonherbal segment of the dietary supplement market is expected to have a steady growth. The North America and Asia Pacific nutraceutical market are expected to have a market share of 39.2 % and 30.4 % in 2017. The United States of America, the largest consumer of functional foods and nutraceuticals, is expected to have moderate growth compared to the dietary supplement market in the region between 2011 and 2017. According to Global Industry Analyst report (2013), 74 % of Americans patronize functional food products such as Greek yogurt, coconut water, and snacking nuts, "free" foods (e.g., gluten free and lactose free), and omega and fatty acid fortified foods because they believe such foods are healthy. Carbonated and non-carbonated energy drinks and shots continue to dominate the functional beverage category. There is also a dramatic increase in availability and options for fruit and vegetable juice drinks. However, the market for dairy and nondairy products fortified with probiotics is very promising for future revenue growth. Currently, dietary supplements are the largest segment of nutraceuticals in the USA, but other segments are rapidly growing to outpace dietary supplements. However, the strength of the dietary supplement market has been due to the presence of sports related and performance enhancing supplements, weight loss and management supplements, and nonherbal supplements (Global Industry Analysts, July 2012). Proteins and peptides are the leading nutraceutical ingredients in the USA and the nutraceutical ingredient market alone is expected to hit $23.7 billion by 2015 (World Nutraceutical Ingredient Industry 2011).

4 Motivators of Functional Foods and Nutraceutical Market

Consumer interest in the relationship between diet and health has increased substantially all over the world. Today, there is much greater recognition that people can reduce the risk of illness and disease and maintain a state of health and well-being through healthy lifestyle and diet. Ongoing support for the important role of foods such as fruits, vegetables, and whole grain cereals in disease prevention as well as dietary antioxidants and combinations of protective substances in plants has helped to provide the drive for further developments in the functional food market in Europe. The changes in population demographics and socioeconomic changes have raised the need for foods with added health benefits. According to a report by the U.S. State Department and United Nations Department of Economic and Social

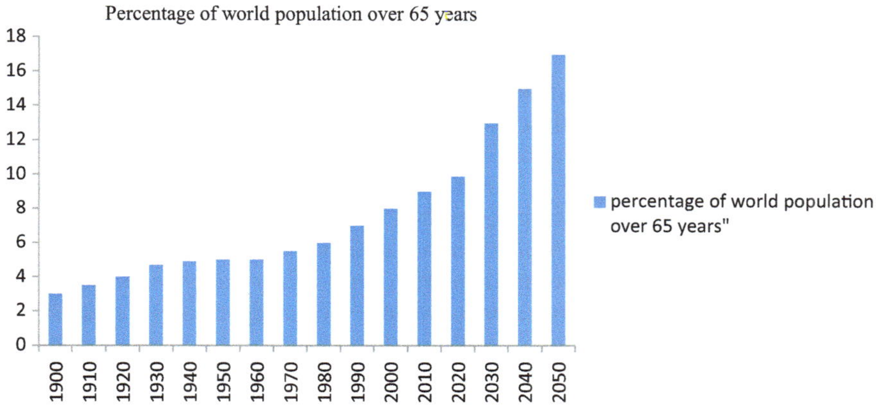

Fig. 5 Percent of world population over 65 years. *Source:* U.S. State Department and United Nations Department of Economic and Social Affairs (http://www.un.org/en/development/desa/population/publications/pdf/ageing/WorldPopulationAgeing2013)

Affairs (2013), about 8 % of people worldwide are over the age of 65 (Fig. 5), and this age group is expected to account for about 13 % of the population by 2030 with most of them residing in developing countries. The increasing life expectancy therefore raises the need for an improved quality of life.

Moreover, the increasing costs of health care have prompted governments, researchers, health professionals, and the food industry to find solutions to how these changes can be managed efficiently. These are strong factors propelling the functional foods market. The rising rates of obesity and diabetes, rising eye-health concerns, increase in cardiovascular disease incidence, transforming consumption patterns in developing markets, rising preference for preventive medicine, rising popularity of nutraceuticals with organic and natural ingredients, and strong demand for multivitamins and other ingredients such as omega 3s and astaxanthin are factors propelling the nutraceutical and functional food market (Zawistowski 2014). Nutraceuticals containing ingredients such as conjugated linoleic acid (CLA) soy, whey, and dietary fibers with claims of weight management, heart care, immunity, and digestive health are gaining popularity.

Product innovations have also promoted consumer patronage of functional foods and nutraceuticals. Consumers are looking for convenience as well as products that offer higher bioavailability of major ingredients. Therefore, the presence of liquid nutraceuticals such as enhanced waters, energy juices, and energy-sports drinks as well as liquid dietary supplements, oral powder and liquid suspensions, oxygenated water, nutrient enhanced vitamin water, ready-to-drink (RTD) beverages, liquid shooters, liquid shots, energy drinks, antioxidant berry juice blends, omega 3 oils, and essential oils offer significant growth opportunities (Bourne Partners 2013).

5 The Intersection of Food and Genes

Diet is one of the environmental factors that our genes are exposed to and their effects on gene expression affect the life of the consumer. Nutrients control the concentration of proteins in various organs by serving as cofactors, substrates, or coenzymes. Studies designed to identify specific effects of diet on phenotypic expression of biochemical components that determine health have resulted in exciting suggestions for dietary interventions designed to modify gene expression. For instance, in a study to understand parsley (*Petroselinum crispum*) as a new and effective nutraceutical in inflammatory bowel disease management, Jia et al. (2014) supplemented colitis-induced mice with parsley. Colonic transcriptome revealed downregulation of inflammatory cytokines, hepatic transcriptome and proteome revealed a downregulation of cancer markers, and plasma metabolite analysis showed shifts in the citric and urea cycle, indicating improved glycolysis and reduced oxidative stress. Also, hepatic transcriptome and metabolome revealed an upregulation of fatty acid synthesis genes, thereby improving body weight loss. This study is useful in shaping the future of nutraceutical and functional food research.

Many humans and animals studies have shown permanent effects of early diet on adult metabolism, cognitive function, and body composition through activation or suppression of gene expression, or turning genes "on" or "off" (Fang et al. 2014; Laker et al. 2013). Such effects can be obvious, such as the effects seen in vitamin deficiency diseases, or more obscure and complex, as occurs in type 2 diabetes, predisposition to obesity, and other chronic diseases. For this reason, the application of nutrigenomics provides understanding of how diet affects gene expression and this knowledge is opening new doors for many nutritional interventions. By evaluating the changes in gene and protein expression as well as metabolic pathways after the administration of dietary nutrients, certain biomarkers could be used to demonstrate the effect of bioactive food components on health.

In a recent study (Hung et al. 2015), apple polyphenols (AP) have been shown to drastically suppress migration, invasion, colony formation, and adhesion of DLD-1 cells. They also reduced expression of tubulin and F-actin and altered the cytoskeleton structure. Taken together, AP significantly inhibited motility of DLD-1 cells via disruption of the interaction between snail and the FAK promoter and consequently diminished tumorigenesis and metastasis of DLD-1 cells (Hung et al. 2015). The results indicate that apple polyphenols could be beneficial in colon cancer treatment through metastasis attenuation. Another study on resveratrol and pterostilbene from plants shows that they suppress the growth-promoting effect of MED28 in cancer cells. MED28 regulates cellular migration and invasion in human breast cancer cells (Huang et al. 2012) and its suppression delayed cell cycle progression and reduced cyclin D1 and p-RelA/p65 expression as well as the transcriptional activity of NFκB (Huang et al. 2015), thereby implying the feasibility of resveratrol and pterostilbene as nutraceuticals in suppressing breast cancer development.

It has been reported that many aspects of metabolism display circadian rhythms that are controlled by peripheral clocks in a tissue-autonomous manner (Tahara and Shibata 2013). Alteration of this clock system has been associated with different illnesses, such as cancer and metabolic syndrome, and this implies that peripheral clocks are essential in maintaining homeostasis and normal body function (Froy 2010). In a recent study, dietary grape seed proanthocyanidin extracts were found to modulate the molecular clock system of human liver cells through the transactivation of *RORα*, resulting in the overexpression of *BMAL1* (Ribas-Latre et al. 2015). This suggests that grape seed proanthocyanidin may have a positive effect in liver glucose and lipid metabolism.

Many industries and the European commission continue to fund projects that unveil the effects of food on health (http://www.cordis.europa.eu/fp7/kbbe/home_en.html); however, merging information about the physiological responses to food with individual genetic information is very important for designing personalized food and diets in the near future. Factors such as the bioavailability of functional food ingredients, appropriate biomarker development for a wider range of functional end points, and the stability of functional food ingredients during manufacturing and passage through the GI tract to reach the target organ remain challenging in functional food and nutraceutical research.

6 International Regulations for Labeling and Health Related Claims

In the USA, functional food and nutraceuticals are regulated by Food and Drugs Authority (FDA) under the authority of the Federal Food, Drug, and Cosmetic Act. Since they are not specifically defined by law, they are regulated under the same statutes as other food and food products. The FDA is also responsible for evaluating the safety and labeling of dietary supplements to ensure that **they** meet all the requirements of Dietary Supplement Health and Education and FDA regulations before marketing (http://www.fda.gov/Food/DietarySupplements/default.htm). By January 2018, food labels in the USA must bear information on updated serving size requirements and new labeling requirements for certain package sizes, greater understanding of nutrition science, and the packages must have refreshed designs as proposed by the FDA (http://www.fda.gov/Food/GuidanceRegulation/GuidanceDocumentsRegulatoryInformation/LabelingNutrition/ucm385663.htm). Manufactures who wish to market new dietary supplements must notify the FDA and add information that shows that the article qualifies as a new dietary supplement and safe under the conditions of use as stipulated in its labeling (http://www.fda.gov/Food/DietarySupplements/NewDietaryIngredientsNotificationProcess/ucm109764.htm) before marketing the new product. Similarly, functional foods are still a "virtual category" in terms of EU or Irish food law but regulated through existing food legislation (https://www.fsai.ie/assets/0/86/204/667b54fe-972c-4c04-a6f8-9a0c5c92f886.pdf). A new food may require a full novel food authorization. In Europe, the European Commission drafts proposals based on the EFSA safety assessment and decides whether or not to accept a new food product. The EU legislation stipulates that only those supplements (vitamins and minerals) on the positive list may be marketed as of August 1st, 2005, while new supplements must undergo a full safety assessment (www.fsai.ie/legislation/food_supp/index.asp). The European Commission makes nutrition labeling mandatory and instructs food manufacturers to provide information on the energy value and 6 nutrients as well as allergens present in the food (http://eur-lex.europa.eu/LexUriServ/LexUriServ.do?uri=OJ:L:2011:304:0018:0063:EN:PDF). According to the FDC Act, Section 201 (g)(1), articles that cure, mitigate, treat, or prevent any disease are drugs. However, the Nutrition Labeling and Education Act (NLEA) of 1990 authorizes FDA to allow certain disease-risk-reduction claims, known as "health claims," to appear in food labeling. Health claims are exempted from drug status, only if all of the applicable requirements for each type of claim are met. Failure to comply with this requirement renders the food mislabeled and illegal (http://www.wilmerhale.com/uploadedFiles/WilmerHale_Shared_Content/Files/Editorial/Publication/SC%20roses.pdf).

7 Challenges in the Functional Food and Nutraceutical Market

Though the functional food and nutraceutical market has existed for several years, its future is hard to predict because of challenges such as country regulations, difficulty in health claim substantiation, and low innovation by food companies. There is mixed global acceptance of functional foods. Though several countries have legislation permitting health claim use and regulation, yet the process has not resulted in permission for the use of such claims (El Sohaimy 2012). Health claim approval for functional foods is difficult in the USA and Japan. Though guidelines for health claim substantiation have been proposed to the European community by The European Food Safety Authority (EFSA), it still seems impossible in Europe. The type of evidence and parameters such as duration of the clinical studies, validation of the biological markers, dose–response curves to define the efficient dosage, and the adverse effects that would support a health claim are not well defined in these regions.

Moreover, for companies that go through the health claim petition and obtain the approval, competing companies can use the claim. Although functional foods and nutraceutical products with exclusive ingredients may be patented, a large majority of products contain "free" ingredients and are therefore easily copied. This situation provides a little competitive advantage to the originator company. Again, since functional foods and nutraceuticals address health care, claims of health benefit must be proven with scientific evidence. This is not a simple task. Biological markers of health improvement or risk reduction are not easily recognized. Also clinical studies often take a long time and hardly give strong evidence of efficient dosage while adverse effects often necessitate supplemental studies. Besides the difficulty in health claim substantiation, finding a good product is not simple. Therefore, the food industry has to consider physiological factors such as bioavailability, molecular interactions that control biological functions, or age-related physiology and many other factors.

Furthermore, the nutraceutical sector still exists more as a concept than a reality because it regroups food supplements and functional foods. Presently, only a few products such as the phytosterol-based margarine, Benecol from Raisio or Nestlé's LC1, and the bifidogenic yogurt are clearly positioned on health with scientific evidence. The vast majority of food companies have not been very innovative because they only "functionalize" conventional products by adding some vitamins, minerals, or herbal extracts, rather than building up new food concepts of ingredients that share common health benefit. Despite a difficult environment however, the market is still expanding.

8 Future Perspectives

Companies now understand more about how nutrients affect people from a healthcare perspective. They are therefore looking at ways both medical treatment and nutrition can be integrated in the medical field to ensure that holistic medical care is provided. As of now, medical care is considered to be the province of drugs, while nutrition is considered to be a product of a healthy living. In the near future, it is expected that much more work on how the two interact and complement one another will be done. Investment in new technology and the application of genetically modified technology within the food industry for medical and health benefits is set to drive further increase in market revenues within the nutraceuticals market. Expanding the body of scientific research which validates the effectiveness and safety of these new products will stimulate further investment in the technology and application. Promising technologies such as nutrigenomics, imaging techniques, and converging technologies are progressively being used in nutrition research. Their huge potential will enhance the development of foods for targeted population groups with defined risk factors or diseases such as obesity, diabetes, allergy, and cardiovascular disease. The creativity of food technology might also contribute to further advances in developing food products that can support optimum health. The increased consumer awareness of functional foods and nutraceuticals will however drive further revenue growth, globally. International growth across the industry is expected to continue as developing countries increase nutraceutical consumption. Also, domestic growth in nutraceutical consumption is expected to continue as novel products and new target segments are introduced by domestic producers, including high growth specialty foods focused on probiotics and heart health. Aging global population and rising healthcare costs have shifted consumer focus to healthier living, preventative care, and secondary source diagnosis or medication (van der Zanden et al. 2014). Continued concerns over "naturalness," increased global regulation, and concerns about safety due to overseas manufacturing could however suppress growth.

References

Abd-ElSalam H-AH, Al-Ghobashy MA, Zaazaa HE, Ibrahim MA (2014) Stability of catechins in green tea nutraceutical products: application of solid phase extraction–thin layer chromatography densitometry. Food Chem 156:94–99

Baek YJ, Lee BH (2009) Probiotics and prebiotics as bioactive components in dairy products. In: Park YW (ed) Bioactive components in milk and dairy products. Wiley-Blackwell, New York, NY, p 449

Balogun KA, Albert CJ, Ford DA, Brown RJ, Cheema SK (2013) Dietary omega-3 polyunsaturated fatty acids alter the fatty acid composition of hepatic and plasma bioactive lipids in C57BL/6 mice: a lipidomic approach. PLoS One 8, e82399. doi:10.1371/journal.pone.0082399

BCC Research (2011) Nutraceuticals: global markets and processing technologies. Report Code: FOD013D

Beecher GR (1999) Phytonutrients role in metabolism: effect on resistance to degenerative processes. Nutr Rev 57:S3–S6

Boeiraa SP, Filhob Carlos B, Del'Fabbrob L, Souza RS, Luiz FFR, Figheraa MR, Cristiano RJ, Mauro Schneider O, Furian AF (2014) Lycopene treatment prevents hematological, reproductive and histopathological damage induced by acute zearalenone administration in male Swiss mice. Exp Toxicol Pathol 66:179–185

Bordbar S, Ebrahimpour A, Abdul HA, Abdul MMY, Anwar F, Saari N (2013) The improvement of the endogenous antioxidant property of stone fish (*Actinopygalecanora*) tissue using enzymatic proteolysis. BioMed Res Int 2013: Article ID 849529, 9 pages

Bourne Partners Sector Report, Nutraceuticals Industry, April 2013

Byun HG, Lee JK, Park HG, Jeon JK, Kim SK (2009) Antioxidant peptides isolated from the marine rotifer, *Brachionus rotundiformis*. Process Biochem 44:842–846

Chandran B, Goel AA (2012) Randomized, pilot study to assess the efficacy and safety of curcumin in patients with active rheumatoid arthritis. Phytother Res 26:1719–25

Charoenphun N, Cheirsilp B, Sirinupong N, Youravong W (2013) Calcium-binding peptides derived from tilapia (Oreochromisniloticus) protein hydrolysate. Eur Food Res Technol 236:57–63

Chen Y, Cao L, Zhong M, Zhang Y, Han C, Li Q, Yang J, Zhou D, Shi W, He B, Liu F, Yu J, Sun Y, Cao Y, Li Y, Li W, Guo D, Cao Z, Yan H (2012) Anti-HIV-1 activity of a new scorpion venom peptide derivative Kn2-7. PLoS One 7, e34947

Cole GM, Frautschy SA (2010) DHA may prevent age-related dementia. J Nutr 140:869–874

Cottin SC, Sanders TA, Hall WL (2011) The differential effects of EPA and DHA on cardiovascular risk factors. Proc Nutr Soc 70:215–231

Das L, Bhaumik E, Raychaudhuri U, Chakraborty R (2012) Role of nutraceuticals in human health. J Food Sci Technol 49:173–183

De Rossi SS, Thoppay J, Dickinson DP, Looney S, Stuart M, Ogbureke KU, Hsu S (2014) A phase II clinical trial of a natural formulation containing tea catechins for xerostomia. Oral Surg Oral Med Oral Pathol Oral Radiol 118:447–454.

Delphine M, Saulnier A, Jennifer Spinler K, Glenn Gibson R, Versalovic J (2009) Mechanisms of probiosis and prebiosis: considerations for enhanced functional foods. Curr Opin Biotechnol 20:135–141

Dewapriya P, Kim SK (2014) Marine microorganisms: an emerging avenue in modern nutraceuticals and functional foods. Food Res Int 56:115–125

Dewulf EM, Cani PD, Claus SP, Fuentes S, Puylaert PG, Neyrinck AM, Bindels LB, de Vos WM, Gibson GR, Thissen JP, Delzenne NM (2013) Insight into the prebiotic concept: lessons from an exploratory, double blind intervention study with inulin-type fructans in obese women. Gut 62:1112–1121

Dong ZB, Liang YR, Fan FY, Ye JH, Zheng XQ, Lu JL (2011) Adsorption behavior of the catechins and caffeine onto polyvinylpolypyrrolidone. J Agric Food Chem 59:4238–4247

El Sohaimy SA (2012) Functional foods and nutraceuticals-modern approach to food science. World Appl Sci J 20:691–708

Fang L, Liang XF, Zhou Y, Guo XZ, He Y, Yi TL, Liu LW, Yuan XC, Tao YX (2014) Programming effects of high-carbohydrate feeding of larvae on adult glucose metabolism in zebrafish, Daniorerio. Br J Nutr 111:808–818

FAO/WHO Report of a Joint FAO/WHO expert consultation on guidelines for the evaluation of probiotics in food. World Health Organization and food and agriculture organization of the United Nations. London, Ontario, Canada, 2002

Freitas AC, Rodrigues D, Rocha-Santos TAP, Gomes AMP, Duarte AC (2012) Marine biotechnology advances towards applications in new functional foods. Biotechnol Adv 30:1506–1515

Froy O (2010) Metabolism and circadian rhythms—Implications for obesity. Endocr Rev 31:1–24

Fukuzawa Y, Kapoor Mahendra P, Yamasaki K, Okubo T, Hottac Y, Juneja LR (2014) Effects of green tea catechins on nonalcoholic steatohepatitis (NASH) patients. J Funct Foods 9:48–59

Galanakis CM (2013) Emerging technologies for the production of nutraceuticals from agricultural by-products: A viewpoint of opportunities and challenges. Food Bioprod Process 91 (4):575–579. doi:10.1016/j.fbp.2013.01.004

Gaullier JM, Halse J, Høye K, Kristiansen K, Fagertun H, Vik H, Gudmundse O (2004) Conjugated linoleic acid supplementation for 1 y reduces body fat mass in healthy overweight humans. Am J Clin Nutr 79:1118–1125

Global Industry Analysts: 10 Key Trends in Food, Nutrition, and Health 2013, New Nutrition Business

Guttler N, Kirila Z, Parahuleva M, Ridvan C, Mehmet B, Neuhof C, Mehmet B, Niemann B, Ali E, Boning A (2012) Omega-3 fatty acids and vitamin D in Cardiology. Cardiol Res Pract 2012: Article ID 729670, 11 pages. doi:10.1155/2012/729670

Hafeez Z, Cakir-Kiefer C, Roux E, Perrin C, Miclo L, Dary-Mourot A (2014) Strategies of producing bioactive peptides from milk proteins to functionalize fermented milk products. Food Res Int 63:71–80

Harbourne N, Eunice M, Jacquier JC, O'Riordan D (2013) Stability of phytochemicals as sources of anti-inflammatory nutraceuticals in beverages—A review. Food Res Int 50:480–486

Hayes M, Ross RP, Fitzgerald GF, Hill C, Stanton C (2006) Casein-derived antimicrobial peptides generated by *Lactobacillus acidophilus* DPC6026. Appl Environ Microbiol 72:2260–2264

Hisano M, Bruschini H, Carlos Nicodemo A, Srougi M (2012) Cranberries and lower urinary tract infection prevention. Clinics 67:661–667

Hsu KC, Li-Chan E, Jao CL (2011) Antiproliferative activity of peptides prepared from enzymatic hydrolysates of tuna dark muscle on human breast cancer cell line MCF7. Food Chem 126:617–622

Huang CY, Chou YH, Hsieh NT, Chen HH, Lee MF (2012) MED28 regulates MEK1-dependent cellular migration in human breast cancer cells. J Cell Physiol 227:3820–3827

Huang CY, Hsieh NT, Pan MH, Li CI, Chou YH, Chiou YS, Lee MF (2015) Resveratrol and pterostilbene regulate MED28 (magicin/EG1) expression and cell growth in MCF7 human breast cancer cells. J Funct Foods 12:158–167

Hung CH, Hung CC, Hsu LS, Kao SH, Wang CJ (2015) Apple polyphenol inhibits colon carcinoma metastasis via disrupting snail binding to focal adhesion kinase. J Funct Foods 12:80–91

Isa Y, Miyakawa Y, Yanagisawa M, Goto T, Kang M-S, Kawada T (2008) 6-Shogaol and 6-gingerol, the pungent of ginger, inhibit TNF-[alpha] mediated downregulation of adiponectin expression via different mechanisms in 3T3–L1 adipocytes. Biochem Biophys Res Commun 373:429–434

Ishimwe N, Daliri E, Lee BH, Fang F, Du G (2015) The perspective on cholesterol-lowering mechanisms of probiotics. Mol Nutr Food Res 59:94–105

Jang A, Jo C, Kang KS, Lee M (2008) Antimicrobial and human cancer cell cytotoxic effect of synthetic angiotensin-converting enzyme (ACE) inhibitory peptides. Food Chem 107:327–336

Jayaraj RL, Tamilselvam K, Manivasagam T, Elangovan N (2013) Neuroprotective effect of CNB-001, a novel pyrazole derivative of curcumin on biochemical and apoptotic markers against rotenone-induced SK-N-SH cellular model of Parkinson's disease. J Mol Neurosci 51:863–870

Jia H, Wanping AW, Hanate M, Takahashi S, Saito K, Tanaka H, Tomita M, Kato H (2014) Multi-faceted integrated omics analysis revealed parsley (*Petroselinum crispum*) as a novel dietary intervention in dextran sodium sulphate induced colitic mice. J Funct Foods 11:438–448

Johanningsmeier SD, Harris GK (2011) Pomegranate as a functional food and nutraceutical source. Ann Rev Food Sci Technol 2:181–201

Jones ML, Martoni CJ, Pietro ED, Simonc RR, Prakash S (2012) Evaluation of clinical safety and tolerance of a *Lactobacillus reuteri* NCIMB 30242 supplement capsule: a randomized control trial. Regul Toxicol Pharmacol 63:313–320

Jones ML, Tomaro-Duchesneau C, Martoni CJ, Prakash S (2013) Cholesterol lowering with bile salt hydrolase-active probiotic bacteria, mechanism of action, clinical evidence, and future direction for heart health applications. Expert Opin Biol Ther 13:631–642

Kannappan R, Chandra Gupta S, Kim JH, Simone R, Aggarwal BB (2011) Neuroprotection by spice-derived nutraceuticals: You are what you eat! Mol Neurobiol 44:142–159

Ken Research (2014). Japan Nutraceuticals Industry Outlook to 2017—Demands of Benefit Specific Nutraceutials to Impel Revenues. http://www.slideshare.net/Ken_Research/japannutraceuticalsindustryresearchreport?related=3&utm_campaign=related&utm_medium=1&utm_source=1

Kerry LI, Lewis JR, Jonathan M, Kun Z, Satvinder S, Thompson PL, Prince RL (2011) Association between yogurt, milk, and cheese consumption and common carotid artery intima-media thickness and cardiovascular disease risk factors in elderly women. Am J Clin Nutr 94:234–239

Keservani RK, Sharma AK, Ahmad F, Baig ME (2014) Nutraceutical and functional food regulations in India. Food Sci Technol 327–342

Kittiphattanabawon P, Benjakul S, Visessanguan W, Shahadi F (2013) Inhibition of angiotensin converting enzyme, human LDL cholesterol and DNA oxidation by hydrolysates from blacktip shark gelatin. LWT—Food Sci Technol 51:177–182

Ko SC, Min CK, Lee JK, Byun HG, Kim SK, Lee SC, Jeon BT, Park PJ, Jung WK, Jeon YJ (2011) Effect of angiotensin I-converting enzyme (ACE) inhibitory peptide purified from enzymatic hydrolysates of *Styelaplicata*. Eur Food Res Technol 233:915–922

Koba K and Yanagita T (2014) Health benefits of conjugated linoleic acid (CLA). Obes Res Clin Pract 8:e525—e532.

Korecka A, de Wouters T, Cultrone A, Lapaque N, Pettersson S, Doré J, Blottière HM, Arulampalam V (2013) ANGPTL4 expression induced by butylate and rosiglitazone in human intestinal epithelial cells utilizes independent pathways. Am J Phys Gastrointest Liver Physiol 304:G1025–G1037. doi:10.1152/ajpgi.00293.2012

Laker RC, Wlodekc ME, Connellya JJ, Zhen Y (2013) Epigenetic origins of metabolic disease: the impact of the maternal condition to the offspring epigenome and later health consequences. Food Sci Hum Wellness 2:1–11

Lee LK, Foo KY (2014) An appraisal of the therapeutic value of lycopene for the chemoprevention of prostate cancer: a nutrigenomic approach. Food Res Int 54:1217–1228

Lian YT, Yang XF, Wang ZH, Yang Y, Yang Y, Shu YW (2013) Curcumin serves as a human kv1.3 blocker to inhibit effector memory T lymphocyte activities. Phytother Res 27:1321–1327

Lin P-P, Hsieh Y-M, Kuo W-W, Lin Y-M, Yeh Y-L, Lin C-C, Tsai F-J, Tsai C-H, Tsai C-C, Huang C-Y (2013) Suppression of TLR-4-related inflammatory pathway and anti-fibrosis effects of

probiotic-fermented purple sweet potato yogurt in hearts of spontaneously hypertensive rats. Chin J Physiol 56:174–83. doi:10.4077/CJP.2013.BAB118
Mahmoud S, Parisa S, Farzin H, Seyed AM (2013) Extraction of lycopene from tomato paste by ursodeoxycholic acid using the selective inclusion complex method. J Food Sci 78: C1680–1685. doi:10.1111/1750-3841.12264
van Baarlen P, Troost F, van der Meer C, Hooiveld G, Mark B, Robert JM, Kleerebezema M (2010) Human mucosal in vivo transcriptome responses to three lactobacilli indicate how probiotics may modulate human cellular pathways. PNAS. doi:10.1073/pnas.1000079107
Martín R, Miquel S, Ulmer J, Kechaou N, Langalla P, Bermúdez-Humarán LG (2013) Role of commensal and probiotic bacteria in human health: a focus on inflammatory bowel disease. Microb Cell Fact 12:71
Mohamed S (2013) Functional foods against metabolic syndrome (obesity, diabetes, hypertension and dyslipidemia) and cardiovasular disease. Trends Food Sci Technol 35:114e–128
Monsanto M, Hooshyar N, Meuldijk J, Zondervan E (2014) Modeling and optimization of green tea precipitation for the recovery of catechins. Sep Purif Technol 129:129–136
Naqash SY, Nazeer RA (2012) In vitro antioxidant and antiproliferative activities of bioactive peptide isolated from Nemipterus Japonicus backbone. Int J Food Prop 15(6):1200–1211. doi:10.1080/10942912.2010.517342
Neagu D, Leopold LF, Thonart P, Destain J, Socaciu C (2014) Enzyme-assisted extraction of carotenoids and phenolic derivatives from tomatoes. Bull UASVM Anim Sci Biotechnol 71:20–26
Ngo DH, Voa TS, Ngob DN, Wijesekarac I, Kim SK (2012) Biological activities and potential health benefits of bioactive peptides derived from marine organisms. Int J Biol Macromol 51:378–383
Ninfali P, Mea G, Giorgini S, Rocchi M, Bacchiocca M (2007) Antioxidant capacity of vegetables, spices and dressings relevant to nutrition. Br J Nutr 93:257–266
Nutraceuticals: a global strategic business report, Global Industry Analysts, July 2012
Palozza PN, Parrone AC, Simone R (2010) Tomato lycopene and inflammatory cascade: basic interactions and clinical implications. Curr Med Chem 17:2547–2563
Parodi PW (2007) A role for milk proteins and their peptides in cancer prevention. Curr Pharm 13:813–828
Power O, Jakeman P, Fitzgerald RJ (2012) Antioxidative peptides: enzymatic production, in vitro and in vivo antioxidant activity and potential applications of milk-derived antioxidative peptides. Amino Acids J 10:1393–1399
Rajagopal N (2014) Indian nutraceutical market is filled with promise. http://www.naturalproductsinsider.com/articles/2014/05/indian-nutraceutical-market-is-filled-with-promis.aspx
Ramchandran L, Shah NP (2011) Yogurt can beneficially affect blood contributors of cardiovascular health status in hypertensive rats. J Food Sci 76:H131–H136
Ribas-Latre A, Baselga-Escudero L, Casanova E, Arola-Arnal A, Salvadó M, Bladé C, Arola L (2015) Dietary proanthocyanidins modulate BMAL1 acetylation, Nampt expression and NAD levels in rat liver. Sci Rep 5:10954–10967
Ryan JT, Ross RP, Declan B, Fitzgerald GF, Stanton C (2011) Bioactive peptides from muscle sources: meat and fish. Nutrients 3:765–791
Saiga A, Okumura T, Makihara T, Katsuda SI, Morimatsu F, Nishimura T (2006) Action mechanism of an angiotensin I-converting enzyme inhibitory peptide derived from chicken breast muscle. J Agric Food Chem 54:942–945
Senevirathne M, Kim SK (2012) Utilization of seafood processing by-products: medicinal applications. Adv Food Nutr Res 65:495–512
Shahidi F (2009) Nutraceuticals and functional foods: Whole versus processed foods. Trends Food Sci Technol 20:376–387
Sheih IC, Fang TJ, Wu TK (2009) Isolation and characterisation of a novel angiotensin I-converting enzyme (ACE) inhibitory peptide from the algae protein waste. Food Chem 115:279–284

Sheril A, Lange K, Amolo T, Grinstead JS, Haakonsson AK, Szalowska E, Koppen A, Mudde K, Haenen D, Al-Lahham S, Roelofsen H, Houtman R, van der Burg B, Mandrup S, Bonvin AM, Kalkhoven E, Müller M, Hooiveld GJ, Kersten S (2013) Short-chain fatty acids stimulate angiopoietin-like 4 synthesis in human colon adenocarcinoma cells by activating peroxisome proliferator-activated receptor γ. Mol Cell Biol 33:1303–1316

Shimizu M (2014) History and current status of functional food regulations in Japan. Food Sci Technol 257–263

Sonmez M, Turk G, Ceribasi AO, Sakin F, Atessahin A (2011) Attenuating effect of lycopene and ellagic acid on 2,3,7,8-tetrachlorodibenzo-p-dioxin-induced spermiotoxicity and testicular apoptosis. Drug Chem Toxicol 34:347–356

Tahara Y, Shibata S (2013) Chronobiology and nutrition. Neuroscience 253:78–88

van der Zanden Lotte DT, van Kleef E, de Wijk René A, van Trijp Hans CM (2014) Knowledge, perceptions and preferences of elderly regarding protein-enriched functional food. Appetite 80:16–22

Veda S, Srinivasan K (2009) Influence of dietary spices—Black pepper, red pepper and ginger on the uptake of b-carotene by rat intestines. J Funct Foods 1:394–398

Vujic G, Jajac KA, Despot SV, Kuzmic VV (2013) Efficacy of orally applied probiotic capsules for bacterial vaginosis and other vaginal infections: a double-blind, randomized, placebo-controlled study. Eur J Obstet Gynecol Reprod Biol 168:75–79

Waliszewski KN, Blasco G (2010) Nutraceutical properties of lycopene. Salud Publica Mex 52 (3):254–265.

Wang LC, Chen SY, Liu R, Wu H (2012) A novel hydrolytic product from flesh of *Mactraveneriformis* and its bioactivities in calcium supplement. J Ocean Univ China 11(3):389–396. doi:10.1007/s11802-012-1960-4

Wassall SR, Stillwell W (2009) Polyunsaturated fatty acid-cholesterol interactions: domain formation in membranes. Biochim Biophys Acta 1788:24–32

Wissam Z, Ghada B, Wassim A, Warid K (2012) Effective extraction of polyphenols and proanthocyanidins from pomegranate's peel. Int J Pharm Sci 4:675–682

World Nutraceutical Ingredient Industry, Freedonia, November 2011. http://www.freedoniagroup.com/brochure/27xx/2799smwe.pdf

Wu D, Bird MR (2010) The interaction of protein and polyphenol species in ready to drink black tea liquor production. J Food Process Eng 33:481–505

Zawistowski J (2014) Regulation of functional foods in selected Asian countries in the pacific rim. Food Sci Technol 419–463

Zhang B, Chen H, Li M, Gu Z, Song Y, Ratledge C, Chen YQ, Zhang Y, Chen W (2013) Genetic engineering of *Yarrowia lipolytica* for enhanced production of trans-10, cis-12 conjugated linoleic acid. Microb Cell Fact 12:70–78

Zhang CR, Dissanayake AA, Kudret KL, Muraleedharan GN (2015) Evaluation of coriander spice as a functional food by using in vitro bioassays. Food Chem 167:24–29

Zhang J, Zhang H, Wang L, Guo X, Wang X, Yao H (2010) Isolation and identification of antioxidative peptides from rice endosperm protein enzymatic hydrolysate by consecutive chromatography and MALDI-TOF/TOF MS/MS. Food Chem 119:226–234

Zhang Y, Yu Y, Li X, Meguro S, Hayashi S, Katashima MY, Asumasu T, Wang J, Li K (2012) Effects of catechin-enriched green tea beverage on visceral fat loss in adults with a high proportion of visceral fat: a double-blind, placebo-controlled, randomized trial. J Funct Foods 4:315–322

http://www.kenresearch.com/healthcare/skin-and-personal-care-industry/india-nutraceutical-market-research-report/484-91.html. Retrieved on 1/08/2014

http://www.researchandmarkets.com/product/4aff5d/nutraceuticals_product_market_global_market. Retrieved on 1/08/2014

Roles of Probiotics on Lifelong Diversifications of Gut Microbiota

Yung-Sheng Loh, Lee-Ching Lew, Boon-Yin Khoo, Nor Azlina Khalil, Chee-Yuen Gan, and Min-Tze Liong

Contents

Abstract A diverse community of microbiota has been proven to inhabit the gastrointestinal tract along its length. This indigenous gut microbiota is important to the host metabolism and health. Although it is widely recognized that the initial contact with mothers during delivery shapes the composition of gut microbiota in newborns, uncertainties arising throughout the human life such as development of

Y.-S. Loh • L.-C. Lew • M.-T. Liong (✉)
Bioprocess Technology Division, School of Industrial Technology, Universiti Sains Malaysia, 11800 Penang, Malaysia
e-mail: yungsheng.loh@gmail.com; mintze.liong@usm.my

B.-Y. Khoo
Institute for Research in Molecular Medicine, Universiti Sains Malaysia, 11800 Penang, Malaysia

N.A. Khalil
Advanced Medical and Dental Institute, Universiti Sains Malaysia, 11800 Penang, Malaysia

C.-Y. Gan
Centre for Advanced Analytical Toxicology Services, Universiti Sains Malaysia, 11800 Penang, Malaysia

M.-T. Liong (ed.), *Beneficial Microorganisms in Food and Nutraceuticals*, Microbiology Monographs 27, DOI 10.1007/978-3-319-23177-8_11

disorders, diseases, and stressful experiences may twist the microbial balance in host. These subsequently lead to the occurrence of aberrant composition of gut microbiota which often favors the proliferation of pathogenic bacteria. While alteration of gut microbiota composition has been suggested to be one of the contributing factors to pathogenesis of diseases, an enormous interest has emerged in the reestablishment of microbial balance in gut via the intervention of probiotics. In this chapter, we provide some evidence on numerous factors that affect the compositions gut microbiota, beginning from infancy to adulthood, and the effectiveness of probiotic therapy in tackling health issues arising from such alterations.

1 Introduction

The human gut harbors a vast range of living microorganisms that forms a complex ecosystem which regulates host metabolism and various physiological functions. Bacteria are by far the most abundant inhabitants of the gut, with a population that outnumbers the human somatic cells by approximately tenfold. From a sterile fetus, infants are immediately colonized by environmental bacteria typically originating from the maternal source during delivery, prior to development and diversification in later life. During the process of gut maturation, initial dominance of the gut by facultative anaerobic bacteria is gradually replaced by strict anaerobes. It is noted that the composition of the gut microbiota is dynamic and constantly changing in response to a wide range of factors (Fig. 1).

Disruption of the balanced gut microbial community has often been linked to the development of various diseases. Increasing evidence has shown that probiotic administration may alleviate certain diseases via restoration of a healthy gut microbial community. Probiotics are live microorganisms which, when administered in adequate amounts, confer health benefit to the host (WHO 2001). While probiotics are commonly used for the maintenance of gut health, recent researches have suggested that the beneficial effects exerted by probiotics can also extend beyond the gut such as the brain. For instance, the involvement of probiotics as a tool to manipulate brain activity has currently been extensively explored for the intention to cure anxiety disorder. While we will discuss on the various factors that

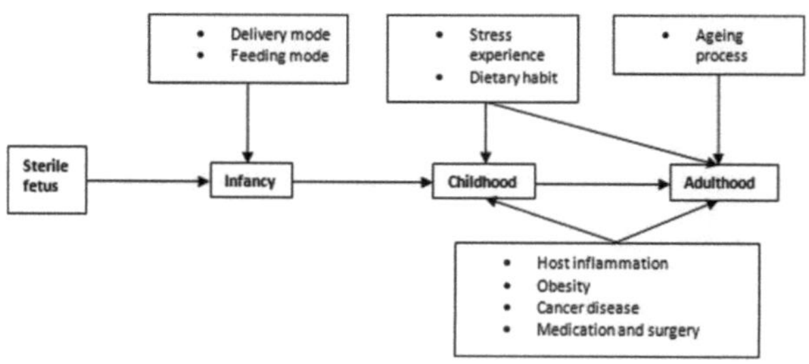

Fig. 1 Overview of factors affecting changes in gut microbiota

lead to the diversification of gut microbiota, this chapter also provides insights into modulating a healthier gut microbiota that is often affected by the abovementioned factors.

2 Infancy

2.1 Cesarean and Vaginally Born Infants

The delivery modes often play a crucial role in the development of the gut microbiota in neonates. Infants delivered vaginally have been reported to have a higher population of bifidobacteria and bacteroides in the gut, whereas infants born through cesarean section (C-section) were often more colonized with *Clostridium difficile*. The differences in gut microbiota composition between these two groups of newborn babies have been reported where vaginally born babies harbor bacterial communities similar to those bacteria found in mother's vagina while the gut of the C-section babies was mainly inhabited by the bacteria which were more commonly found on the skin (Dominguez-Bello et al. 2010).

A much more diversified intestinal microbiota composition was observed in vaginally delivered infants with predominant species such as *B. longum* and *B. catenulatum*. The significant presence of bifidobacteria helps promote stronger immune responses in vaginally born infants, and this has been associated with the lower prevalence of atopic diseases as compared to infants delivered via C-section. In addition, increased allergic events have also been much associated with delivery via C-section. This is largely attributed to the fact that C-section children were deprived of massive bacterial load from vaginal delivery. However, recent study has shown that the increased prevalence of allergies in C-section infants can be reversed via probiotic supplementation containing bifidobacteria (Kuitunen et al. 2009). Ig-E sensitization which contributes to allergic diseases was also reported to be greatly reduced after probiotic intervention.

While probiotic colonization in the newborn from maternal source is reported to be more common in vaginally born infants, oral administration of probiotic can also result in colonization of the probiotic strain in the intestine of C-section infants. Probiotic supplementation to pregnant women reduced cumulative incidence of atopic dermatitis in infants during the first 2 years of life (Dotterud et al. 2010). Similarly, infants with high cord blood IgE concentration have also been reported to have benefited from maternal probiotic supplementation (Rautava et al. 2002).

2.2 Breast-Feeding and Infant Milk Formula

Infant milk formulas were supplemented with prebiotics in order to mimic the functional and beneficial effects of human breast milk. Infant formula supplemented with galacto- and fructo-oligosaccharides has been demonstrated to induce an intestinal ecosystem with diversified bifidobacteria species. On the other hand, a standard infant formula is more likely to promote an adult-like bifidobacteria distribution which consists mostly of *B. adolescentis* and *B. catenulatum* (Haarman and Knol 2005). It is noted that an adult-like bifidobacteria flora establishment in early age is associated with the development of allergic diseases (He et al. 2001a, b).

Recently, infant formula is increasingly supplemented with probiotics. Thus, numerous studies have been initiated to investigate on the effects of the probiotic supplementation on the growth and health status of infants. Supplementation of milk formula with *Bifidobacterium lactis* and *Streptococcus thermophilus* has been reported to result in adequate growth, reduced incidence of colic, and lower frequency of antibiotic use in infants (Saavedra et al. 2004). Supplementation of viable *Bifidobacterium lactis* in infant formula effectively decreased the incidence of gastrointestinal infection in infants (Chouraqui et al. 2004). Other study has also demonstrated that milk formula supplemented with *Lactobacillus reuteri* improved the morbidity parameters as well as reduced the occurrence of diarrhea (Weizman et al. 2005).

Contrary to the traditional belief that breast milk is sterile, recent evidence has shown that breast milk is a reservoir of bacteria; thus, it plays an important role in shaping the neonatal gut microbiota. It is recently accepted that the vertical transfer of viable microorganisms from mother's milk to the infants has contributed to the early establishment of gut microbiota. Increased awareness on the protective effects of breast milk against infectious disease in infants has led to the interest in isolation of bacteria with probiotic potential. For instances, *Lactobacillus salivarius* CECT5713 and *Lactobacillus fermentum* CECT5716 which were isolated from human milk enhanced host immune responses via the induction of cytokines and the activation of natural killer cells as well as regulatory T-cells (Pérez-Cano et al. 2010). In another study, incorporation of *L. fermentum* isolated from human breast milk into infant formula significantly reduced the occurrence of gastrointestinal infection and upper respiratory tract infection in 6–12 months infants (Maldonado et al. 2012).

3 Childhood and Adulthood

3.1 Nutrition and Dietary Habit

Dietary habit plays a role in shaping the microbial ecology of the human gut. Growing evidence has unraveled the interrelationship between nutritional profiles and the intestinal microbiota composition. Different foods relatively contribute to different total energy input in the human body, affecting microbial diversity in the gut and along different regions of the gastrointestinal tract. A day of switching from low fat to high fat diet could cause an immediate alteration of the gut microbiome.

It has been hypothesized that coevolution of the gut microbiota could occur in order for them to adapt to the dietary habits of the hosts. This may allow the bacteria to maximize energy output from dietary intakes, and at the same time providing the hosts an access to additional energy needed for metabolic processes. This idea was derived upon the detection of increased short-chain fatty acids (SCFAs) in fecal samples of populations consuming diets rich in fiber. The presence of plant polysaccharide-degrading bacteria in this population has enabled the extraction of extra calories from the energy-rich food, via the production of SCFAs (De Filippo et al. 2010).

Other than the role as an additional energy source, the importance of SCFAs has been well proven for their protective role against gut inflammation as well as other diseases. SCFAs are organic fatty acids which arise from bacterial fermentation of proteins, peptide, polysaccharides, oligosaccharides, and glycoprotein precursors in the colon. Numerous studies have shown that probiotics enhance gut concentration of major SCFAs such as acetate, butyrate, and propionate (Hijova and Chmelarova 2007), leading to the modulation of colonic microbiota due to changes in pH. Oral consumption of *Lactobacillus plantarum* P-8 increased fecal concentration of acetic and propionic acids which was accompanied by an increased population of bifidobacteria and a decrease in opportunistic pathogens such as *Desulfovibrio* (Wang et al. 2014).

SCFA specifically butyrate can stimulate the proliferation and differentiation of colonocytes. Butyrate possesses anti-inflammatory properties given its inhibition activity on histone deacetylase which consequently suppress the activation of nuclear factor-kappa B (NF-κB) (Yin et al. 2001). Improper regulation of NF-κB has been related to improper immune development, autoimmune disorder, cancer, and inflammatory diseases. In addition, butyrate also improves intestinal defense barrier and decreases epithelial permeability via stimulated production of mucin as well as tight junction proteins.

A high consumption of alcohol has been associated with leaky gut and liver injuries, leading to the development of diseases such as alcoholic steatohepatitis and alcoholic fatty liver, accompanied by a higher level of plasma endotoxin (Gramenzi et al. 2006). The excessive generation of endotoxin has been associated with the overgrowth of gram-negative bacteria, and the continuous production of endotoxin further supports the proliferation of pathogenic bacteria over the beneficial microorganisms (Rao et al. 2004). Chronic alcohol abuse has also been demonstrated to promote bacterial overgrowth, especially gram-positive aerobic cocci in the upper gastrointestinal tract. Alcohol consumption impaired host immune system by damaging gut-associated lymphoid tissue (GALT), and subsequently rendering the intestines to be compromised upon colonization of *Salmonella typhimurium* (Sibley and Jerrells 2000).

Probiotic therapy consisting of *Bifidobacterium bifidum* and *Lactobacillus plantarum* 8PA3 has been demonstrated to restore the gut microbiota profile by significantly increasing the number of lactobacilli and bifidobacteria in patients with alcoholic liver disease (Kirpich et al. 2008). These probiotics had also significantly reduced plasma level of aspartate aminotransferase, γ-glutamyl transpeptidase, and alanine aminotransferase. High levels of these enzymes are usually detected in subjects with mild alcoholic hepatitis. Another study has also reported that probiotic such as *L. rhamnosus* GG ameliorated alcohol-induced gut leakiness and liver injury (Forsyth et al. 2009). The improvement in both gut and liver was associated with reduced markers of oxidative stress and inflammation and a reduction in transport of pro-inflammatory bacterial products across the gut to the liver. Another study has also reported that *Lactobacillus acidophilus* and *Bifidobacterium longum* effectively protected individuals from alcohol-induced gastric and liver injury by decreasing blood level of alcohol via increased metabolism of alcohol in the stomach and liver (Qing and Wang 2008).

3.2 Stress and Behavior

The gut–brain axis, a bidirectional interaction between the brain and the gut which includes the neural, endocrine, and immune pathways, has been proposed to regulate homeostasis during the events of stress. Early exposure to stress in life may exacerbate an individual's susceptibility to diseases in later life. Disruption in the balance of intestinal microbiota such as increasing population of coliforms has been associated with the consequences of stress. Reduction in bifidobacteria and lactobacilli was often observed in animals upon exposure to acute stress, thus compromising the intestinal barrier function and subsequently immunity (Barreau et al. 2004).

Animals that were exposed to social stressors showed increased circulating cytokines, accompanied by reduced phyla of Bacteroidetes accompanied by increased phyla of Firmicutes (Fig. 2a); a decrease in genera of *Coprococcus*, *Pseudobutyrivibrio*, and *Dorea* was observed, accompanied by increased genera of *Clostridium* (Fig. 2b) (Bailey et al. 2011). DGGE analysis of fecal samples from animals experiencing maternal stress showed a more disrupted gut microbiota with an increased coliform population (O'Mahony et al. 2009). Moreover, damaged noradrenergic neurons as a consequence of traumatic experience and subsequent release of norepinephrine into the systemic circulation may also shift the indigenous microorganism population, favoring the growth of gram-negative bacteria.

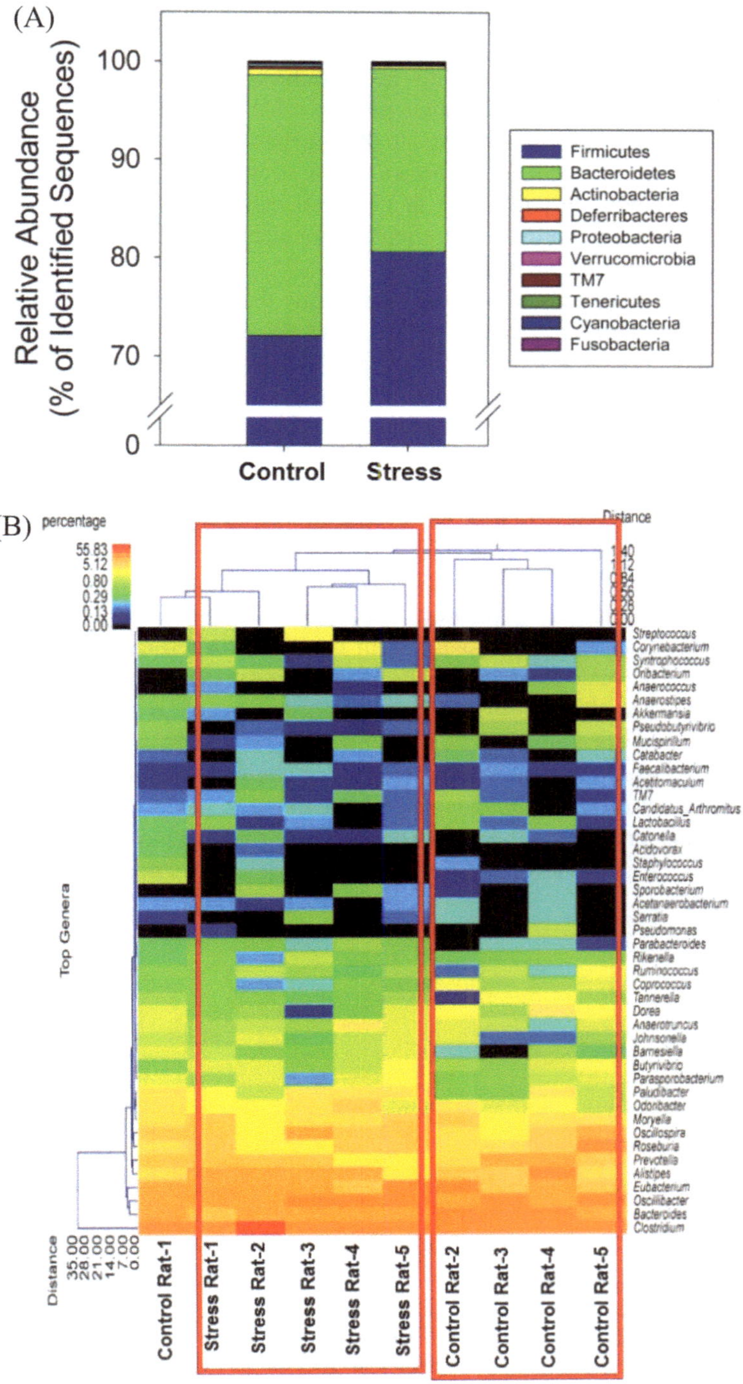

Fig. 2 Cecal microbiota of mice that were stressed via exposure to an aggressive male mouse for 2 h/day for 6 days. (**a**) Cecal microbiota of stressed and non-stressed (control) mice comprising bacteria from 10 divisions. Mice that were subjected to stress had higher levels of Firmicutes

Increased colonic motility has also been reported in humans upon exposure to stressful conditions. Elevated level of motilin, which improves peristalsis, was observed in stressful patients with irritable bowel syndrome. In addition to increased peristaltic motility, another notable effect involved the detachment of adherent bacteria in the gut, attributed to the degenerated intestinal mucosa and loss of goblet cells. Increased level of cortisol during stress also leads to apoptosis of the gut epithelial cells, thus resulting in intestinal cellular damage. The combined effects may eventually provide a desired environment for opportunistic pathogenic bacteria to colonize the GI tract.

There is growing body of literature suggesting that gut microbiota composition also affects the central mediated system such as alteration of mood and behavior as well as modulation of brain neurochemistry. An increasing number of studies have been dedicated to search for putative probiotics as therapeutic solution to tackle both central and peripheral elements of anxiety-related disorder. Study showed that treatment of probiotic *Lactobacillus farciminis* to rats submitted to stress procedure has prevented stress-induced gut hyperpermeability, hypothalamic–pituitary–adrenal axis stress response, as well as neuroinflammation by means of reduction of luminal lipopolysaccharides concentration (Ait-Belgnaoui et al. 2012). Ingestion of 5×10^9 cfu/ml of probiotic *L. rhamnosus* in mice could alleviate stress-related disorder as evidenced by reduced corticosterone level and altered gamma-aminobutyric acid (GABA) expression in the brain (Bravo et al. 2011). Similar researches also revealed an influential role of probiotic in reversal of abnormalities in behavior via modulation of central neurochemical such as restoration of basal noradrenaline levels and normalization of central serotonergic system (Desbonnet et al. 2008, 2010). In clinical trial, probiotic formulation consisting of *Lactobacillus helveticus* and *Bifidobacterium longum* has presented an anxiolytic effect on human healthy volunteers (Messaoudi et al. 2011). These findings have further suggested that probiotics may have a broader therapeutic application than previously considered.

3.3 Aging

Generally, aging was often accompanied by an increase in inflammatory status. Aging process deeply influenced the immune system and increased its vulnerability to illnesses. Malnutrition has been reported as one of the reasons for the lower

Fig. 2 (continued) ($p = 0.08$), but lower levels of Bacteroidetes ($p = 0.11$) compared to the control mice. (**b**) Heat map of relative abundance of different genera from stressed and control mice ($n = 5$ per group), showing a lower population of *Coprococcus*, *Pseudobutyrivibrio*, and *Dorea* and higher population of *Clostridium* from stressed mice compared to the non-stressed control. Reprinted from Bailey et al. (2011), with permission from Elsevier (License number: 3632820761432)

immune response in the elderly. Concomitantly, a shift in intestinal microbiota composition was also frequently reported in the elderly. Analysis of human fecal samples showed that the bacteroides counts decreased with age, while *E. coli* and *Enterococci* spp. increased with age (Enck et al. 2009).

Antibiotic therapy has been reported to magnify the effect of dwindling population of bacteroides in the elderly. The reduction in bacterial diversity is more apparent in the elderly as compared to young adults. A decline in species diversity of bifidobacteria was more prevalent in the gut of elderly, where a wide range of bifidobacteria species eventually narrowed to one or two dominant species such as *B. adolescentis* and *B. longum* (He et al. 2001a, b). It has also been suggested that most of the known species with anti-inflammatory properties were markedly decreased in the elderly.

Since aging process may lead to a weaker immune function, the elderly are therefore at higher risks of infectious and noninfectious diseases. Probiotic intervention could represent as an effective means in restoring the cellular immunity by improving the function of intestinal microflora in elderly. Consumption of probiotic supplementation consisting of *B. lactis* HN019 enhanced the phagocytic capacity of mononuclear and polymorphonuclear phagocytes and the tumoricidal activity of natural killer cells in elderly subjects (Gill et al. 2001). Probiotic *B. animalis* has been reported to increase longevity in animal host via suppression of chronic low-grade inflammation by increasing the polyamines' concentration in colon (Matsumoto et al. 2011). Probiotic treatment using *Lactobacillus rhamnosus* HN001 and *Lactobacillus acidophilus* NCFM on elderly subjects has showed a significantly lower count of *Clostridium difficile* in feces (Lahtinen et al. 2012). Although no changes in immune markers were observed in this study, the reduction in population of *Clostridium difficile* could be beneficial to the elderly by reducing the risk of diseases ranging from trivial diarrhea to life-threatening colitis.

Aging was closely associated with increased oxidative damage which is induced by imbalance between pro-oxidants and antioxidants in cells. The elevated oxidative stress in aging organism was postulated to be one of the major factors contributing to senescence. While overexpression of antioxidative enzymes retards age-associated oxidative damage, the reduction in level of reactive oxygen species extends the life span of organisms. Supplementation of *Lactobacillus rhamnosus* IMC 501 and *Lactobacillus paracasei* IMC 502 exerted strong antioxidant activity by increasing plasma level of antioxidant and reducing reactive oxygen metabolites' level (Martarelli et al. 2011). Consequently, the reactive oxygen species neutralization effect of probiotics may enhance the prevention of age-inflicted cancer and neurodegenerative diseases in the elderly.

3.4 Host Inflammation

A dense population of microbiota in the gut has been associated with increased resistance against pathogenic bacteria, via competing with pathogens for space and nutrients as well as inhibition via the actions of metabolic by-products. However, resistance to pathogenic attacks is often compromised in the events of gut inflammation. Infection of *Salmonella enteric* subspecies 1 serovar Typhimurium (S. Tm) has been found to trigger severe inflammation and colitis, leading to decreased resistance by gut microbiota. In contrast, inflammation was not apparent with mutant S. Tm which is avirulent, where colonization resistance of the gut microbiota was not altered (Stecher et al. 2007). These findings suggested that inflammation shifted the balance of gut microbiota in favor of pathogens. Besides, gut inflammation also reduces bacterial diversity such as the reduction of *Firmicutes* and increases intestinal permeability, thus allowing translocation of pathogenic microorganisms and their metabolic products across the intestinal epithelium.

The administration of probiotic *L. casei* strain Shirota has improved murine chronic inflammatory bowel disease by downregulating the pro-inflammatory agents such as IL-6 and IFN-γ (Matsumoto et al. 2005). This beneficial health effects exerted by probiotics are generally achieved via the interaction of bacterial metabolites, cell wall components, or DNA with epithelial and gut-associated immune cells. P40, a soluble protein derived from probiotic *Lactobacillus rhamnosus* GG, ameliorated intestinal inflammation through the activation of epidermal growth factor receptor (EGFR) which subsequently inhibited cytokine-induced apoptosis in intestinal cells (Yan et al. 2011). Cell wall component of *L. casei* alleviated skin inflammation in murine models, via the suppression of $CD8^+$ type-1 cytotoxic T-cells and regulation of $CD4^+$ T-cells (Chapat et al. 2004). DNA isolated from probiotic VSL-3 and *Escherichia coli* (DH5α) possessed immune-stimulatory effect, attenuating the severity of experimental colitis. These DNAs were reported to be biologically active and involved in the activation of toll-like receptor (TLR) pathway (Rachmilewitz et al. 2004). Lactic acid bacteria (LAB) such as *Lactobacillus salivarius* are also reported to produce antimicrobial peptides known as bacteriocin, to inhibit growth of pathogenic microorganisms (Messaoudi et al. 2013). The antagonistic activity of the bacteriocin against pathogens may suggest the potential use of probiotic as oral, topical, antibiotics or disinfectants as well as in food processing and preservation.

3.5 Obesity

Gut microbiota has been associated with the regulation of energy balance and body weight gain of hosts; obese humans and animals have been shown to harbor microbiota that are more capable of harvesting energy from food, accompanied by enhanced absorption of monosaccharides, higher hepatic production of triglycerides, and adipogenesis. A higher count of methanogenic *Archaea* was detected in obese individuals rather than their normal weight counterparts; the coexistence between these H_2-utilizing methanogens and H_2-producing bacteria enabled higher energy extraction and uptake in obese individuals (Zhang et al. 2009).

The difference in dietary nutrient load is one of the major factors behind the rapid changes in gut bacterial composition. In the event of higher caloric intake by lean individuals, a greater stool energy loss and increased energy harvest were observed, accompanied by a 20 % increase in *Firmicutes* and a corresponding decrease in *Bacteroidetes* (Jumpertz et al. 2011). This observation was supported where a lower stool energy excretion was more apparent in obese individuals compared to lean participants.

The ability of certain microbes to harvest and store excessive energy posed complications to pregnant women as it may cause excessive weight gain; *Bacteroides* concentration before pregnancy is closely linked to weight gain in women during pregnancy (Collado et al. 2008). Overweight pregnant women presented a higher bacterial count of *Bacteroides* before pregnancy, and increased count throughout pregnancy. Overweight pregnant women often gave birth to newborn with higher birth weight due to the high energy supplied to fetus during pregnancy. Consequently, overweight infants have a higher risk of developing obesity during adolescence.

Obesity is also often accompanied by low-grade systemic inflammations and metabolic disorders, typically from high fat diet-induced obesity. Experiments using rats with different phenotypes such as the obesity-prone and obesity-resistance rats showed that obesity can induce ileal inflammation via increased activation of toll-like receptor (TLR4) and decreased intestinal alkaline phosphatase which are needed for the detoxification of lipopolysaccharides (LPS) (de La Serre et al. 2010). LPS has been associated with the onset of most metabolic diseases, where increased plasma level of LPS has led to increased abundance of gut *Clostridiales*.

Given that obesity is one of the current major public health issues, various efforts have been made to improve current intervention or strategies for a better control of obesity. Recent discovery on the implication of gut microbiota in energy homeostasis may provide a novel target in the prevention and treatment of obesity disorder. Early gut microbiota intervention via probiotic such as *L. rhamnosus* has been reported to modify the growth pattern of children by moderating excessive weight gain in early ages (Luoto et al. 2010). Since weight reduction is defined as a state of negative energy balance, it is therefore proposed that probiotic intervention may contribute to prevention of weight gain by reducing food intake, stimulating energy expenditure, and inhibiting energy absorption. Probiotic VSL#3 administered to murine model has induced hormone GLP-1 release which subsequently resulted in reduced food intake (Yadav et al. 2013). Modulation of gut microbiota by probiotic therapy may also help prevent excessive energy harvest in obese individuals. A study involving vancomycin treatment on obese mice showed that *Firmicutes* counts were proportionate with weight gain and energy harvest or absorption (Murphy et al. 2013).

Modification of gut microbiota upon probiotic consumption also influences lipid homeostasis. A shift in microbial composition after the consumption of yogurt containing *Lactobacillus amylovorus* and *Lactobacillus fermentum* by 28 overweight participants led to significant reduction in body fat mass (Omar et al. 2013), possibly via alteration of body adiposity through alteration of gut microbiota composition. *Lactobacillus paracasei* ST11 modulated activity of nerve innervating brown adipose tissue and enhanced lipolysis, leading to reduced body weight in the rats (Tanida et al. 2008). *L. reuteri* induced expression of carnitine palmitoyltransferase 1a (Cpt1a) enzymes (Fåk and Bäckhed 2012). Cpt1a is essential for fatty acid oxidation, where a deficiency in this enzyme may lead to accumulation of fats in cells and damage the liver, heart, and brain.

3.6 Cancer Diseases

Intestinal microbiota has been identified to play a role in colorectal carcinogenesis. Investigation on the correlationship between gut microbiota and colorectal cancer development revealed changes of microbial genes along the adenoma (benign polyps) to carcinoma (cancer) sequence (Fig. 3) (Feng et al. 2015), suggesting that the gut microbiome and their interactions contribute to the development of colorectal cancer.

It has been postulated that the constant exposure of mucosa to bacterial metabolites such as phenols, indoles, hydrogen sulfide, reactive oxygen species, and other carcinogens are associated with the etiology of tumor and cancer diseases. Increased in diversity of *Clostridium leptum* and *C. coccoides* was more evident in patients with colorectal cancer as compared to the control (Scanlan et al. 2008). *Clostridium* is one of the most commonly recognized factors involved in the metabolism of phenolic compounds, and could culminate in higher production of toxic or carcinogenic metabolites which are detrimental to the host. Colorectal cancer patients also often harbor elevated Bacteroides/Prevotella populations, accompanied by elevated amount of IL-17 producing cells in the mucosa layer, suggesting that the aberrancy in the mucosal microbiota is closely associated with the disruption in immune response (Sobhani et al. 2011).

The administration of probiotics has appeared to be promising in reducing the risks and incidences of colon, liver, and bladder cancers, mainly to the carcinogenic-inhibitive activities and reduction in cancer cells promoting enzymes. *Bifidobacteria adolescentis* SPM0212 exhibited antiproliferative effects on human colon cancer cell lines by inducing macrophage activation and increasing the production of tumor necrosis factor-α (TNF-α) and nitric oxide (NO) (Lee et al. 2008). Polysaccharides isolated from *Lactobacillus acidophilus* 606 and *Lactobacillus casei* ATCC 393 induced apoptosis on cancer cells while modulating circulatory oxidative stress that protected the cells against carcinogen-induced damages (Choi et al. 2006). Bifidobacteria also produce anticarcinogenic metabolites that interfere with cytochrome P450s-dependent monooxygenase system which could subsequently lead to the production of carcinogens (Liong 2008).

Mutagens such as benzo[a]pyrene, amino acid pyrolysates, N-nitroso compounds, and aflatoxin B are generally carcinogenic due to its ability to induce genetic mutation and damage. *Bifidobacterium longum* exhibited antimutagenic activity by increasing the activity of glutathione transferase (Challa et al. 1997),

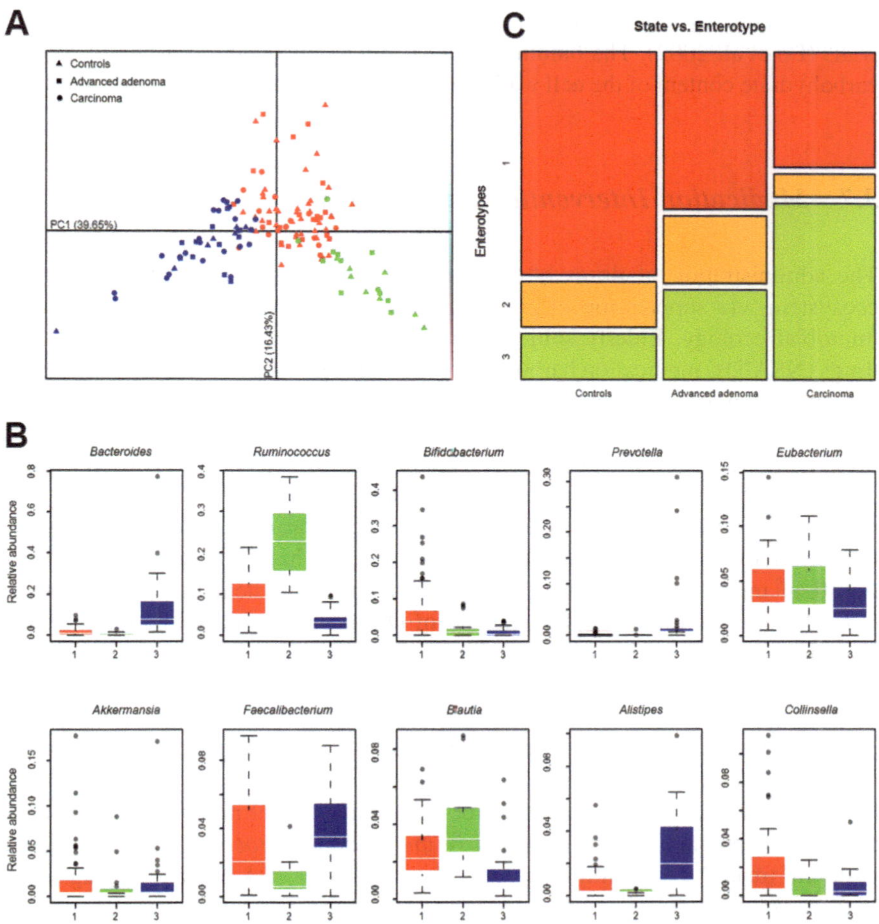

Fig. 3 Metagenome-wide association study on stool samples from healthy subjects and advanced adenoma and carcinoma patients. Microbial genes were categorized into three different communities: *Red*, community type 1; *green*, community type 2; *blue*, community type 3. (**a**) Principal component analysis for the stool samples at the genus level. (**b**) Relative abundances of the top 10 most abundant genera in the three community types. (**c**) Distribution of the healthy control, advanced adenoma, and carcinoma samples in the community types. Reprinted from Feng et al. (2015), with permission from Elsevier (License number: 3632880210485)

which reduces the cytotoxicity and genotoxicity of mutagens through the catalysis of reaction between sulfhydryl compounds and the mutagens to form less harmful conjugates. Consumption of yogurt containing *Bifidobacterium lactis* LKM512 reduced mutagenicity levels in human, accompanied by increased levels of spermidine in the gut (Matsumoto and Benno 2004). Spermidine is vital for cell proliferation, differentiation, and regulation of enzymatic activity as well as synthesis and stabilization of DNA, RNA, and proteins. Viable cells of *Bifidobacterium lactis* Bb-12 and *B. longum* CCRC 14634 have also been shown to possess high antimutagenic activity against mutagens via binding of mutagens to bacterial cell walls (Lo et al. 2002). The binding ability is often dependent on bacterial species, carbohydrate content of the cell wall, and the chemical complexity of the mutagens.

3.7 Medication Intervention and Surgery

The administration of drugs often causes disturbances in the normal intestinal ecosystem, via suppression of gut microbial growths, leading to an imbalanced microbial ecology. Elderly subjects on a regular nonsteroidal anti-inflammatory drugs (NSAIDs) medication harbored lower counts of *Lactobacillus* and *Collinsella* spp. in the gut as compared to those without NSAIDs (Mäkivuokko et al. 2010).

Antibiotic treatments are widely associated with reduced colonization resistance of certain intestinal microorganisms, leading to the overgrowth of resistant pathogens. While the overall taxonomic composition of intestinal microbiota remained nearly unchanged after an antibiotic treatment, some of the taxa could not be recovered to its pretreatment state for as long as half a year. Assessment of intestinal microbiota composition using denaturing gradient gel electrophoresis (DGGE) showed a reduced total number of DGGE bands, indicating lesser microbial biodiversity in the gut upon antibiotic treatment (Iapichino et al. 2008). Disturbance of the gut microbial balance may subsequently increase the risk of developing diarrhea. Antibiotic-associated diarrhea (AAD) is common after an antibiotic withdrawal or upon further referral to a different prescription of antibiotic. Probiotics such as *Saccharomyces boulardii* and *Lactobacillus rhamnosus* GG reduce the risk of AAD, via restoration of a stable bowel microflora (McFarland 2006). Concomitant administration of *B. lactis* and *Streptococcus thermophilus* in children treated with antibiotic also showed lesser risk of developing AAD (Szajewska et al. 2006).

Surgical intervention intended to improve metabolism and inflammation in human can also contribute to the compositional modulation of gut microbiota. Batriatric surgery offers morbidly obese patient a definitive approach for long-term weight loss and reverses the effect of adverse metabolic effects. However, a substantial shift in the main gut phyla is often observed upon metabolic surgery, where lower levels of *Firmicutes* and *Bacteroidetes* and higher levels of Proteobacteria were detected (Li et al. 2011). Surgical trauma and perioperative infection control also contributed to the fluctuations of gut ecosystems, attributed to altered redox state, pH, and norepinephrine release that affect gut gene regulations and bacterial survival and competency, in addition to increased virulence of pathogenic bacteria.

Pouchitis is a nonspecific inflammation of ileal pouch which occurs in patients with ulcerative colitis after colectomy surgery. While the etiology of pouchitis remains unclear, bacterial overgrowth is deduced to play an important role given the efficacy of antibiotic treatment in pouchitis. The use of probiotic combination consisting of bifidobacteria, lactobacilli, and streptococci, following pouch surgery for ulcerative colitis, presented a positive preventive effect on the onset and relapses of pouchitis, attributed to normalization of immunological markers such as cytokine IL-10 and tumor necrosis factor-α (Ulisse et al. 2001).

Similar beneficial effects of probiotic were also observed in surgical patients who have undergone Roux-en-Y gastric bypass (RNYGB) procedure. While RNYGB offers a promising therapy for morbid obesity, the surgical procedure also disrupts the gut microbiota. Postoperative administration of lactobacilli in RNYGB surgery patients showed a restored intestinal microbial balance supported by a significant reduction in bacterial overgrowth measured by the indicative hydrogen gas level (Woodard et al. 2009) and accompanied by a greater weight loss in RNYGB surgery patients compared to control.

Bacterial translocation and postoperative morbidity have been reported in patients who had undergone major abdominal surgeries. Physical injuries on the intestinal mucosa inflicted during surgery may have disrupted the gut barrier, enabling passage of bacteria from the gastrointestinal tract to the sterile tissues such as internal organs and mesenteric lymph nodes (MLN). Increased intestinal permeability and bacterial translocation can however be reversed with probiotic treatment. Administration of *Lactobacillus rhamnosus* strain R0011 and *Lactobacillus helveticus* strain R0052 has been reported to reduce translocation of bacteria to distant organ via their adherence to the epithelial surface, thus limiting the binding of pathogenic bacteria to the intestinal epithelium (Zareie et al. 2006). It was also suggested that the interaction of probiotic with indigenous cells and host immune cells modulated the immune responses and therefore improved intestinal barrier integrity.

4 Conclusions

Prevalence of various diseases is closely linked to the occurrence of aberrant composition of gut microbiota. Accumulating evidence has indicated that the influence of gut microbiota extends beyond the gut, most probably via immunological and neuroendocrine pathways. Thus, the intervention of gut bacterial colonization via administration of probiotics has been advocated as a new therapeutic approach to alleviate a wide range of disorders. Understanding factors that cause the declining population of certain gut bacterial species may allow us to reintroduce suitable beneficial microbes for the reestablishment of an effective functional gut.

Acknowledgment This work was supported by the FRGS grant (203/PTEKIND/6711372) and MyBrain scholarship provided by the Ministry of Higher Education Malaysia and the USM IReC grant (1002/PTEKIND/910406) provided by Universiti Sains Malaysia.

Conflict of Interest No.

References

Bailey MT, Dowd SE, Galley JD, Hufnagle AR, Allen RG, Lyte M (2011) Exposure to a social stressor alters the structure of the intestinal microbiota: implications for stressor-induced immunomodulation. Brain Behav Immun 25(3):397–407

Barreau F, Ferrier L, Fioramonti J, Bueno L (2004) Neonatal maternal deprivation triggers long term alterations in colonic epithelial barrier and mucosal immunity in rats. Gut 53(4):501–506

Ait-Belgnaoui A, Durand H, Cartier C, Chaumaz G, Eutamene H, Ferrier L, Houdeau E, Fioramonti J, Bueno L, Theodorou V (2012) Prevention of gut leakiness by a probiotic treatment leads to attenuated HPA response to an acute psychological stress in rats. Psychoneuroendocrinology 37(11):1885–1895

Bravo JA, Forsythe P, Chew MV, Escaravage E, Savignac HM, Dinan TG, Bienenstock J, Cryan JF (2011) Ingestion of *Lactobacillus* strain regulates emotional behavior and central GABA receptor expression in a mouse via the vagus nerve. Proc Natl Acad Sci U S A 108 (38):16050–16055

Challa A, Rao DR, Chawan CB, Shackelford L (1997) *Bifidobacterium longum* and lactulose suppress azoxymethane-induced colonic aberrant crypt foci in rats. Carcinogenesis 18 (3):517–521

Chapat L, Chemin K, Dubois B, Bourdet-Sicard R, Kaiserlian D (2004) *Lactobacillus casei* reduces CD8+ T cell-mediated skin inflammation. Eur J Immunol 34(9):2520–2528

Choi SS, Kim Y, Han KS, You S, Oh S, Kim SH (2006) Effects of *Lactobacillus* strains on cancer cell proliferation and oxidative stress in vitro. Lett Appl Microbiol 42(5):452–458

Chouraqui JP, Van Egroo LD, Fichot MC (2004) Acidified milk formula supplemented with *Bifidobacterium lactis*: impact on infant diarrhea in residential care settings. J Pediatr Gastroenterol Nutr 38(3):288–292

Collado MC, Isolauri E, Laitinen K, Salminen S (2008) Distinct composition of gut microbiota during pregnancy in overweight and normal-weight women. Am J Clin Nutr 88(4):894–899

De Filippo C, Cavalieri D, Di Paola M, Ramazzotti M, Poullet JB, Massart S, Collini S, Pieraccini G, Lionetti P (2010) Impact of diet in shaping gut microbiota revealed by a comparative study in children from Europe and rural Africa. Proc Natl Acad Sci U S A 107 (33):14691–14696

De La Serre CB, Ellis CL, Lee J, Hartman AL, Rutledge JC, Raybould HE et al (2010) Propensity to high-fat diet-induced obesity in rats is associated with changes in the gut microbiota and gut inflammation. Am J Physiol Gastrointest Liver Physiol 299(2):G440–G448

Desbonnet L, Garrett L, Clarke G, Bienenstock J, Dinan TG (2008) The probiotic *Bifidobacterium infantis*: an assessment of potential antidepressant properties in the rat. J Psychiatr Res 43 (2):164–174

Desbonnet L, Garrett L, Clarke G, Kiely B, Cryan JF, Dinan TG (2010) Effects of the probiotic *Bifidobacterium infantis* in the maternal separation model of depression. Neuroscience 170 (4):1179–1188

Dominguez-Bello MG, Costello EK, Contreras M, Magris M, Hidalgo G, Fierer N, Knight R (2010) Delivery mode shapes the acquisition and structure of the initial microbiota across multiple body habitats in newborns. Proc Natl Acad Sci U S A 107(26):11971–11975

Dotterud CK, Storrø O, Johnsen R, Øien T (2010) Probiotics in pregnant women to prevent allergic disease: a randomized, double-blind trial. Br J Dermatol 163(3):616–623

Enck P, Zimmermann K, Rusch K, Schwiertz A, Klosterhalfen S, Frick JS (2009) The effects of ageing on the colonic bacterial microflora in adults. Z Gastroenterol 47(7):653

Fåk F, Bäckhed F (2012) *Lactobacillus reuteri* prevents diet-induced obesity, but not atherosclerosis, in a strain dependent fashion in Apoe−/− mice. PLoS One 7(10), e46837

FAO/WHO (2001) Report of a joint FAO/WHO expert consultation on evaluation of health and nutritional properties of probiotics in food including powder milk with live lactic acid bacteria, Córdoba, Argentina

Feng Q, Liang S, Jia H, Stadlmayr A, Tang L, Lan Z, Zhang D, Xia H, Xu X, Jie Z, Su L, Li X, Li J, Xiao L, Huber-Schönauer U, Niederseer D, Xu X, Al-Aama JF, Yang H, Wang J, Kristiansen K, Arumugam M, Tilg H, Datz C, Wang J (2015) Gut microbiome development along the colorectal adenoma–carcinoma sequence. Nat Commun 6

Forsyth CB, Farhadi A, Jakate SM, Tang Y, Shaikh M, Keshavarzian A (2009) *Lactobacillus* GG treatment ameliorates alcohol-induced intestinal oxidative stress, gut leakiness, and liver injury in a rat model of alcoholic steatohepatitis. Alcohol 43(2):163–172

Gill HS, Rutherfurd KJ, Cross ML, Gopal PK (2001) Enhancement of immunity in the elderly by dietary supplementation with the probiotic *Bifidobacterium lactis* HN019. Am J Clin Nutr 74 (6):833–839

Gramenzi A, Caputo F, Biselli M, Kuria F, Loggi E, Andreone P, Bernardi M (2006) Review article: alcoholic liver disease–pathophysiological aspects and risk factors. Aliment Pharmacol Ther 24(8):1151–1161

Haarman M, Knol J (2005) Quantitative real-time PCR assays to identify and quantify fecal *Bifidobacterium* species in infants receiving a prebiotic infant formula. Appl Environ Microbiol 71(5):2318–2324

He F, Ouwehand AC, Isolauri E, Hashimoto H, Benno Y, Salminen S (2001a) Comparison of mucosal adhesion and species identification of bifidobacteria isolated from healthy and allergic infants. FEMS Immunol Med Microbiol 30(1):43–47

He F, Ouwehand AC, Isolauri E, Hosoda M, Benno Y, Salminen S (2001b) Differences in composition and mucosal adhesion of bifidobacteria isolated from healthy adults and healthy seniors. Curr Microbiol 43(5):351–354

Hijova E, Chmelarova A (2007) Short chain fatty acids and colonic health. Bratisl Lek Listy 108 (8):354

Iapichino G, Callegari ML, Marzorati S, Cigada M, Corbella D, Ferrari S, Morelli L (2008) Impact of antibiotics on the gut microbiota of critically ill patients. J Med Microbiol 57(8):1007–1014

Jumpertz R, Le DS, Turnbaugh PJ, Trinidad C, Bogardus C, Gordon JI, Krakoff J (2011) Energy-balance studies reveal associations between gut microbes, caloric load, and nutrient absorption in humans. Am J Clin Nutr 94(1):58–65

Kirpich IA, Solovieva NV, Leikhter SN, Shidakova NA, Lebedeva OV, Sidorov PI, Bazhukova TA, Soloviev AG, Barve SS, Cave M (2008) Probiotics restore bowel flora and improve liver enzymes in human alcohol-induced liver injury: a pilot study. Alcohol 42(8):675–682

Kuitunen M, Kukkonen K, Juntunen-Backman K, Korpela R, Poussa T, Tuure T, Haatela T, Savilahta E (2009) Probiotics prevent IgE-associated allergy until age 5 years in cesarean-delivered children but not in the total cohort. J Allergy Clin Immunol 123(2):335–341

Lahtinen SJ, Forssten S, Aakko J, Granlund L, Rautonen N, Salminen S, Viitanen M, Ouwehand AC (2012) Probiotic cheese containing *Lactobacillus rhamnosus* HN001 and *Lactobacillus acidophilus* NCFM® modifies subpopulations of fecal lactobacilli and *Clostridium difficile* in the elderly. Age 34(1):133–143

Lee DK, Jang S, Kim MJ, Kim JH, Chung MJ, Kim KJ, Ha NJ (2008) Anti-proliferative effects of *Bifidobacterium adolescentis* SPM0212 extract on human colon cancer cell lines. BMC Cancer 8(1):310

Li JV, Ashrafian H, Bueter M, Kinross J, Sands C, le Roux CW, Bloom SR, Darzi A, Athanasiou T, Marchesi JR, Nicholson JK, Holmes E (2011) Metabolic surgery profoundly influences gut microbial–host metabolic cross-talk. Gut 60(9):1214–1223

Liong MT (2008) Roles of probiotics and prebiotics in colon cancer prevention: postulated mechanisms and in-vivo evidence. Int J Mol Sci 9(5):854–863

Lo PR, Chou CC, Tsai YH (2002) Antimutagenic activity of several probiotic bifidobacteria against benzo [a] pyrene. J Biosci Bioeng 94(2):148–153

Luoto R, Kalliomäki M, Laitinen K, Isolauri E (2010) The impact of perinatal probiotic intervention on the development of overweight and obesity: follow-up study from birth to 10 years. Int J Obes (Lond) 34(10):1531–1537

Mäkivuokko H, Tiihonen K, Tynkkynen S, Paulin L, Rautonen N (2010) The effect of age and non-steroidal anti-inflammatory drugs on human intestinal microbiota composition. Br J Nutr 103(2):227

Maldonado J, Cañabate F, Sempere L, Vela F, Sánchez AR, Narbona E, López-Huertas E, Geerlings A, Valero AD, Olivares M, Lara-Villoslada F (2012) Human milk probiotic *Lactobacillus fermentum* CECT5716 reduces the incidence of gastrointestinal and upper respiratory tract infections in infants. J Pediatr Gastroenterol Nutr 54(1):55–61

Martarelli D, Verdenelli MC, Scuri S, Cocchioni M, Silvi S, Cecchini C, Pompei P (2011) Effect of a probiotic intake on oxidant and antioxidant parameters in plasma of athletes during intense exercise training. Curr Microbiol 62(6):1689–1696

Matsumoto M, Benno Y (2004) Consumption of *Bifidobacterium lactis* LKM512 yogurt reduces gut mutagenicity by increasing gut polyamine contents in healthy adult subjects. Mutat Res Fundam Mol Mech Mutagen 568(2):147–153

Matsumoto M, Kurihara S, Kibe R, Ashida H, Benno Y (2011) Longevity in mice is promoted by probiotic-induced suppression of colonic senescence dependent on upregulation of gut bacterial polyamine production. PLoS One 6(8), e23652

Matsumoto S, Hara T, Hori T, Mitsuyama K, Nagaoka M, Tomiyasu N, Suzuki A, Sata M (2005) Probiotic *Lactobacillus*-induced improvement in murine chronic inflammatory bowel disease is associated with the down-regulation of pro-inflammatory cytokines in lamina propria mononuclear cells. Clin Exp Immunol 140(3):417–426

McFarland LV (2006) Meta-analysis of probiotics for the prevention of antibiotic associated diarrhea and the treatment of *Clostridium difficile* disease. Am J Gastroenterol 101(4):812–822

Messaoudi M, Lalonde R, Violle N, Javelot H, Desor D, Nejdi A, Bisson JF, Rougeot C, Pichelin M, Cazaubiel M, Cazaubiel JM (2011) Assessment of psychotropic-like properties of a probiotic formulation (*Lactobacillus helveticus* R0052 and *Bifidobacterium longum* R0175) in rats and human subjects. Br J Nutr 105(05):755–764

Messaoudi S, Manai M, Kergourlay G, Prévost H, Connil N, Chobert JM, Dousset X (2013) *Lactobacillus salivarius*: bacteriocin and probiotic activity. Food Microbiol 36(2):296–304

Murphy EF, Cotter PD, Hogan A, O'Sullivan O, Joyce A, Fouhy F, Clarke SF, Marques TM, O'Toole PW, Stanton C, Quigley EMM, Daly C, Ross PR, O'Doherty RM, Shanahan F (2013) Divergent metabolic outcomes arising from targeted manipulation of the gut microbiota in diet-induced obesity. Gut 62(2):220–226

O'Mahony SM, Marchesi JR, Scully P, Codling C, Ceolho AM, Quigley EM, Cryan JF, Dinan TG (2009) Early life stress alters behavior, immunity, and microbiota in rats: implications for irritable bowel syndrome and psychiatric illnesses. Biol Psychiatry 65(3):263–267

Omar JM, Chan YM, Jones ML, Prakash S, Jones PJ (2013) *Lactobacillus fermentum* and *Lactobacillus amylovorus* as probiotics alter body adiposity and gut microflora in healthy persons. J Funct Foods 5(1):116–123

Pérez-Cano FJ, Dong H, Yaqoob P (2010) In vitro immunomodulatory activity of *Lactobacillus fermentum* CECT5716 and *Lactobacillus salivarius* CECT5713: two probiotic strains isolated from human breast milk. Immunobiology 215(12):996–1004

Qing L, Wang T (2008) Lactic acid bacteria prevent alcohol-induced steatohepatitis in rats by acting on the pathways of alcohol metabolism. Clin Exp Med 8(4):187–191

Rachmilewitz D, Katakura K, Karmeli F, Hayashi T, Reinus C, Rudensky B, Akira S, Takeda K, Lee J, Takabayashi K, Raz E (2004) Toll-like receptor 9 signaling mediates the anti-inflammatory effects of probiotics in murine experimental colitis. Gastroenterology 126 (2):520–528

Rao R, Seth A, Sheth P (2004) Recent advances in alcoholic liver disease I. Role of intestinal permeability and endotoxemia in alcoholic liver disease. Am J Physiol Gastrointest Liver Physiol 286(6):G881–G884

Rautava S, Kalliomäki M, Isolauri E (2002) Probiotics during pregnancy and breast-feeding might confer immunomodulatory protection against atopic disease in the infant. J Allergy Clin Immunol 109(1):119–121

Saavedra JM, Abi-Hanna A, Moore N, Yolken RH (2004) Long-term consumption of infant formulas containing live probiotic bacteria: tolerance and safety. Am J Clin Nutr 79 (2):261–267

Scanlan PD, Shanahan F, Clune Y, Collins JK, O'Sullivan GC, O'Riordan M, Holmes E, Wang Y, Marchesi JR (2008) Culture-independent analysis of the gut microbiota in colorectal cancer and polyposis. Environ Microbiol 10(3):789–798

Sibley D, Jerrells TR (2000) Alcohol consumption by C57BL/6 mice is associated with depletion of lymphoid cells from the gut-associated lymphoid tissues and altered resistance to oral infections with *Salmonella typhimurium*. J Infect Dis 182(2):482–489

Sobhani I, Tap J, Roudot-Thoraval F, Roperch JP, Letulle S, Langella P, Corthier G, Van Nhieu JT, Furet JP (2011) Microbial dysbiosis in colorectal cancer (CRC) patients. PLoS One 6(1), e16393

Stecher B, Robbiani R, Walker AW, Westendorf AM, Barthel M, Kremer M (2007) *Salmonella enterica* serovar typhimurium exploits inflammation to compete with the intestinal microbiota. PLoS Biol 5(10), e244

Szajewska H, Ruszczyński M, Radzikowski A (2006) Probiotics in the prevention of antibiotic-associated diarrhea in children: a meta-analysis of randomized controlled trials. J Pediatr 149 (3):367–372

Tanida M, Shen J, Maeda K, Horii Y, Yamano T, Fukushima Y, Nagai K (2008) High-fat diet-induced obesity is attenuated by probiotic strain *Lactobacillus paracasei* ST11 (NCC2461) in rats. Obes Res Clin Pract 2(3):159–169

Ulisse S, Gionchetti P, D'Alò S, Russo FP, Pesce I, Ricci G, Rizello F, Helwig U, Cifone MG, Campieri M, De Simone C (2001) Expression of cytokines, inducible nitric oxide synthase, and matrix metalloproteinases in pouchitis: effects of probiotic treatment. Am J Gastroenterol 96 (9):2691–2699

Wang L, Zhang J, Guo Z, Kwok L, Ma C, Zhang W, Lv Q, Huang W, Zhang H (2014) Effect of oral consumption of probiotic *Lactobacillus planatarum* P-8 on fecal microbiota, SIgA, SCFAs, and TBAs of adults of different ages. Nutrition 30(7):776–783

Weizman Z, Asli G, Alsheikh A (2005) Effect of a probiotic infant formula on infections in child care centers: comparison of two probiotic agents. Pediatrics 115(1):5–9

Woodard GA, Encarnacion B, Downey JR, Peraza J, Chong K, Hernandez-Boussard T, Morton JM (2009) Probiotics improve outcomes after Roux-en-Y gastric bypass surgery: a prospective randomized trial. J Gastrointest Surg 13(7):1198–1204

Yadav H, Lee JH, Lloyd J, Walter P, Rane SG (2013) Beneficial metabolic effects of a probiotic via butyrate-induced GLP-1 hormone secretion. J Biol Chem 288(35):25088–25097

Yan F, Cao H, Cover TL, Washington MK, Shi Y, Liu L, Chaturvedi R, Peek RM Jr, Wilson KT, Polk DB (2011) Colon-specific delivery of a probiotic-derived soluble protein ameliorates intestinal inflammation in mice through an EGFR-dependent mechanism. J Clin Invest 121 (6):2242

Yin L, Laevsky G, Giardina C (2001) Butyrate suppression of colonocyte NF-κB activation and cellular proteasome activity. J Biol Chem 276(48):44641–44646

Zareie M, Johnson-Henry K, Jury J, Yang PC, Ngan BY, McKay DM, Soderholm JD, Perdue MH, Sherman PM (2006) Probiotics prevent bacterial translocation and improve intestinal barrier function in rats following chronic psychological stress. Gut 55(11):1553–1560

Zhang H, DiBaise JK, Zuccolo A, Kudrna D, Braidotti M, Yu Y, Parameswaran P, Crowell MD, Wing R, Rittman BE, Krajmalnik-Brown R (2009) Human gut microbiota in obesity and after gastric bypass. Proc Natl Acad Sci U S A 106(7):2365–2370

Food Colorant from Microorganisms

Marlia Singgih and Elin Julianti

Contents

1 Colorant-Producing Microorganisms

Traditional processed food in Asia is famous worldwide. For many years, some fermentative food-grade color pigments produced by microorganisms have been used to enhance their appearance. In Asia, mostly in China, India, Japan, and Indonesia, although natural colorants extracted from vegetables and fruits, such as saffron, turmeric, and suji leaf, have been widely used as food colorants, microbial color pigment is now intensely studied to produce a high-standard, high-yield, and stable food colorant, for example, more stable against light and heat, and with a wider range of pH than the natural food colorants. Some examples of the pigment molecules are carotenoids, melanin, flavin, quinones, monascins, and xanthophylls. Colorant-producing microorganisms, such as *Monascus*,

M. Singgih (✉) • E. Julianti
School of Pharmacy, Bandung Institute of Technology, Jalan Ganesha 10, Bandung 40132, Indonesia
e-mail: marlia@fa.itb.ac.id; elin_julianti@fa.itb.ac.id

M.-T. Liong (ed.), *Beneficial Microorganisms in Food and Nutraceuticals*,
Microbiology Monographs 27, DOI 10.1007/978-3-319-23177-8_12

Table 1 Microbial pigment and their color (Joshi et al. 2003)

Microorganism	Color pigments
Bacteria	
Achromobacter	Creamy
Bacillus	Brown
Brevibacterium sp.	Orange, yellow
Corynebacterium	Grayish, creamy
Pseudomonas	Yellow
Rhodococcus	Bluish red
Streptomyces	Yellow, red, blue
Molds	
Aspergillus sp.	Orange, red
Blakeslea	Cream
Helminthosporium	Red, bronze, maroon
Monascus	Red, Yellow, orange
Penicillium	Orange, yellow
Yeast	
Cryptococcus	Red
Phaffia	Red
Rhodotorula	Red
Yarrowia	Brown
Algae	
Dunaliella	Red

Rhodotorula, *Bacillus*, and *Achromobacter*, can produce a large number of color pigments (Table 1). They grow mostly in soil and some of them grow on coast or in deep sea.

Some reviews have been published about microbial pigment, for example, by Malik et al. (2012), Joshi et al. (2003), and Dufossé (2006).

The trend toward greater use in natural colorants compared to synthetic colorants for food, drugs, and cosmetics will continue. Some of these pigments from microorganisms may also play a role in supporting biological activity, such as food supplements, probiotic, and anticancer activity.

1.1 Bacteria

Some bacteria have been used as colorants or food additives in traditional food in Asia. Use of several bacterial pigments as food colorant was reported. For example, zeaxanthin (yellow) was isolated from *Flavobacterium* sp. and *Paracoccus zeaxanthinifaciens*, carotenoids (yellow) were isolated from *Streptomyces* sp., canthaxanthin (dark red) was isolated from photosynthetic bacterium *Bradyrhizobium* sp., and *Halobacterium* sp., astaxanthin (pink-red) was isolated from *Agrobacterium aurantiacum*, *Paracoccus carotinifaciens*, and *Halobacterium salinarum* (Venil et al. 2013; Malik et al. 2012). Other carotenoid-producing bacteria are *Corynebacterium michiganense*, *Micrococcus roseus*, *Brevibacterium* spp., *Bradyrhizobium* spp., *Gordonia jacobaea*, and *Dietzia natronolimnaea* (Mata-Gómez et al. 2014). Barnett and Hageman (1983) reported isolation of a brown pigment from *Bacillus subtilis*. A blue color pigment, named as λ-actinorhodin, which was isolated from cultured soil bacteria *Streptomyces coelicolor*, was reported to have excellent stability to light, heat, and food additives in common use and resistant to oxidant and reducers under acidic conditions and to reducers under alkaline conditions. In addition, this pigment is nontoxic; therefore, it became useful as a potential food additive (Venil et al. 2013; Zhang et al. 2006). Other than providing attractive colors, the bacterial colorants are environmentally friendly and they provide health benefits of probiotics in food products (Nagpal et al. 2011).

1.2 Fungi

Fungi are the most well-known microorganisms that have been exploited to obtain their metabolites and used as natural food colorants due to their safety. *Monascus purpureus*, *Neurospora sitophila*, *Blakeslea trispora*, and *Phycomyces blakesleeanus* are some common fungi that have been used in Asian countries. The color pigments from *Monascus* have been very well known since the eighteenth century. It is used to prepare red angkak fermented rice. This angkak rice was reported to be used for treatment of dengue fever and as a cholesterol-lowering agent (Diansyah et al. 2013). Although *Monascus* pigments have not been approved to be used in the USA and Europe, it has long been used as food colorant in Asia, especially in China and Indonesia.

The influence of different substrates on pigment production was reported by Lin and Lizuka (1982). Other than rice, some alternative substrates such as bread, oat, corn, and wheat grain could be used for *Monascus* fermentation in concern of pigment production. Some suitable conditions of *Monascus* culture were reported to be the factors of producing high-quality fermentation product (Priatni et al. 2014).

Specially in Indonesia, a traditional fermented peanut-kernel cake, called "oncom," has a very bright color ranging from red to orange and yellow. It is a source of vegetable-originated protein, inoculated by a fungus, *Neurospora* sp. (Priatni 2014).

1.3 Yeast

Some yeasts are good source of microbial pigments. Carotenoids from some of the red yeasts like *Rhodotorula* and *Phaffia rhodozyma* have attracted commercial interest as a natural pigment for foods (Latha and Jeevaratnam 2010). From genus *Rhodotorula*, there are some species such as *Rhodotorula glutinis*,

Rhodotorula minuta, *Rhodotorula mucilaginosa*, *Rhodotorula acheniorum*, and *Rhodotorula graminis* known as carotenoid producer (Mata-Gómez et al. 2014). The other genera of red yeasts that have potential as natural alternative sources of carotenoids are *Rhodosporidium*, *Sporidiobolus*, *Sporobolomyces*, *Cystofilobasidium*, and *Kockovaella* (Čertík et al. 2009). *Rhodosporidium paludigenum* was reported as a new potent source of carotenoids (Yimyoo et al. 2011). Kim et al. (2012) reported isolation carotenoid from *Rhodosporidium babjevae* JI-1 which was isolated from citrus fruit peel. Focusing on genus *Sporobolomyces*, there are some species recognized as carotenoid producers such as *Sporobolomyces roseus*, *Sporobolomyces salmonicolor*, and *Sporobolomyces patagonicus* (Mata-Gómez et al. 2014; Maldonade et al. 2008). Malik et al. (2012) reported that the red basidiomycetous yeast *Xanthophyllomyces dendrorhous* produced astaxanthin and it is one of the main sources for natural astaxanthin. The main carotenoid from *Rhodotorula* is β-carotene. The other pigments that were reported as food colorant from the yeast *Rhodotorula* and of industrial interest were torularhodin and torulene, the orange red keto-carotenoids for which at present no commercially exploitable plant or animal sources exist. Yeasts of the genera *Rhodotorula* are able to synthesize those pigments, but the low production rate of pigment in these microorganisms limits its industrial application. Therefore, the strategies to improve the strain of *Rhodotorula mucilaginosa* for obtaining hyperpigmentation mutants by ultraviolet-B radiation have been studied (Moliné et al. 2012).

1.4 Microalgae

Microalgae are one of the potential sources of colorants. Several species of microalgae produce high concentrations of carotenoids such as β-carotene, astaxanthin, and canthaxanthin. Their most important uses are as natural food colorants. *Dunaliella salina* is the best organism producing β-carotene among the algae and other organisms since it can produce β-carotene up to 14 % of its dry weight, and therefore, it was first proposed for commercial production of β-carotene (Spolaore et al. 2006). *Dunaliella salina* is a green unicellular alga and bi-flagellate and produces β-carotene in a high salt environment (Emeish 2012).

Another marine microalga which is achieving commercial success in colorant production is a green microalga, *Haematococcus pluvialis*. *Haematococcus* produces high-value carotenoid, namely astaxanthin. *Haematococcus* generally contains 1.5–3 % astaxanthin. Microalgae synthesize astaxanthin and concentrate it into the food chain through zooplankton and crustaceans, which are prey for salmon, trout, and other aquatic animals (Lorenz and Cysewski 2000). *Haematococcus* microalgae meal has been used as a natural red food colorant and as a pigment for fish feeds especially for salmon and trout farming (Dufossé et al. 2005). Astaxanthin has been reported to have biological activity as antioxidant, anti-inflammatory, antihypertensive, and neuroprotective (Hussein et al. 2006).

Phycocyanin (blue) and Phycoerythrin (red) are collectively known as phycobiliprotein. They are also main natural pigments that are commercially produced from microalgae. Both phycoerythrin and phycocyanin are water soluble and can be use as natural colorants in food, cosmetics, and pharmaceuticals. The microalgae as sources for those colorant are red microalga genus *Phorpyridium* (Dufossé et al. 2005) and blue-green microalga genus *Spirulina* sp. (Cesar de Carvalho et al. 2008).

2 Metabolism of Colorant

Microorganism has been reported to produce a variety of color pigments. Therefore, they become one of the potential sources of natural colorants for food, cosmetics, and pharmaceuticals. Color pigments are secondary metabolites and are produced mostly after the growth phase of a bacterial, algal, or fungal culture. The amount of color pigments produced is directly dependent on biomass accumulation (Matthews and Wurtzel 2007). Enhancement of biomass production by controlling the environment and the nutrition of organism favors the accumulation of metabolic color pigments. The following section focuses on the metabolism of carotenoid and *Monascus* pigments which are the major color pigments used as food colorant.

2.1 Biosynthetic Pathway of Colorant

2.1.1 *Carotenoids*

Carotenoids are natural pigments that are primarily synthesized within plants, algae, and some species of fungi and bacteria. Animals are not able to synthesize carotenoids; therefore, they must be supplied in the diet (Lorenz and Cysewski 2000). The carotenoid pigments are classified into carotenes and xanthophylls. Some carotenes only have carbon and hydrogen on their chemical structure such as β-carotene and torulene, whereas xanthophylls also contain oxygen such as astaxanthin and canthaxanthin (Mata-Gómez et al. 2014).

The carotenoids are biosynthesized through isoprenoid pathway from the basic C_5-terpenoid precursor, isopentenyl diphosphate (IPP). The entire biosynthesis takes place in the chloroplasts (in green tissues) or chromoplasts (in yellow to red tissues) encoded by nucleus genes. The biosynthesis of all natural carotenoids begins with the enzymatic assembly of a C_{30} or C_{40} backbone. In C_{40} carotenoids backbone, the isoprenoid chain is built up from mevalonic acid (MVA) and is catalyzed by prenyltransferases, to the C_{20} level, as geranylgeranyl diphosphate. Two molecules of this are joined tail to tail to give 15-cis phytoene as the first product with the C_{40} carotenoid skeleton, which is catalyzed by the phytoene synthase (PSY) (Ötles and Çagindi 2007). C_{40} carotenoids are mostly made in plants and microbial species. C_{30} carotenoid pathways starting with the condensation of two molecules of farnesyl diphosphate ($C_{15}PP$) to form (15Z)-4,4-′-diapophytoene (also called dehydrosqualene) are much less widespread (Umeno et al. 2005). Different types and levels of modification of C_{40} or C_{30} backbone by carotenoid biosynthetic enzymes such as isoprenyl diphosphate synthases, carotenoid synthases, desaturases, cyclases, and others specific transformations lead to synthesize a large number of different products (Umeno et al. 2005; Ötles and Çagindi 2007). The biosynthesis pathway of carotenoid can be organized into a tree-like hierarchy, Fig. 1 (Umeno et al. 2005).

2.1.2 Biosynthesis of Monascus Pigments

Monascus pigments are classified as azaphilone metabolites. Based on the color of metabolites, *Monascus* pigments consist of yellow pigments (ankaflavin, monascin), orange pigments (rubropunctatin, monascorubrin), and red pigments (monascorubramine and rubropunctamine) (Mostafa and Abbady 2014). These pigments have chromophore groups such as hexaketides and fatty acids such as hexanoic acid (C6) or octanoic acid (C8). The derivative compounds are obtained by reduction, oxidation, or reaction of the *Monascus* pigments with other products, especially with a variety of amino acids (Pastrana et al. 1995; Blanc et al. 1997; Lakrod et al. 2000). Other pigments that had been isolated from *Monascus* spp. were two yellow components with furanoisophthalide skeleton, xanthomonasin A and xanthomonasin B; six yellow coumarin derivatives, namely monankarins A–F, and four orange pigments, namely monapilol A–D, Fig. 2 (Feng et al. 2012).

Monascus also produces statin derivative compounds, such as monacolin K (mevinolin), which have biological activities such as antihypercholesterolemia. Statin derivative compound was isolated for the first time from *Aspergillus terreus* which is known as lovastatin, Fig. 3 (Blanc et al. 1998; Hai 1998; Keane 1999).

Other than pigments and monacolin K, *Monascus* also produces monascidin A which has antibacterial activity. This compound has been further characterized as citrinin (Fig. 3), a mycotoxin that was isolated for the first time from *Penicillium citrinum* and also was produced from other fungi, especially from Aspergilaceae (Blanc et al. 1995; Hai 1998). Citrinin was reported as nephrotoxic, teratogenic, and carcinogenic agents; therefore, citrinin is not safe for using as medicines (Gremmels and dan Vilar 1998).

The *Monascus* pigments are biosynthesized through polyketides pathway as other secondary metabolites, but the pathway remains unclear. Biosynthesis of polyketides pathway is specific process metabolism of fungi especially for Ascomycetes. In this biosynthesis pathway, a condensation reaction occurs between one molecule of acetyl CoA with one or more molecules of malonyl CoA. It is followed by a series of reduction, dehydration, methylation, and cyclization reaction to produce polyketide metabolites. All these reactions are catalyzed by polyketide

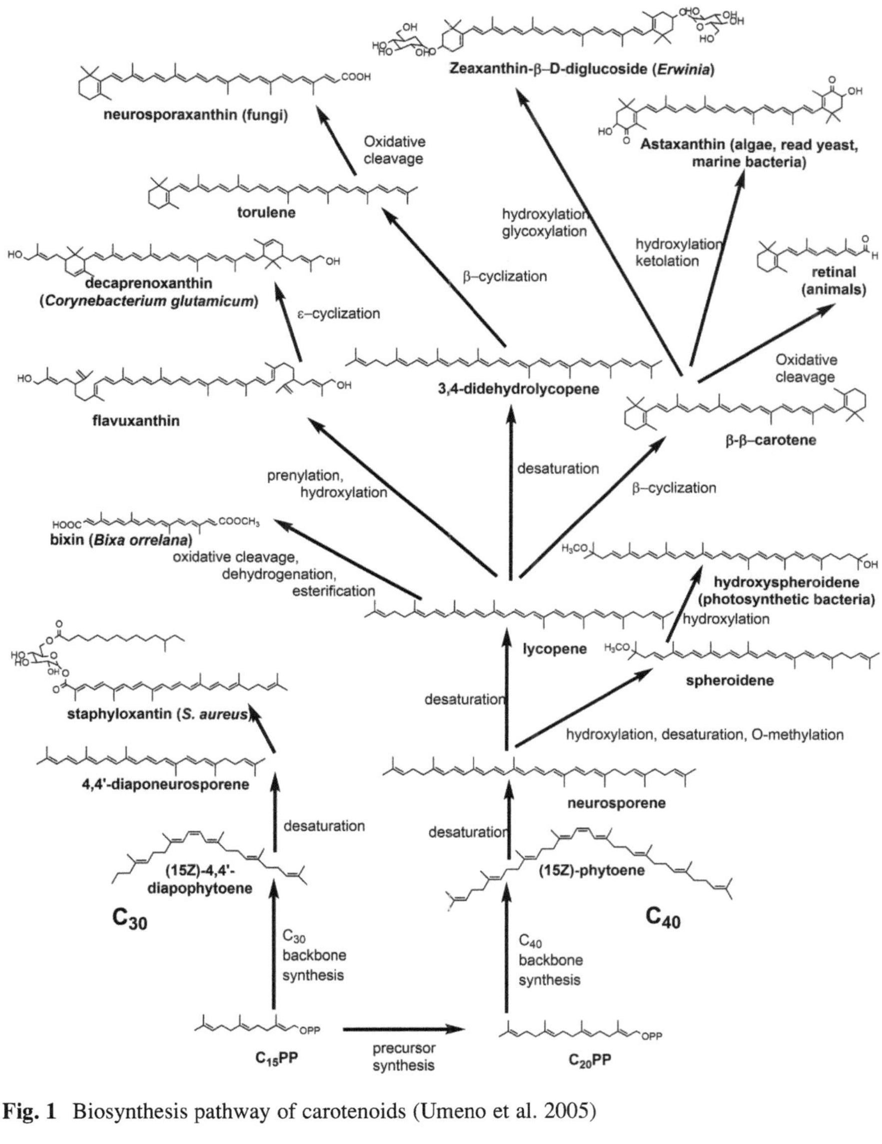

Fig. 1 Biosynthesis pathway of carotenoids (Umeno et al. 2005)

biosynthesis synthase (PKS), a multifunctional enzyme encoded by PKS genes (Pfeifer and Khosla 2001).

The study of the biosynthesis pathway of *Monascus* red pigments and monascidin A in *M. ruber* showed that both metabolites were produced from a precursor of tetraketides. This precursor was produced from a condensation reaction between one molecule of acetyl-CoA with three molecules of malonyl CoA, followed by one molecule of acetyl-CoA. In some species of *Penicillium* and *Aspergillus*, citrinin was produced from precursor of pentaketides without pigment

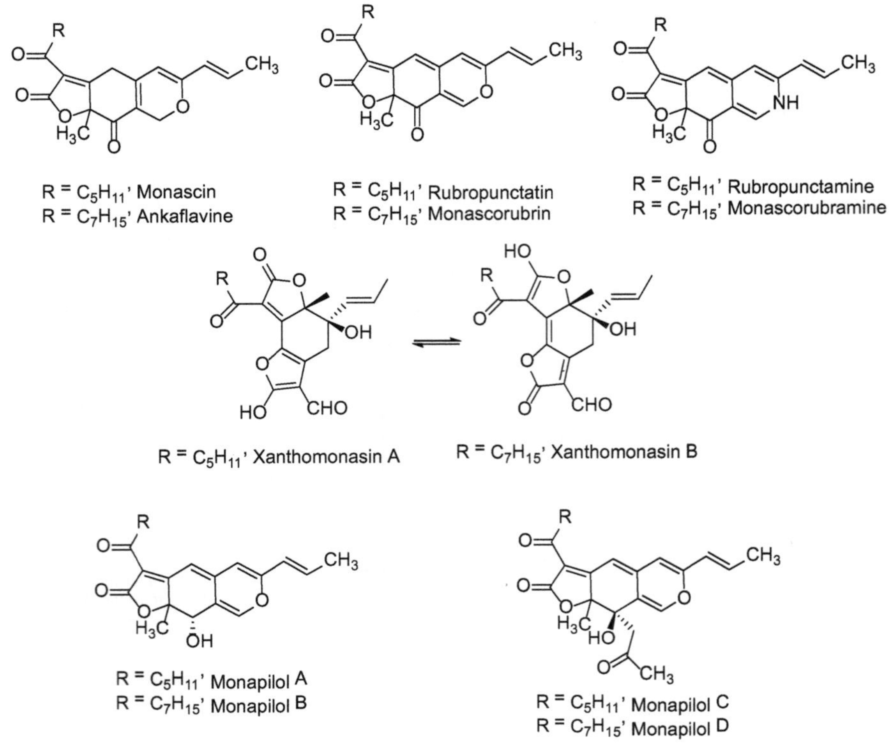

Fig. 2 Chemical structure of some pigments produced by *Monascus* sp. (Feng et al. 2012)

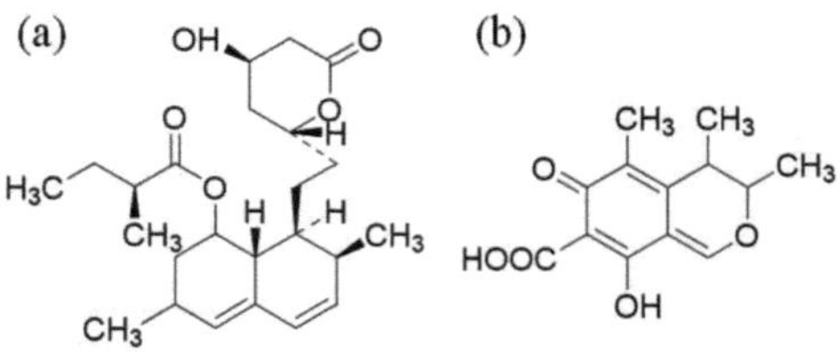

Fig. 3 Chemical structure of lovastatin (**a**) and monascidin A (**b**) (Blanc et al. 1998)

production. Figure 4 illustrated the biosynthesis of red pigment and citrinin (Hajjaj et al. 1999).

2.2 *Factors Affecting of Pigment Microbial Productions*

There are many factors that need to be considered when growing microorganism and producing secondary metabolite pigments. Microorganisms require supply of

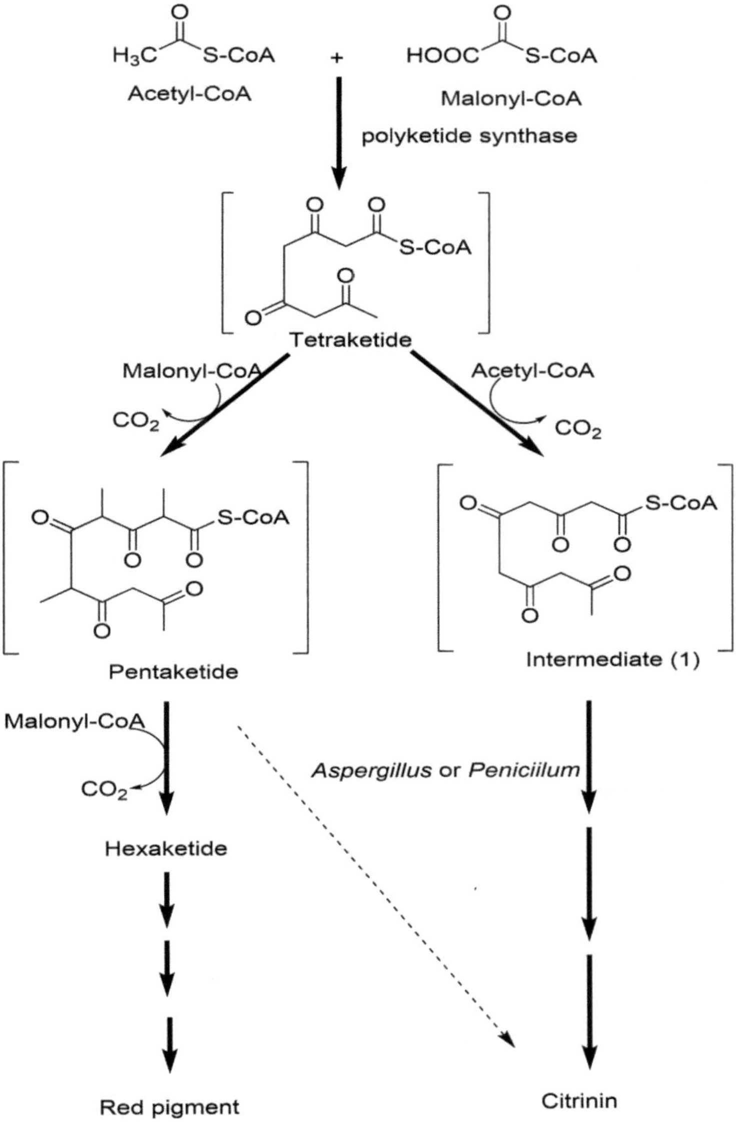

Fig. 4 Biosynthesis pathway of red pigment and citrinin in *Monascus ruber*. The biosynthesis pathway of citrinin in *Aspergillus* and *Penicillium* indicated by dashed arrow. All the reactions in this biosynthesis pathway are catalyzed by polyketide synthases/PKS (Hajjaj et al. 1999)

water, carbon, nitrogen, and minerals. Certain physical conditions, such as temperature of incubation and pH of the medium, will also affect the microbial growth and secondary metabolites' production.

2.2.1 Carbon Source

Type of carbon sources, such as polysaccharides, oligosaccharides, and various monosaccharides, are influencing the growth of mycelium and pigment production. It was reported that glucose and its oligo- and polysaccharides were better carbon source than maltose and fructose for the growth of *Monascus* sp. TTWMB 6042 and the production of *Monascus* pigment (Lin and Demain 1991; Lee et al. 2001). The type of carbon also influences the constituent of *Monascus* pigments. A study reported that using maltose and glucose in *Monascus purpureus* fermentation produced a very dark brown pigment, whereas using sucrose would produce light red pigments (Joshi et al. 2003). For pigment production from *Rhodotorula glutinis* var. glutinis, the highest carotenoid production was obtained in the medium containing sucrose (El-Banna et al. 2012). Similar result was reported by Latha et al. (2005) that the maximum growth and pigmentation of *Rhodotorula glutinis* DFR-PDY were achieved by using glucose and fructose as carbon sources. Attri and Joshi (2005) reported optimization of apple pomace-based medium and fermentation conditions for pigment production by *Micrococcus* sp. The addition of fructose to the apple pomace medium increased the yield of biomass and carotenoid production. For *Sarcin*a sp. that also was reported using apple pomace medium, the maximum yield of biomass was observed in medium with glucose followed by fructose, but the carotenoid production was statistically equal in glucose and fructose (Joshi et al. 2011).

2.2.2 Nitrogen Source

The production of microbial pigment is also influenced by nitrogen source. Inorganic ammonic compounds, such as ammonium chloride, ammonium nitrate, and organic nitrogen sources such as urea, peptone, monosodium glutamate, and amino acids, are good nitrogen sources for both growth and pigment production of *Monascus* spp. It was reported that the ammonium chloride is the best nitrogen source for pigment production, followed by ammonium nitrate and glutamate. Amino acids such as leucine, lysine, valine, and methionine, especially leucine, were reported to have inhibitory effects on the formation of water-soluble *Monascus* pigment (Joshi et al. 2003; Feng et al. 2012). The production of carotenoids from *Rhodotorula glutinis* var and *R. glutinis* showed the highest yield of β-carotene when ammonium nitrate was utilized as nitrogen source (El-Banna et al. 2012). According to Latha et al. (2005), among the various nitrogens which was used in optimization of growth condition of *Rhodotorula glutinis* DFR-PDY, maximum growth and pigmentation were observed in sodium nitrate. For *Sarcina* sp., the highest yield of biomass and carotenoids production was achieved when potassium nitrate was used followed by sodium nitrate and ammonium sulfate (Joshi et al. 2011).

2.2.3 Minerals

Among the other elements, minerals also have a great effect on the growth of microorganisms and the production of microbial pigments. For pigment production from *Monascus purpureus* (wild and mutant NIIS), it was reported that 2×10^{-3} M and 3×10^{-3} M of Zn^{2+} in liquid culture media could stop the growth and pigment production, whereas on solid media, for the mutant *M. purpureus* N11S, the pigment production was promoted by 5×10^{-5} M of Zn^{2+}. However, the mycelial growth was significantly inhibited by 5×10^{-5} M of Zn^{2+}. The effect of other metal ions such as Fe^{2+} and Mn^{2+} on *M. purpureus* pigment production also had been studied. Fe^{2+} exhibited the strongest stimulatory effect on red pigment production compared to Zn^{2+} and Mn^{2+} (Lee et al. 2001). For pigment production from a mutant of *Rhodotorula glutinis*, it was shown that divalent cation salts, especially Mg^{2+}, have a stimulatory effect on the production of β-carotene (Bhosale and Gadre 2001).

2.2.4 Temperature

Temperature of incubation plays important roles in the growth of microorganisms and the production of microbial pigments. The optimum temperature for growth and pigments production depends on the type of microorganisms. For example, *Monascus* spp. are cultured at 25–30 °C for pigment production and *Pseudomonas aeruginosa* at 35–36 °C for growth and its pigment production (Joshi et al. 2003; Feng et al. 2012). According to Zhou et al. (2009) and Liu et al. (2007), *Monascus* spp. could also grow at higher temperature such as 31 °C and 32 °C, respectively. *Penicillium oxalicum* was reported to produce a red colorant at optimum temperature 27–29 °C (Dufossé et al. 2005). The yeast *Rhodotorula glutinis* was reported to produce carotenoid pigments maximum at the temperature 29–32 °C (Latha et al. 2005), whereas the optimum temperature for *R. rubra* has been reported at 22 °C (Martin et al. 1993). The temperature of the growth medium had also an effect on the growth and carotenoid production of marine bacteria, *Micrococcus* sp. The biomass production increased remarkably with raising the temperature up to 37 °C, while the carotenoids' production rate enhanced strictly up to 30 °C. The cell biomass production reduced sharply at 45 °C due to the denaturation of the enzyme system of microorganism at higher temperatures (Ibrahim 2008). For the pigment production of *Micrococcus* sp., Bhat et al. (2013) and Attri and Joshi (2005) reported the maximum production at 35 °C. Further increase in temperature showed a decrease in the growth of bacteria and its pigment production. Production of pigment by *Sarcina* sp. also occurred at temperature 35 °C (Joshi et al. 2011).

2.2.5 pH

The growth of microorganisms and the production of microbial pigments are influenced by the pH of culture medium. Each microorganism has a specific pH tolerance range for growth and pigment production. The optimum pH for *Monascus* spp. is 5.5–6.5 (Joshi et al. 2003; Feng et al. 2012). According to Chen and Johns (1993), the highest red pigment production from *Monascus purpureus* was obtained using a glucose-peptone medium at pH 6.5, while yellow pigment ankaflavin synthesis was favored at pH 4.0. Orozco and Kilikian (2008) reported that the highest total red pigments from *Monascus purpureus* were achieved at pH 5.5. Another fungus, *Penicillium oxalicum* produced a red pigment and reached the optimum production at pH 5.6–6.2 (Dufossé et al. 2005). Latha et al. (2005)reported that a red yeast *Rhodotorula glutinis* was able to grow and produce carotenoid pigments over a wide range of pH from 2.5 to 2.9 and the optimum condition at pH 5.5. The lycopene production is optimum in neutral to slight alkaline condition, whereas the synthesis of β-carotene is optimum in acidic conditions (Joshi et al. 2003). Bhat et al. (2013) reported the effect of pH on pigmentation of bacteria *Micrococcus* sp. The pigment production increased with the increase of pH. Maximum pigment production was observed at pH 9 and at pH 4; none of the cultures observed growth. For *Micrococcus flavus*, it was reported that pH 7.5 was optimum condition for maximum pigment production (Deb and Madhugiri 2012). Joshi et al (2011) also reported that among different pH values from 3.0 to 6.0, a pH of 5.5 was an optimum pH for growth and carotenoids' pigment production of *Sarcina* sp.

2.2.6 Type of Fermentation

The type of fermentation such as solid or submerged fermentation influences the production of microbial pigments. Bacteria may grow in solid substrate but usually show better development in liquid media. The use of photoautotrophic microalgae culture such as *Spirulina* in solid state fermentation (SSF) is limited because light has a poor penetration on the substrate. Fungi are the most adequate microorganisms for pigment production in SSF (Cesar de Carvalho et al. 2008). For example, in *Monascus purpureus* fermentation, solid culture was reported to give a better yield compared to submerged fermentation though media composition; pH agitation also influences the pigment production (Joshi et al. 2003). A study reported that submerged fermentation of *Monascus purpureus* with optimal condition produced pigments that could be comparable to the solid culture, whereas the yield of citrinin was increased more than 100 times. This study indicated that the type of fermentation could be a primary factor that influenced the citrinin production and a secondary factor for the pigment production (Zhang et al. 2013).

3 Examples of Microbial Colorant Production

3.1 Monascus purpureus

Monascus purpureus has been used as a starter culture in the production of red mold rice for food colorant. *Monascus purpureus* was isolated by Went in 1895 from red mold rice that was obtained from Java, Indonesia. Red mold rice has traditionally been used in Chinese food as a preservative, food coloring and flavoring agent, and diet supplement, in addition to its use as a food medicine for improving digestion and enhancing blood circulation (Hsieh et al. 2008).

There are two major processes for *Monascus* pigment production, namely solid state fermentation (SSF) and liquid-state fermentation (LSF). The SSF method is a traditional processing method to produce *Monascus* pigment in China. The starter of *Monascus* strain is inoculated into the steamed rice that is spread on the big trays and cultured in an air-, moisture-, and temperature-controlled room for about 20 days (Dufossé et al. 2005; Feng et al. 2012). In Indonesia, the rice usually was placed into the glass jars or bottle and then sterilized by steam. Inoculation was carried out by mixing 10 % (v/v) of starter culture of *Monascus* sp. (2×24 h) aseptically. The inoculated rice in the jars was shaken until homogeneous and incubated for 14 days in incubator at 30 °C. In the first 7 days of inoculation, the rice was moisturized by using sugar solution in every 24 h. The shaking was performed once a day until incubation time was completed. The rice grains gradually turn red in color. Each rice grain becomes bright red in its core and reddish purple on the outside when they are fully cultured for about 14 days. On the 14th day, the product was harvested and dried in oven at 50 °C for 24 h (Singgih et al. 2014b). The SSF product, namely red mold rice, also known as hongqu (Chinese), red yeast rice, red fermented rice, red Kojic rice, and angkak, can be directly used as food colorants (Liu et al. 2010) or as materials for *Monascus* pigment extraction, while for LSF products, the pigment must be extracted before using as colorant. *Monascus* pigment production by SSF has more advantages compared to LSF processing method, such as simpler technique, less capital investment, lower levels of end product inhibition and catabolite repression, lower amount of waste output, better product recovery, and higher yield (Joshi et al. 2003; Lee et al. 2002). Some different sources of SSF also have been developed, for example, on rhizome of *Dioscorea hispida*, called "Gadung," tofu waste (Fig. 5) gave different rays of red pigments (Priatni et al. 2014).

Other publications on effects of different solid substrates of *Monascus* were elaborated using local sources of rice, manihot, etc. (Dikshit and Tallapragada 2011).

On the other hand, LSF process also gives an attractive alternative because the process is easier to manage, and has shorter cultivation time, lower production costs, and higher product quality (Silveira et al. 2013). *Monascus purpureus* grown in submerged culture has been investigated compared to solid culture. The submerged fermentation in potato dextrose broth (PDB) media gave maximum red

(a) (b) (c)

Fig. 5 *Monascus* sp. on different substrates gained different rays of *red* color; (**a**) on rice, (**b**) on tofu waste, (**c**) on gadung (*Dioscorea hispida*) (Priatni et al. 2014)

pigment production at 16th days of incubation at 30 °C and pH of 5.5 (Dikshit and Tallapragada 2011).

Besides their function as food colorant, *Monascus* pigments were reported to have effects as antibacterial agents, as reported by Kim et al. (2006). Two hydrophobic and two hydrophilic derivatives exhibited high antimicrobial activity. The derivates were combination of the pigments with L-Tyr and L-Phe, L-Glu and L-Asn.

It also have been reported that extract of *Monascus purpureus* showed significant effect on thrombopoietin (TPO) level in dengue-infected patient (Diansyah et al. 2013). In this research, the patients who get dengue infection received standard therapy with and without red yeast rice and the TPO levels were analyzed using single-blind randomized controlled trial. The *Monascus* extract was suggested to increase platelet counts presumably through effects on increasing megakaryopoiesis and thrombopoiesis in bone marrow or anti-inflammatory effects of metabolites.

3.2 Neurospora *sp.*

Neurospora, the Ascomycetes fungi that mostly grow in tropical and subtropical countries, grew and sporulated quickly on the surface of fire scorched vegetation. In Europe at least five species were identified, i.e., *N. crassa*, *N. discreta*, *N. intermedia*, *N. sitophila*, and *N. tetrasperma* (Jacobson et al. 2006). In Asian countries, *Neurospora* sp. is a famous fungus used in traditional food. For example, in West Java, Indonesia, Oncom or fermented peanut cake is a traditional fermented food with orange to red color spores on the surface of the cake. The food is produced by inoculating a β-carotene pigment producer, *Neurospora sitophila*, on peanut waste, and it is incubated for 3 days. The fermentation yields a red, orange, and pink color, depending on strains used. Instead of peanuts, many substrates can be used, such as rice cake and tofu (bean curd). There are two kinds of Oncom according to their color appearance, i.e., Red oncom and Black oncom. The difference is due to the fungal strain and substrate used. The red Oncom is produced by using *Neurospora sitophila*, whereas the black Oncom is produced by using *Rhizopus oligosporus*. The substrates used for solid fermentation can vary, mostly waste of cereals or any plants which are rich in carbohydrates and amino acids, such as peanuts, soybeans, manihot, or coconut (Siswono 2002).

Seventy-one cultures have been isolated by Ho et al. (1978) and a series of pigments proved to be similar to the color that was produced in carrot and apricots. Yellow to orange pigments are mostly produced by *Neurospora crassa*. *Neurospora sitophila* produces pink to orange pigments on their conidia. The pigments produced by fungi are secondary metabolites with different chemical structures. The range of color produced by one isolate can be changed when the substrate is different. The growth rate of the fungi will also be varied when the substrate is changed. The mechanism of different substrates on the changing of their secondary metabolite structure in morphology is yet to be investigated.

In Indonesia, *Neurospora sitophilia* and *Neurospora intermedia* are the fungi species that are very important in "oncom" fermentation process. *Neurospora intermedia* belongs to a group of Ascomycetes, with 8 ascus spores, mycelia, and conidia representing color from yellow to reddish-orange (Perkins and Turner 1988). Biosynthesis of carotenoids in the mold usually involves 3-hydroxy-3-methyl glutaryl CoA (HMG-CoA) synthase enzyme and acetyl-CoA as a main precursor. This precursor will be converted to 3-hydroxy-3-methyl glutaryl CoA (HMG CoA), catalyzed by the enzyme. The HMG-CoA is then converted into mevalonic acid (MVA). MVA, a precursor in terpenoid biosynthesis pathway, is further converted into isopentenyl pyrophosphate (IPP) by a series of reactions including phosphorylation. Metal ions are very important for microbes' metabolism, such as potassium (K^+), magnesium (Mg^{2+}), sodium (Na^+), calcium (Ca^{2+}), zinc (Zn^{2+}), and some other transition elements. It has been reported that 10 mM Mg^{2+} and above can increase phytoen production (Mitzka and Schanabel 1985). Phytoen is desaturated to produce licopen; licopen is a precursor of cyclic carotenoid. *Neurospora intermedia* is a mold that produces spores, mycelia, and conidia with orange or yellow-orange color (Perkins and Turner 1988). Pigments that are produced by *Neurospora* can be various, from yellow to orange reddish, in color. The pigments are carotenoid with α-, β-, and γ-form of carotenes. The carotenoids from Ascomycetes (except for Pezizalesi) are similar to the carotenoids that are produced by Phycomycetes (Goodwin 1980). A research on the evaluation of the role of some metal ions as cofactor in carotenogenesis of *Neurospora intermedia* N-1 mold in "oncom" production had been carried out (Singgih et al. 2014a). Solid substrate Fermentation of *Neurospora sitophila* or *Neurospora intermedia* was made by inoculating 1 mL of spores' suspension of the fungi onto 200 g of peanut kernel and added with 200 mL of sterile water containing 0.15 % of NH_4NO_3, 0.1 % of MgO_47H_2O, and 0.25 % of KH_2PO_4. The culture was incubated at 30 °C for 3 days. The fungi grow over the peanut cake and its spores produce bright orange colored pigments, as seen in Fig. 6.

On day 1, the hypha of the fungi was colorless. It became reddish in the following days and finally it showed orange color. The growth of the fungi, *Neurospora sitophila* on PDA, microscopically showed that its hypha brings the conidia up, which have a smooth texture, are septate, and form a structure like chain. The conidia is singly connected with a protein hialin, dispersed rapidly, to

Fig. 6 Bright *Orange* color from spores of oncom culture *Neurospora intermedia* (Priatni 2014)

form ellipse shapes or cylinder or globose or subglobose shapes (Samson et al. 2002).

Priatni (2014) reported that optimization production of carotenoids by *Neurospora intermedia* on waste solid tofu gave at least five carotenoid compounds, i.e., lycopene, neurosporene, γ-carotene, β-carotene, and phytoene. Identification of these carotenoids isolated from oncom has been carried out by using the HPLC equipped with a photodiode array. All compounds were hydrophobic molecules, which can improve their solubility in aqueous phase by conjugating them with some copolymer, so they can be used widely as food and cosmetic colorants. This study also proved that biosynthesis pathway in *Neurospora intermedia* is similar to *Neurospora crassa*.

The fermentation of *Neurospora crassa* on tapioca by product and waste tofu has been carried out also to produce the alternative poultry feed with high content of β-carotene. In order to achieve optimum physical and chemical stability and bioavailability of carotenoids, innovative processes for their production with modern methods of encapsulation technology have been developed and investigated (Ribeiro et al. 2010).

To produce colorant, some efforts have been done. Microencapsulation of carotene extracts from *Neurospora* sp. spores has been carried out by using protein base as shell material. Encapsulation carotenoid extract with sodium caseinate as shell material gives the highest microencapsulation efficiency, total carotenes, and carotene retention values, compared to soy protein isolate and milk protein isolate (Pahlevi et al. 2008).

Another encapsulation process of carotenoids is from *N. intermedia* N-1 by using copolymer of gelatin and maltodextrin. In this study, the average EY of carotenoids powder obtained by using spray dryer was 48 %. The stability of carotenoids powder can be maintained at low humidity and dark storage. Encapsulated carotenoids from *N. intermedia* N-1 were stable when they are stored in brown glass at RH between 20 % and 30 % (Priatni 2014).

References

Attri D, Joshi VK (2005) Optimization of apple pomace based medium and fermentation conditions for pigment production by *Micrococcus* species. J Sci Ind Res 64:598–601

Barnett TA, Hageman JH (1983) Characterization of a brown pigment from *Bacillus subtilis* cultures. Can J Microbiol 29(3):309–315

Bhat SV, Khan SS, Amin T (2013) Isolation and characterization of pigment producing bacteria from various foods for their possible use as biocolours. Int J Recent Sci Res 4(10):1605–1609

Bhosale PB, Gadre RV (2001) Production of β-carotene by a mutant of *Rhodotorula glutinis*. Appl Microbiol Biotechnol 55:423–427

Blanc PJ, Loret MO, Goma G (1995) Production of citrinin by various species of *Monascus*. Biotechnol Lett 17(7):291–294

Blanc PJ, Loret MO, Goma G (1997) Pigments and citrinin production during cultures of *Monascus* in liquid and solid media. In: Roussos S, Lonsane BK, Raimbault M, Gonzalez GV (eds) Advance in solid state fermentation. Springer, Netherlands, pp 393–406

Blanc PJ, Hajjaj H, Loret MO, Goma G (1998) Control of production of citrinin by *Monascus*. Proceeding of symposium of *Monascus* culture and applications, June, Toulouse, France

Chen M-H, Johns R (1993) Effect of pH and nitrogen source on pigment production by *Monascus purpureus*. Appl Microbiol Biotechnol 40:132–138

Čertík M, Hanusová V, Breierová E, Márová I, Rapta P (2009) Biotechnological production and properties of carotenoid pigments. In: Hou CT, Shaw J-F (eds) Biocatalysis and agricultural biotechnology. CRC Press, Boca Raton, FL, pp 356–375

Cesar de Carvalho J, Soccol CR, Babitha S, Pandey A, Woiciechowski AL (2008) Production of pigments. In: Pandey A, Soccol CR, Larroche C (eds) Current developments in solid-state fermentation. Springer/Asiatech Publisher, New Delhi, India, pp 337–355

Deb P, Madhugiri JM (2012) Optimization of apple-pomace based medium for pigment production by *Micrococcus flavus*. Bioscan 7(1):57–60

Dufossé L, Galaupa P, Yaron A, Arad SM, Blanc P, Murthy KNC, Ravishankar GA (2005) Microorganisms and microalgae as sources of pigments for food use: a scientific oddity or an industrial reality? Trends Food Sci Technol 16:389–406

Dufossé L (2006) Microbial production of food grade pigments. Food Technol Biotechnol 44 (3):313–321

Diansyah MN, Suharto, Triyono EA (2013) Effect of red yeast rice (*Monascus Purpureus*) extract to the thrombopoietin level in dengue infected patients. Folia Med Indones 49(4):220–225

Dikshit R, Tallapragada P (2011) *Monascus purpureus*: a potential source for natural pigment production. J Microbiol Biotech Res 1(4):164–174

El-Banna AA, El-Razek AMA, El-Mahdy AR (2012) Some factors affecting the production of carotenoids by *Rhodotorula glutinis* var. glutinis. Food Nutr Sci 3:64–71

Emeish S (2012) Production of natural β-carotene from *Dunaliella* living in the Dead sea. JJEES 4:23–27

Feng Y, Shao Y, Chen F (2012) *Monascus* pigments. Appl Microbiol Biotechnol 96:1421–1440

Goodwin TW (1980) Fungi. In: Goodwin TW (ed) The biochemistry of the carotenoids, vol 1. Springer, Netherlands, pp 257–290

Gremmels JF, dan Vilar MS (1998) Pharmacological and toxicological aspects of *Monascus* and the role of citrinin. Proceeding of symposium of *Monascus* culture and applications, June, Toulouse, France

Hai Z (1998) Production of monacolin by *Monascus purpureus* on rice solid state fermentation. Proceeding of symposium of *Monascus* culture and applications, June, Toulouse, France

Hajjaj H, Klaebe A, Loret MO, Gom G, Blanc PJ, Francois J (1999) Biosynthesis pathway of citrinin in the filamentous fungi *Monascus ruber* as revealed by ^{13}C nuclear magnetic resonance. Appl Environ Microbiol 65(1):311–314

Ho CC, Chen YF, Koh TH, Mak HP (1978) Improved production of extracellular amyloglucosidase in sorbose resistant mutants in *Neurospora*. Proceedings of the first ASEAN workshop on solid substrate fermentation, ASEAN sub-committee on protein, May, Bandung, Indonesia
Hsieh Y-HP, Pao S, Li J (2008) Traditional Chinese fermented foods. In: Farnworth ER (ed) Handbook of fermented functional foods, 2nd edn. CRC, Boca Raton, FL, pp 433–474
Hussein G, Sankawa U, Goto H, Matsumoto K, Watanabe H (2006) Astaxanthin, a carotenoid with potential in human health and nutrition. J Nat Prod 69(3):443–449
Ibrahim AS (2008) Production of carotenoids by a newly isolated marine *Micrococcus* sp. Biotechnology 7:469–474
Jacobson DJ, Dettman JR, Adams RI, Boesl C, Sultana TS, Roenneberg MM, Duarte M, Marques I, Ushakova A, Carnerio P, Videira A, Navarro-Sampedro L, Olmedo M, Chorrochano LM, Taylor JW (2006) New findings of *Neurospora* in Europe and comparisons of diversity in temperate climates on continental scales. Mycologia 98(4):550–559
Joshi VK, Attri D, Bala A, Bhushan S (2003) Microbial pigments. Indian J Biotechnol 2:362–369
Joshi VK, Attri D, Rana NS (2011) Optimization of apple pomace based medium and fermentation conditions for pigment production by *Sarcina* sp. Indian J Nat Prod Resour 2(4):421–427
Keane M (1999) The red yeast rice cholesterol solution. Adams Media, Avon, MA, pp 1–90
Kim C, Jung H, Kim H, Shin CS (2006) Effect of *Monascus* pigment derivatives on the electrophoretic mobility of bacteria and the cell adsorption and antibacterial activities of pigments. J Colloids Surf B 47:153–159
Kim JI, Lee NK, Yeo I-C, Ryu YJ, Park HS, Kim BY, Kim HY, Hahm YT (2012) Isolation of Carotenoid-producing yeast, *Rhodosporidium babjevae* JI-1, and evaluation of cell extract toxicity against rat hepatic cells. J Korean Soc Appl Biol Chem 55:137–140
Lakrod K, Chaisrisook C, Yongsmith B, Skinner DZ (2000) RAPD analysis of genetic variation within a collection of *Monascus* spp. isolated from red rice (angkak) and sofu. Mycol Res 104 (4):403–408
Latha BV, Jeevaratnam K, Murali HS, Manja KS (2005) Influence of growth factors on carotenoid pigmentation of *Rhodotorula glutinis* DFR-PDY from natural source. Indian J Biotechnol 4:353–357
Latha BV, Jeevaratnam K (2010) Purification and characterization of the pigments from *Rhodotorula glutinis* DFR-PDY isolated from natural source. Glob J Biotechnol Biochem 5 (3):166–174
Lee BK, Park NH, Piao HY, Chung WJ (2001) Production of red pigments by *Monascus purpureus* in submerged culture. Biotechnol Bioprocess Eng 6:341–346
Lee BK, Piao HY, Chung WJ (2002) Production of red pigments by *Monascus purpureus* in solid-state culture. Biotechnol Bioprocess Eng 7:21–25
Lin TF, Demain AL (1991) Effect of nutrition of *Monascus* sp. on formation of red pigments. Appl Microbiol Biotechnol 36:70–75
Lin CF, Lizuka H (1982) Production of extracellular pigment by a mutant of M. kaoling sp. Appl Environ Microbiol 43:671–676
Liu ZX, Du JH, Wang XX, Ma M (2007) Study on submerged fermentation conditions of a strain *Monascus anka* sp. producing pigment and glucoamylase. Food Ferment Ind 33:77–81
Liu DC, Wu SW, Tan FJ (2010) Effects of addition of anka rice on the qualities of low-nitrite Chinese sausages. Food Chem 118:245–25
Lorenz R, Cysewski R (2000) Commercial potential for *Haematococcus* microalgae as a natural source of astaxanthin. TIBTECH 18:160–166
Maldonade IR, Rodriguez-Amaya DB, Scamparini ARP (2008) Carotenoids of yeasts isolated from the Brazilian ecosystem. Food Chem 107(1):145–150
Malik K, Tokkas J, Goyal S (2012) Microbial pigments: a review. Int J Microbial Res Technol 1 (4):361–365
Martin AM, Lu C, Patel TR (1993) Growth parameters for the yeast *Rhodotorula rubra* grown in pear extracts. J Ferment Bioeng 76:321–325

Mata-Gómez LC, Montañez JC, Méndez-Zavala A, Aguilar CN (2014) Biotechnology production of carotenoids by yeast: an overview. Microb Cell Fact 13:1–11

Matthews PD, Wurtzel ET (2007) Biotechnology of food colorant production. In: Socaciu C (ed) Food colorants chemical and functional properties. CRC, Boca Raton, FL, pp 348–398

Mitzka U, Schanabel (1985) Carotenogenic enzymes from *Neurospora*. Pure Appl Chem 57:667–669

Moliné M, Libkind D, van Broock M (2012) Production of torularhodin, torulene, and β-carotene by *Rhodotorula* yeasts. Methods Mol Biol 898:275–283

Mostafa ME, Abbady MS (2014) Secondary metabolites and bioactivity of the *Monascus* pigments review article. Glob J Biotechnol Biochem 9(1):01–13

Nagpal N, Munjal N, Chatterjee S (2011) Microbial pigments with health benefits—a mini review. Trends Biosci 4:157–160

Orozco SFB, Kilikian BV (2008) Effect of pH on citrinin and red pigments production by *Monascus purpureus* CCT3802. World J Microbiol Biotechnol 24:263–268

Ötles S, Çagindi O (2007) Carotenoids as natural colorants. In: Socaciu C (ed) Food colorants chemical and functional properties. CRC, Boca Raton, FL, pp 51–70

Pahlevi YW, Estiasih T, Saparianti E (2008) Microencapsulation of carotene extracts from *Neurospora* sp. spores with protein based encapsulation using spray drying method. J Teknologi Pertanian 1:31–39

Pastrana L, Blanc PJ, Santerre AL, Loret MO, Goma G (1995) Production of red pigments by *Monascus ruber* in synthetic media with a strictly controlled nitrogen source. Process Biochem 30(4):333–341

Perkins D, Turner BC (1988) *Neurospora* from natural population. Exp Mycol 12:19–131

Pfeifer B, Khosla C (2001) Biosynthesis of polyketides in heterologous hosts. Microbiol Mol Biol Rev 65(1):106–118

Priatni S (2014) Review: potential production of carotenoids from *Neurospora*, Nusantara bioscience, ISEA. J Biol Sci 6(1):65–70

Priatni S, Damayanti S, Saraswaty V, Ratnaningrum D, Singgih M (2014) The utilization of solid substrates on *Monascus* fermentation for anticholesterol agent production. Procedia Chem 9:34–39

Ribeiro HS, Schuchmann HP, Engel R, Walz E, Briviba K (2010) Encapsulation of carotenoids. In: Zuidam NJ, Nedovic VA (eds) Encapsulation technologies for active food ingredients and food processing. Springer, New York, NY

Samson RA et al (2002) Introduction to food and airborne fungi, 6th edn. CBS, Utrecht. ISBN 90-70351-42-0

Silveira ST, Daroit DJ, Sant'Anna V, Brandelli A (2013) Stability modeling of red pigments produced by *Monascus purpureus* in submerged cultivations with sugarcane bagasse. Food Bioprocess Technol 6:1007–1014

Singgih M, Ardiansyah SA, Damayanti S, Priatni S (2014a) Production of carotenoid pigment from *Neurospora intermedia* N-1: the influence of Mg^{2+}, Fe^{2+} and Cu^{2+} ion cofactor. Proceeding of international conference on beneficial microbes ICOBM, May, Penang, Malaysia

Singgih M, Saraswaty V, Ratnaningrum D, Priatni S, Damayanti S (2014b) The influence of temperature and ethanol concentration in monacolin K extraction from *Monascus* fermented rice. Procedia Chem 9:242–247

Siswono (2002) Oncom, menutup kekurangan energi dan protein, Gizi.net. Indonesian Nutrition Network

Spolaore P, Joannis-Cassan C, Duran E (2006) Commercial applications of microalgae. J Biosci Bioeng 101(2):87–96

Umeno D, Tobias AV, Arnold FH (2005) Diversifying carotenoid biosynthetic pathways by directed evolution. Microbiol Mol Biol Rev 69(1):51–78

Venil CK, Zakaria ZA, Ahmad WA (2013) Bacterial pigments and their applications. Process Biochem 48:1065–1079

Yimyoo T, Yongmanitchai W, Limtong S (2011) Carotenoid production by *Rhodosporidium paludigenum* DMKU3-LPK4 using glycerol as the carbon source. Kasetsart J (Nat Sci) 45:90–100

Zhang H, Zhan J, Su K, Zhang Y (2006) A kind of potential food additive produced by *Streptomyces coelicolor*: characteristics of blue pigment and identification of a novel compound γ-actinorhodin. Food Chem 95:186–92

Zhang L, Li Z, Dai B, Zhang W, Yuan Y (2013) Effect of submerged and solid-state fermentation on pigment and citrinin production by *Monascus purpureus*. Acta Biol Hung 64(3):385–394

Zhou B, Wang J, Pu Y, Zhu M, Liu S, Liang S (2009) Optimization of culture medium for yellow pigments production with *Monascus anka* mutant using response surface methodology. Eur Food Res Technol 228:895–901

Index

M.-T. Liong (ed.), *Beneficial Microorganisms in Food and Nutraceuticals*, Microbiology Monographs 27, DOI 10.1007/978-3-319-23177-8

MIX
Papier aus verantwortungsvollen Quellen
Paper from responsible sources
FSC® C105338

If you have any concerns about our products,
you can contact us on
ProductSafety@springernature.com

In case Publisher is established outside the EU,
the EU authorized representative is:
Springer Nature Customer Service Center GmbH
Europaplatz 3, 69115 Heidelberg, Germany

Printed by Libri Plureos GmbH
in Hamburg, Germany